Physics of Condensed Matter

Edited by G. Busch, Zürich

In association with
W. Baltensperger · H. Gränicher · W. Känzig · F. Laves
J. Muller · J. L. Olsen · A. Thellung · H. Thomas

With the permanent cooperation of
R. Blinc, Ljubljana · W. Buckel, Karlsruhe
M. H. Cohen, Chicago · J. G. Daunt, Columbus
J. Friedel, Orsay · H. Fröhlich, Liverpool
H. Haken, Stuttgart · K. H. Hellwege, Darmstadt
A. Herpin, Saclay · V. Hovi, Turku
G. Leibfried, Jülich · P. O. Löwdin, Uppsala
W. Low, Jerusalem · K. A. Müller, Zürich-Rüschlikon
L. Néel, Grenoble · S. Nikitine, Strasbourg
H. Raether, Hamburg · G. W. Rathenau, Eindhoven
N. Riehl, München · R. A. Smith, Edinburgh
H. Welker, Erlangen · H. Witte, Darmstadt
K. Yosida, Tokyo

Managing Editor: S. Strässler

Volume 17 · 1973/74

Springer-Verlag Berlin Heidelberg GmbH

Printers: Konrad Triltsch, Graphischer Betrieb, Würzburg

© by Springer-Verlag Berlin Heidelberg 1974
Originally published by Springer-Verlag Berlin Heidelberg New York in 1974.
Softcover reprint of the hardcover 1st edition 1974
ISBN 978-3-662-38713-9   ISBN 978-3-662-39595-0 (eBook)
DOI 10.1007/978-3-662-39595-0

Contents

Phys. cond. Matter 17, 1—10 (1973)

Schallgeschwindigkeitsmessungen in Schmelzen aus den Systemen Al-Mg, Al-Sn, Mg-Sn und Fe-C*

Ulrich Maier und Siegfried Steeb

Max-Planck-Institut für Metallforschung
Institut für Sondermetalle, Stuttgart

Eingegangen am 23. Januar 1973

Sound Velocity in Molten Alloys of the Systems Al-Mg, Al-Sn, Mg-Sn, *and* Fe-C

The concentration dependency of sound velocity was measured in molten alloys of the three systems Al-Mg, Al-Sn, and Mg-Sn in the temperature region between liquidus and 1000 °C. Furthermore it was measured in molten alloys of the system Fe-C in the temperature range between liquidus temperature T_L and $T_L + 150$ °C. From the velocities the compressibilities were calculated, whose concentration dependency revealed changes in the structure of the melts.

The structure of the three systems Al-Mg, Al-Sn, and Mg-Sn in the molten state was already investigated by X-Ray diffraction. These results were compared with the results of velocity measurements. Inhomogeneities in the melts yield increased values of compressibility compared with the values to be expected for statistical distribution of the atoms. This result was used for the discussion of the structure of molten Fe-C-alloys:

The conclusion can be drawn, that within the melts of the system Fe-C above $3^w/_0$ C inhomogeneities are existent.

From the values of sound velocity in the molten elements the temperature dependency of the structure factors $I(0)$ was calculated and compared with experimental values.

1. Einleitung

In vorliegender Arbeit sollen Schallgeschwindigkeitsmessungen an geschmolzenen Legierungen beschrieben werden. Diese wurden in den Schmelzen von je einem Löslichkeits- (Al-Mg [1]), Entmischungs- (Al-Sn [2]) und Verbindungssystem (Mg-Sn [3]) durchgeführt. Es wird aufgezeigt, inwieweit mit dieser Methode strukturelle Besonderheiten in den Schmelzen dieser drei verschiedenen Typen festgestellt werden können.

Mit den daraus gewonnenen Erkenntnissen wird gezeigt, daß auch bei Schmelzen aus dem System Fe-C der Verlauf der Schallgeschwindigkeit und Kompressibilität in Abhängigkeit von der Konzentration und Temperatur Rückschlüsse auf strukturelle Besonderheiten zuläßt.

2. Grundlagen

Für die adiabatische Schallgeschwindigkeit gilt:

$$u_{ad}^2 = -\frac{V}{\varrho}\left(\frac{\partial p}{\partial V}\right)_{ad} \tag{1}$$

mit V = Molvolumen, ϱ = Dichte und p = Druck.

---

* Teil der Dissertation von U. Maier, Universität Stuttgart, 1972.

Der Zusammenhang mit der adiabatischen Kompressibilität ist durch die Beziehung

$$\beta_{\mathrm{ad}} = -\frac{1}{V} \cdot \left(\frac{\partial V}{\partial p}\right)_{\mathrm{ad}} = \frac{1}{\varrho \cdot u_{\mathrm{ad}}^2} \tag{2}$$

gegeben.

Für die Verknüpfung mit den entsprechenden isothermen Größen gelten folgende Zusammenhänge:

$$\beta_{\mathrm{is}} = \beta_{\mathrm{ad}} + \frac{T\,\alpha^2}{\varrho\,c_p} \tag{3}$$

$$\varkappa = c_p/c_V = \beta_{\mathrm{is}}/\beta_{\mathrm{ad}} \tag{4}$$

$$u_{\mathrm{ad}}^2 = \varkappa \cdot u_{\mathrm{is}}^2 , \tag{5}$$

wobei $\alpha = 1/V(\partial V/\partial T)_p$ der Ausdehnungskoeffizient, T die Temperatur in °K und c_p sowie c_V die spezifischen Wärmen sind.

Die Schallgeschwindigkeit in einer Substanz läßt sich nur dann nach Gl. (1) eindeutig bestimmen, wenn eine für die betreffende Substanz gültige Zustandsgleichung der Form $p = p(V, T)$ vorliegt. Nach Ascarelli [4] folgt danach für die Temperaturabhängigkeit der Schallgeschwindigkeit u von Elementschmelzen die folgende Beziehung:

$$u(T) = \left\{\frac{1}{M} \cdot \frac{c_p}{c_V} \cdot k \cdot T \left[\frac{(1+2\eta)^2}{(1-\eta)^4} + \frac{2}{3} \cdot \frac{z \cdot E_{\mathrm{F}}}{k \cdot T} - B \cdot \frac{4}{9 \cdot k \cdot T \cdot v_s^{1/3}}\right]\right\}^{1/2} \tag{6}$$

mit $M\ $ = Atomgewicht

$\quad k\ \ $ = Boltzmann-Konstante

$\quad \eta\ \ $ = Packungsdichte

$\quad z\ \ $ = Zahl der Valenzelektronen pro Atom

$\quad E_{\mathrm{F}}$ = Fermi-Energie

$\quad B\ \ $ = Konstante

$\quad v_s\ $ = Volumen der Proben bei der Temperatur T.

Aus der Beziehung (6) ergibt sich ein geringfügig nichtlinearer Verlauf der Schallgeschwindigkeit in Abhängigkeit von der Temperatur. Mit steigender Temperatur wird der nach diesem Modell berechnete Wert des Temperaturkoeffizienten etwas negativer.

Aus den gemessenen Werten der Schallgeschwindigkeit läßt sich nach Guinier und Fournet [5] über die isotherme Kompressibilität β_T die für $s = (4\pi \sin\theta)/\lambda = 0$ bei Beugungsexperimenten zu erwartende Interferenzfunktion $I(0)$ berechnen:

$$I(0) = \varrho_0 \cdot k \cdot T \cdot \beta_T \tag{7}$$

mit ϱ_0 = mittlere Atomdichte

$\quad 2\theta$ = Winkel zwischen Primärstrahl und abgebeugtem Strahl

$\quad \lambda\ $ = Wellenlänge der für das Beugungsexperiment verwendeten Strahlung.

Nach Omini [6] gilt für die Verknüpfung der Schallgeschwindigkeit mit $I(0)$:

$$\frac{1}{I(0)} = 1 + \frac{1}{F(y)} \tag{8}$$

mit $F(y)$ = Funktion von m, u, k und T.

m = Masse eines Atoms.

3. Experimentelle Durchführung und Ergebnisse

Zur Messung der Schallgeschwindigkeit von Schmelzen bei Temperaturen bis zu 1400 °C wurde das Impuls-Echo-Verfahren nach Seemann und Klein [7] benutzt, und zwar bei konstanter Frequenz von 3 MHz. Dabei wird der von dem bei Raumtemperatur arbeitenden Schwingquarz ausgehende Schall über einen Al_2O_3-Koppelstempel in die Schmelze übertragen. Die akustische Ankoppelung dieses Stempels an die Schmelze wurde durch dünne Schichten aus geeignetem Flußmittel bewirkt, so z.B. für die Schmelzen der Systeme Al-Sn, Al-Mg und Mg-Sn durch ($MgCl_2$ + KCl), beim System Fe-C dagegen durch $BaCl_2$.

Die Messung erfolgte mit fallender Temperatur von 1000 °C bzw. 1400 °C bis zur Liquidustemperatur in Schritten von $\Delta T = 10°$. Bei einzelnen Legierungen wurde zur Kontrolle auch bei steigender Temperatur gemessen. Eine über die Fehlergrenze hinausgehende Abweichung wurde dabei nicht gefunden.

Für jedes der vier zu untersuchenden Systeme wurden außer den reinen Randkomponenten (ausgenommen Kohlenstoff) jeweils ungefähr zehn über den gesamten Konzentrationsbereich verteilte Legierungen hergestellt. Die Reinheit der Ausgangssubstanzen betrug 99,99% für Al, Sn und Fe; 99,98% für Mg und 99,95% für C.

Im folgenden werden die Ergebnisse für die vier untersuchten Systeme dargestellt.

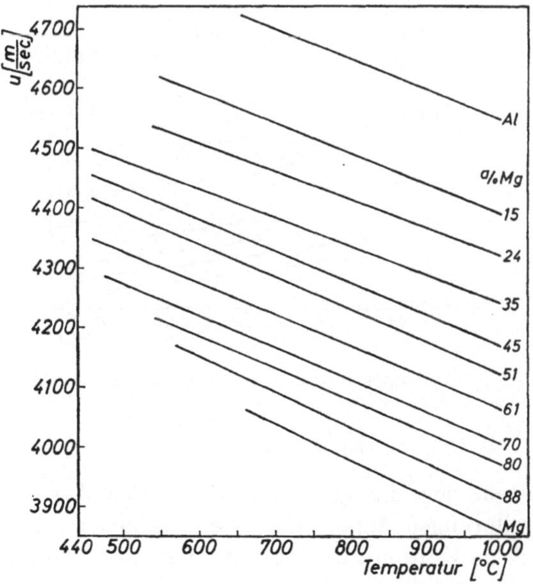

Fig. 1. System Al-Mg: Schallgeschwindigkeit in Abhängigkeit von der Temperatur

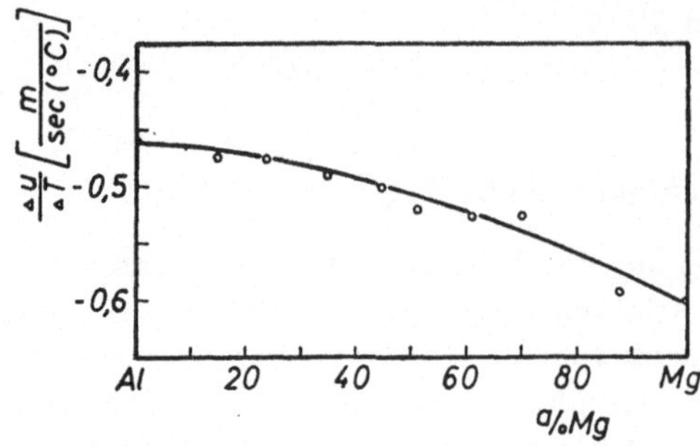

Fig. 2. System Al-Mg: Temperaturkoeffizient der Schallgeschwindigkeit
in Abhängigkeit von der Konzentration

a) *System* Al-Mg

In diesem System sind im geschmolzenen Zustand die Atome beider Atomsorten statistisch verteilt [1]. Fig. 1 zeigt die Temperaturabhängigkeit der Schallgeschwindigkeit für verschiedene Konzentrationen, Fig. 2 die Konzentrationsabhängigkeit des Temperaturkoeffizienten der Schallgeschwindigkeit. Es zeigt
sich, daß für die Schmelzen dieses Löslichkeitssystems die Werte von $\Delta u/\Delta T$ der
Legierungen zwischen denen der reinen Elemente liegen und daß ein kontinuier-
licher Abfall vom Wert des Aluminiums zu dem des Magnesiums besteht.

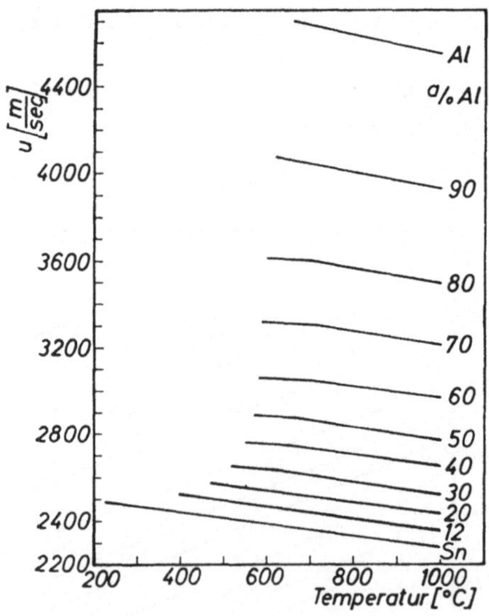

Fig. 3. System Al-Sn: Schallgeschwindigkeit in Abhängigkeit von der Temperatur

b) *System* Al-Sn

In diesem System liegt im geschmolzenen Zustand Entmischung vor mit
Bildung von kugelförmigen Inhomogenitäten, bestehend aus Sn-Atomen [2].
In Fig. 3 setzt sich im Konzentrationsbereich von 30 bis 80 At.% Al der Verlauf
der Schallgeschwindigkeit über der Temperatur jeweils aus zwei Geraden zusammen, und zwar ausgehend von der Liquidus-Temperatur bis etwa 150° darüber

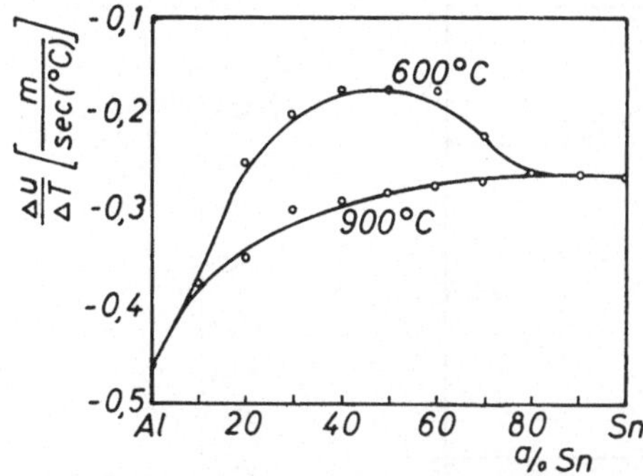

Fig. 4. System Al-Sn: Temperaturkoeffizient der Schallgeschwindigkeit in Abhängigkeit von der Konzentration

aus einer flacher verlaufenden und bei höheren Temperaturen aus einer etwas steiler verlaufenden Geraden. Dieses Verhalten ist etwas deutlicher in Abb. 4 zu sehen, wo die obere Kurve den Verlauf des Temperaturkoeffizienten bei 600 °C und der untere Teil denjenigen bei 900 °C zeigt.

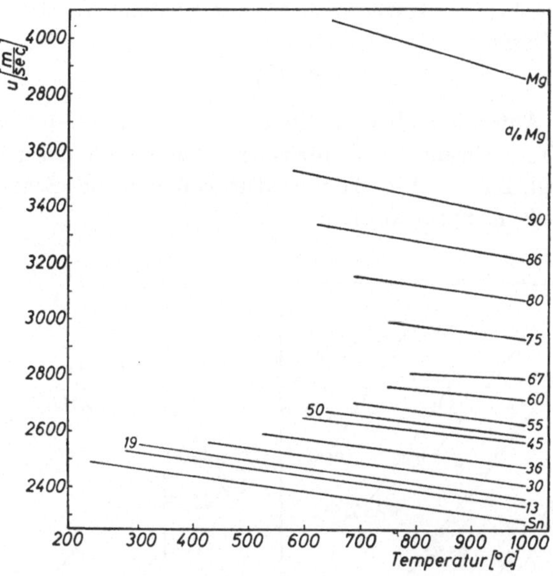

Fig. 5. System Mg-Sn: Schallgeschwindigkeit in Abhängigkeit von der Temperatur

c) System Mg-Sn

In den Schmelzen dieses Systems bilden sich Inhomogenitäten aus, in denen die intermetallische Phase Mg_2Sn vorgebildet ist [3]. Die Schallgeschwindigkeit über der Temperatur (vgl. Fig. 5) bzw. der Temperaturkoeffizient der Schallgeschwindigkeit über der Konzentration (vgl. Fig. 6) zeigen ein ähnliches Verhalten wie beim System Al-Sn. In Fig. 6 fällt das ausgeprägte Maximum an der Stelle der intermetallischen Verbindung Mg_2Sn auf. In Fig. 7 werden für zwei

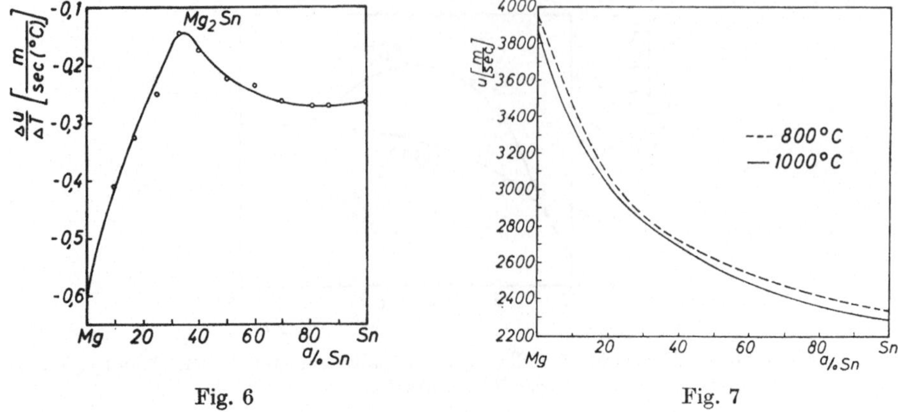

Fig. 6   Fig. 7

Fig. 6. System Mg-Sn: Temperaturkoeffizient der Schallgeschwindigkeit in Abhängigkeit von der Konzentration

Fig. 7. System Mg-Sn: Isothermen der Schallgeschwindigkeit

Temperaturen die Isothermen der Schallgeschwindigkeit für dieses System gezeigt. Die entsprechenden Isothermen für die Systeme Al-Mg und Al-Sn zeigen einen ähnlichen Verlauf.

d) System Fe-C

Um weiteren Aufschluß über die Struktur der Schmelzen (vgl. [8]) in diesem System zu erhalten, wurde die Temperaturabhängigkeit der Schallgeschwindigkeit gemessen (vgl. Fig. 8). Der Temperaturkoeffizient der Schallgeschwindigkeit ist offenbar für alle Konzentrationen etwa derselbe.

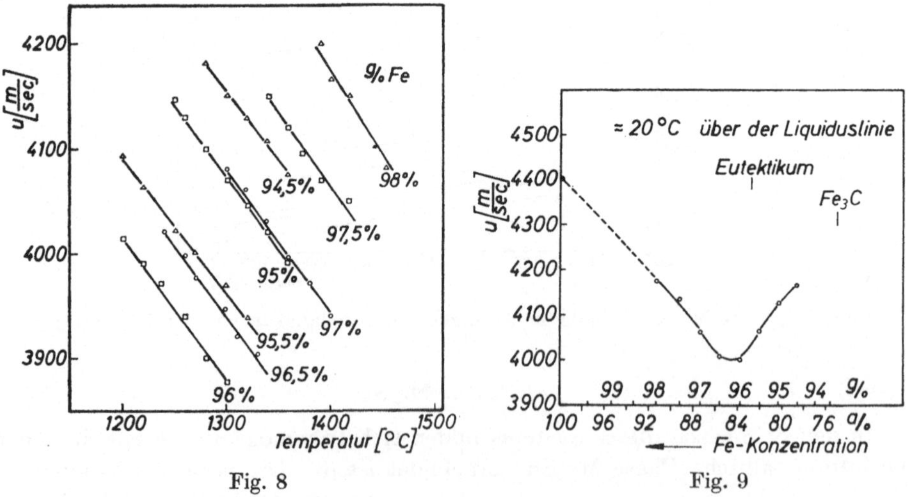

Fig. 8   Fig. 9

Fig. 8. System Fe-C: Schallgeschwindigkeit in Abhängigkeit von der Temperatur

Fig. 9. System Fe-C: Konzentrationsabhängigkeit der Schallgeschwindigkeit, gemessen etwa 20° über Liquidus

In Fig. 9 ist die Schallgeschwindigkeit über der Konzentration aufgetragen. Diese Werte wurden bei ungefähr 20° über der Liquiduslinie gemessen. Vom Wert des reinen Eisens, den Filippov *et al.* [9] mit einer anderen Meßmethode gemessen hatten, ist ein nahezu linearer Verlauf bis zu dem ausgeprägten Minimum bei 3,7 Gew.% C zu erkennen, von dem aus die Schallgeschwindigkeit mit steigendem Kohlenstoffgehalt wieder ansteigt.

4. Diskussion

Der Vergleich der hier experimentell bestimmten Schallgeschwindigkeitswerte für die geschmolzenen Elemente Al, Mg, Sn und Fe mit den nach [4] berechneten Werten gibt eine Übereinstimmung innerhalb einer Abweichung von etwa 10% [10].

Der Verlauf der isothermen Kompressibilität für jeweils konstante Temperatur über der Konzentration ist für die Systeme Al-Mg, Al-Sn, Mg-Sn bzw. Fe-C in den Fig. 10, 11, 12 bzw. 13 dargestellt. Berechnet wurden diese Werte mittels Gl. (2) und (3) aus den gemessenen Schallgeschwindigkeitswerten.

Fig. 10 zeigt zwei Isothermen der Kompressibilität für Schmelzen aus dem System Al-Mg. Beide Kurven weisen ungefähr denselben Verlauf zwischen den Werten der beiden Randkomponenten auf. Sie hängen etwas durch, zeigen jedoch keine Anomalien. Dies steht im Einklang damit, daß in diesem System mit verschiedenen Untersuchungsmethoden im gesamten Konzentrationsbereich ebenfalls keine Anomalien gefunden wurden. Die einzige Ausnahme bildet der Verlauf der Viskositätsisothermen [11].

Im System Al-Sn zeigt die Isotherme der Kompressibilität bei 1000 °C (vgl. Fig. 11) einen linearen Verlauf über dem gesamten Konzentrationsbereich, während die 650 °C-Isotherme einen leicht nach oben gekrümmten Verlauf aufweist, was bedeutet, daß die Kompressibilität bei dieser Temperatur nahezu im ganzen Konzentrationsbereich einen etwas höheren Wert besitzt als es dem statistischen Mittel entspricht. Im Unterschied zum System Al-Mg zeigen die beiden Isothermen einen unterschiedlichen Verlauf. Dies deutet offenbar darauf

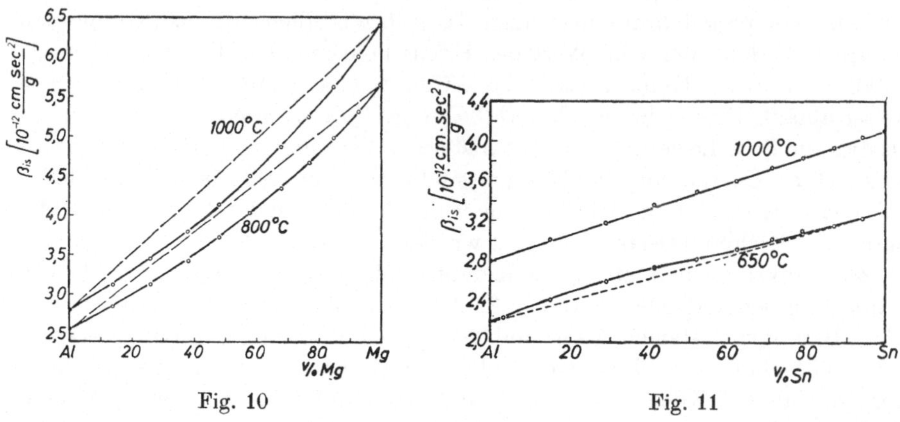

Fig. 10   Fig. 11

Fig. 10. System Al-Mg: Konzentrationsabhängigkeit der isothermen Kompressibilität

Fig. 11. System Al-Sn: Konzentrationsabhängigkeit der isothermen Kompressibilität

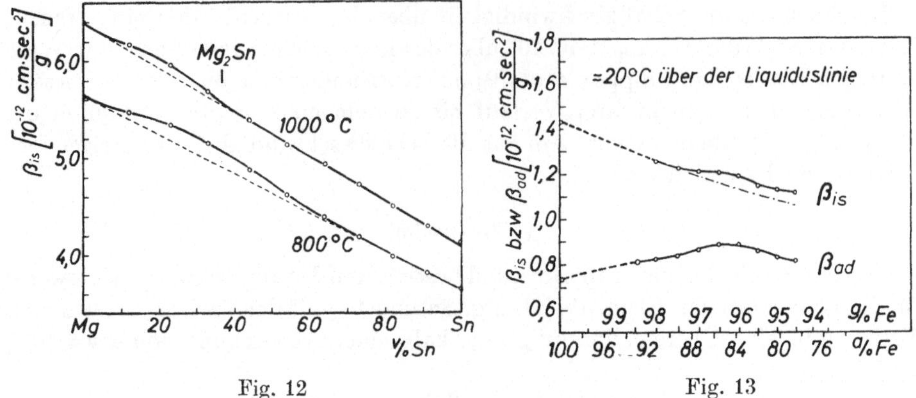

Fig. 12. System Mg-Sn: Konzentrationsabhängigkeit der isothermen Kompressibilität

Fig. 13. System Fe-C: Konzentrationsabhängigkeit der isothermen Kompressibilität

hin, daß sich die Struktur der Schmelze zwischen den beiden Temperaturen verändert: Die bei 650 °C durch Strukturuntersuchungen [2] nachgewiesenen Inhomogenitäten verschwinden bei Temperaturerhöhung und die Schmelze weist bei 1000 °C statistische Verteilung der Atome verschiedener Sorten auf.

Ein ähnliches Verhalten wie beim System Al-Sn zeigen die Isothermen der Kompressibilität im System Mg-Sn (vgl. Fig. 12). Die 1000 °C-Isotherme weist im Konzentrationsbereich von 5 bis 45 Vol.% Sn eine leichte Überhöhung und die 800 °C-Isotherme im Konzentrationsbereich von 5 bis 65 Vol.% eine stärkere Überhöhung gegenüber der statistischen Geraden auf, wobei in beiden Fällen die stärkste Überhöhung bei etwa 33 At.% Sn auftritt. Auch bei diesem System äußern sich die strukturell belegten Mg₂Sn-Agglomerate [3] durch eine Erhöhung der Kompressibilität, die bei höheren Temperaturen entsprechend dem Übergang in eine statistische Verteilung verschwindet.

Für das System Fe-C wurde in Fig. 13 sowohl die isotherme als auch die adiabatische Kompressibilität eingetragen. Danach zeigt die isotherme Kompressibilität einen Verlauf, der vom Wert des Eisens kontinuierlich bis etwa 3 Gew.% C abfällt, von dieser Konzentration bis 4 Gew.% C konstant bleibt und von dort weiter abfällt. Dies bedeutet, daß ab 3 Gew.% C die Kompressibilität größer ist, als dies für Fe-C-Legierungen mit statistischer Verteilung der Atome zu erwarten wäre. Diese Überhöhung im Kompressibilitätsverlauf tritt in Fig. 13 dadurch deutlich in Erscheinung, daß die zum reinen Fe hin gestrichelt gezeichnete Linie durch die strichpunktierte verlängert wurde.

Zur strukturellen Deutung der Kompressibilitätskurven in den Fig. 10 bis 13 wurde dargelegt, daß offenbar das Auftreten einer Inhomogenität in den Schmelzen, wie z.B. in den Systemen Al-Sn und Mg-Sn, sich in einer Überhöhung der Kompressibilitätskurven äußert. In Analogie dazu kann aus Fig. 13 geschlossen werden, daß bei Konzentrationen von 3 bis mindestens 5,5 Gew.% C in den Fe-C-Schmelzen Inhomogenitäten vorliegen.

Die Frage, ob es sich bei diesen Inhomogenitäten überwiegend um Zementit (Fe₃C) oder um Graphit handelt, kann zur Zeit noch nicht entschieden werden.

Ein ähnliches Verhalten der Schallgeschwindigkeit wie die Legierungen im Bereich der Kompressibilitätsüberhöhung zeigen die Elemente Wismut [12] und Antimon [12]. Zur Erklärung des Kompressibilitätsverlaufes über der Temperatur bei geschmolzenem Antimon werden in [12] zwei Strukturen in der Schmelze angenommen, die gleichzeitig nebeneinander existent sind. Die eine ist analog zu der im festen Antimon vorhandenen Struktur aufgebaut, während die andere dichtest gepackt ist. Mit steigender Temperatur nimmt nun der Anteil mit der weniger dicht gepackten, erstgenannten Struktur ab, wodurch das Verschwinden des Überschußanteiles der isothermen Kompressibilität erklärt werden kann.

5. Anhang

In Fig. 14 sind die nach der Boltzmannschen Fundamental-Gleichung [6] bzw. nach Guinier und Fournet [5] aus Schallgeschwindigkeitsmessungen berechneten Werte der Interferenzfunktion $I(0)$ in Abhängigkeit von der Temperatur für die Elemente Al, Mg und Sn als durchgezogene Kurven eingetragen. Außerdem sind noch ein von Hezel [2] experimentell bestimmter Wert für Aluminium bei 665 °C und zwei von Höhler [13] gemessene Werte für Zinn bei 330 °C und 730 °C eingezeichnet. Der gemessene Wert für Aluminium liegt um ungefähr 50% und die Werte für Zinn um etwa 15% über den nach Guinier berechneten Kurven. Die Temperaturabhängigkeit von $I(0)$ bei Zinn stimmt mit den berechneten Werten näherungsweise überein. Die möglichen Ursachen für die Abweichung der gemessenen von den aus der Schallgeschwindigkeit berechneten Werten wurden schon von Hezel [2] diskutiert und sind hauptsächlich durch experimentelle Gegebenheiten bei der Durchführung von Röntgen-Kleinwinkelbeugungsexperimenten bedingt.

Wir danken dem Verein Deutscher Gießereifachleute e. V. für die finanzielle Unterstützung dieser Arbeit sowie Herrn Professor H. J. Seemann, Saarbrücken, und der Fraunhofer-Gesellschaft für die Bereitstellung von Experimentiergerät.

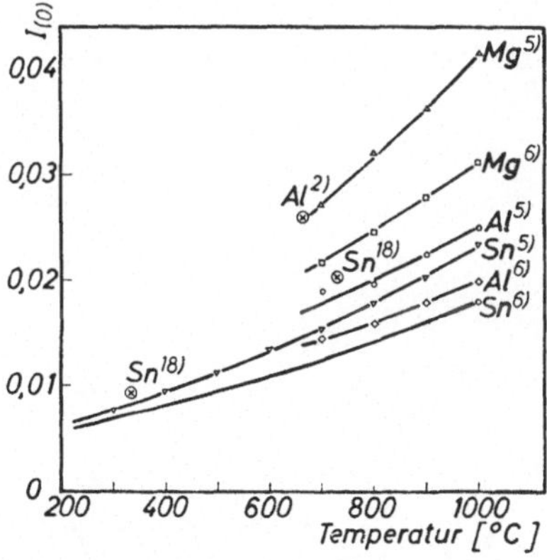

Fig. 14. Berechnete Werte von $I(0)$ für verschiedene Elemente als Funktion der Temperatur. Experimentelle Werte sind mit $\otimes$ gekennzeichnet. Das Zitat 18) in dieser Abbildung ist als 13) zu lesen

Literatur

1. Steeb, S., Woerner, S.: Z. Metallkde. **56**, 771 (1965)
2. Hezel, R., Steeb, S.: Z. Naturforsch. **25a**, 1085 (1970)
   Steeb, S.: Z. Naturforsch. **25a**, 740 (1970)
   Hezel, R., Steeb, S.: Phys. kondens. Materie **14**, 314, 307 (1972)
3. Steeb, S., Entress, H.: Z. Metallkde. **57**, 803 (1966)
4. Ascarelli, P.: Phys. Rev. **173**, 271 (1968)
5. Guinier, A., Fournet, G.: Small-Angle Scattering of X-Rays. New York: John Wiley and Sons, Inc. 1955
6. Omini, M.: Il Nuovo Cimento, Ser. X, **63 B**, 385 (1969)
7. Seemann, H. J., Klein, F. K.: Z. angew. Phys. **19**, 368 (1965)
8. Maier, U., Steeb, S.: Phys. kondens. Materie **16**, 11 (1973)
9. Filippov, S. J., Kazakov, N. B., Pronin, L. A.: Izv. vyssich ucebnych zaved. cern. Met. **9**, 8 (1966)
10. Maier, U.: Dissertation, Universität Stuttgart 1972
11. Gebhardt, E., Detering, K.: Z. Metallkde. **50**, 379 (1959)
12. Gitis, M. B., Mikhailov, I. G.: Soviet Phys.-Acoust. **11**, No. 4, 372 (1966)
13. Hoehler, J.: private Mitteilung

Privat-Dozent Dr. S. Steeb
Max-Planck-Institut
für Metallforschung
Institut für Sondermetalle
D-7000 Stuttgart 1
Seestraße 92
Bundesrepublik Deutschland

Phys. cond. Matter 17, 11—16 (1973)

Röntgen- und Neutronenbeugungsuntersuchungen an Schmelzen des Systems Fe-C*

Ulrich Maier und Siegfried Steeb

Max-Planck-Institut für Metallforschung,
Institut für Sondermetalle, Stuttgart

Eingegangen am 23. Januar 1973

X-Ray and Neutron Wide-Angle Diffraction in Molten Alloys of the System Fe-C

By means of X-Ray- and neutron wide-angle diffraction in molten alloys of the system Fe-C intensity curves were obtained, from which atomic distribution functions could be deduced. From these results, together with results of ultrasound-velocity-measurements, follows the structural behaviour of molten Fe-C alloys in dependency of C-concentration. Increasing C concentration (up to $1.8^w/_0$) causes a higher packing density of the melts because of increasing nearest neighbours distance and number. Above $1.8^w/_0$ C the distance and above $3^w/_0$ up to $5.5^w/_0$ the number of nearest neighbours remains constant. (The upper limit of the concentration range under investigation was $5.5^w/_0$ C.)

Furthermore, the existence of inhomogeneities in melts above $3.5^w/_0$ C can be deduced from ultrasound velocity measurements.

Einleitung

Die Struktur von reinem geschmolzenem Eisen wurde mittels Röntgenbeugung im Jahre 1966 [1] und mittels Neutronenbeugung im Jahre 1970 [2] bestimmt. In letzter Zeit wurden Eigenschaftsmessungen im System Fe-C bekannt, nämlich Viskositätsmessungen [3] und Messungen der Schallgeschwindigkeit [4]. Um weiteren Einblick in das strukturelle Verhalten von geschmolzenen Legierungen im System Fe-C zu bekommen, wird in vorliegender Arbeit über Röntgen- und Neutronenbeugungsuntersuchungen berichtet.

Theoretische Grundlagen

Die kohärent gestreute Intensität im Falle der Röntgen- bzw. Neutronenbeugung wird aus den Experimenten nach den in [5] angegebenen Verfahren erhalten. Dabei wird für das Streuvermögen eines Atomes im ersten Fall der winkelabhängige Atomformfaktor, im zweiten Fall der winkelunabhängige Streuquerschnitt verwendet. Aus der kohärent gestreuten Intensität folgen die Interferenzfunktionen und die verkürzten Interferenzfunktionen sowie hieraus durch Fourierumkehr die Paarkorrelations- bzw. Atomverteilungsfunktionen. Die bei der Neutronenbeugungsmethode anzuwendenden Korrekturen sind ausführlich in [5] beschrieben.

Experimentelle Durchführung und Ergebnisse

Bei der Untersuchung der Schmelzen aus dem System Fe-C mußten sowohl an der Heizvorrichtung der bereits beschriebenen Röntgenbeugungsanlage [6]

---

* Teil der Dissertation von U. Maier, Universität Stuttgart, 1972.

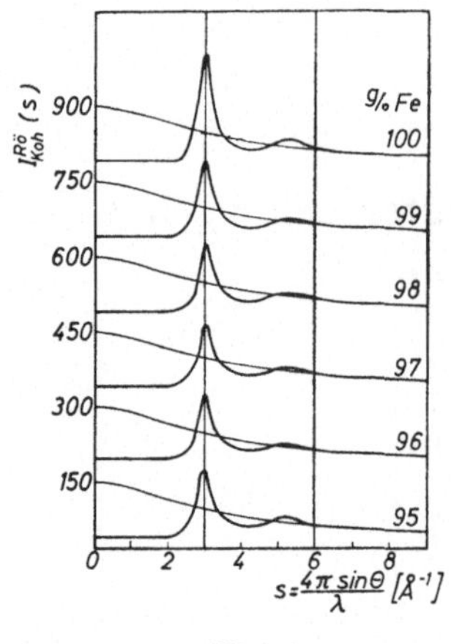

Fig. 1

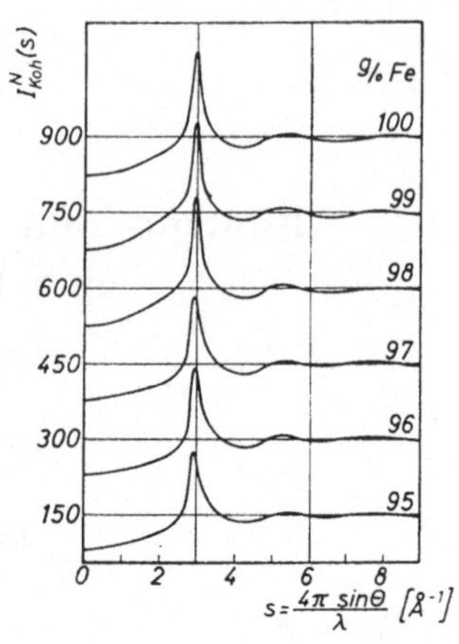

Fig.2

Fig. 1. System Fe-C:Kohärent gestreute Röntgenstrahlenintensität für sechs verschiedene Legierungen

Fig. 2. System Fe-C: Kohärent gestreute Neutronenstrahlenintensität für sechs verschiedene Legierungen

(θ-θ Diffraktometer) als auch an derjenigen der Neutronenbeugungsanlage [5] (Projekt 14 am Forschungsreaktor 2, Kernforschungszentrum Karlsruhe) besondere Vorkehrungen getroffen werden, um die hier benötigten Temperaturen von bis zu 1600 °C erreichen zu können. Als Ausgangssubstanzen dienten Eisen (99,99%) und Kohlenstoff (99,95%), aus denen Proben in einem Vakuuminduktionsofen erschmolzen wurden. Als Tiegelmaterial wurde Al_2O_3 verwendet. Die Messungen erfolgten unter Argon von 10 Torr bei der Röntgenbeugung und von 400 Torr bei der Neutronenbeugung.

1. Intensitätskurven

Es wurden die Intensitätskurven von reinem Eisen und neun bzw. zehn Legierungen gemessen. Die Kohlenstoffkonzentrationen betrugen dabei 0; 0,5; 1; 1,5; 2; 2,5; 3; 3,5; 4; 4,5; 5 Gew.% und bei den Neutronenbeugungsuntersuchungen zusätzlich noch 5,5 Gew.% C. Die Meßtemperaturen lagen zwischen 30° und 40° über dem jeweiligen Schmelzpunkt der Legierungen. Fig. 1 zeigt die aus Röntgenbeugungsexperimenten erhaltenen Intensitätskurven $I_{koh}^{R\ddot{o}}(s)$, die mit dem in [6] beschriebenen Verfahren aus den gemessenen Intensitätsverteilungen berechnet wurden. Außerdem sind die nach Atomprozenten gemittelten Quadrate der Atomformfaktoren eingezeichnet. Unterschiede im Verlauf der einzelnen Kurven treten in dieser Darstellung kaum auf.

In Fig. 2 sind die kohärent gestreuten Neutronenintensitäten $I_{koh}^{N}(s)$ dargestellt, die nach [5] aus den gemessenen Intensitätskurven berechnet wurden. Entsprechend den f^2-Kurven bei der Röntgenbeugung (vgl. Fig. 1) sind hier die nach Atomprozenten gemittelten Quadrate der kohärenten Neutronenstreuamplituden b_{koh} aufgezeichnet. Der Zusammenhang zwischen den aus [7] ent-

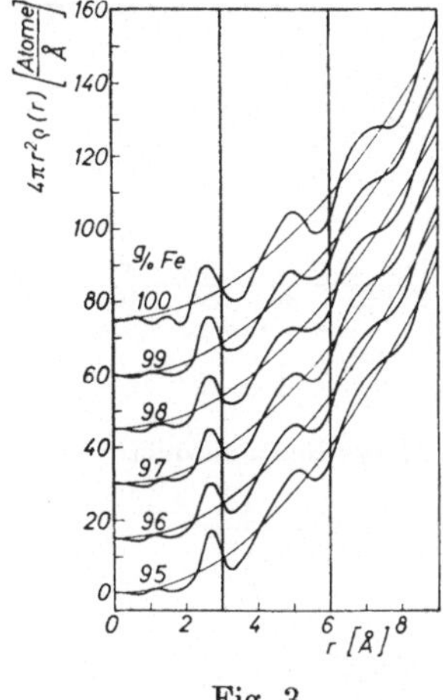

Fig. 3

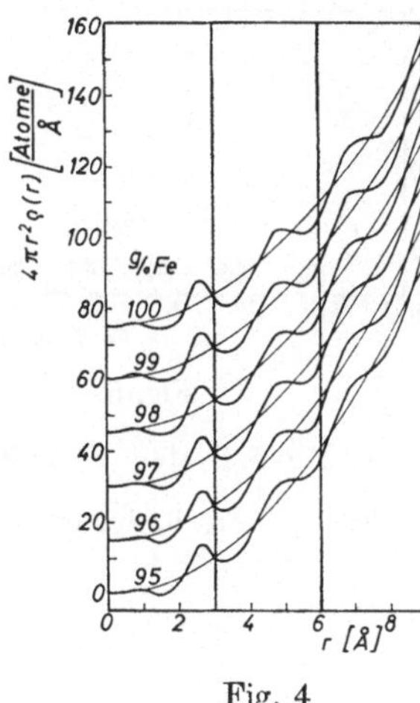

Fig. 4

Fig. 3. System Fe-C: Atomverteilungskurven (Röntgenbeugung)

Fig. 4. System Fe-C: Atomverteilungskurven (Neutronenbeugung)

nommenen Wirkungsquerschnitten σ und der Streuamplitude b ist durch die Beziehung

$$\sigma = 4\pi\,b^2 \qquad [\text{barn}]$$

gegeben.

2. Atomverteilungskurven

Durch Fouriertransformation wurden die mit den beiden Strahlenarten erhaltenen Intensitätskurven der Fig. 1 und 2 in Atomverteilungskurven transformiert. In Fig. 3 sind die so aus Fig. 1 erhaltenen Kurven wiedergegeben. Fig. 4 zeigt die entsprechenden Kurven aus den Neutronenbeugungsexperimenten. In den beiden Figuren oszillieren die Atomverteilungskurven um die Parabeln $4\pi r^2 \varrho_0$. Diesen Kurven können zwei wichtige Bestimmungsstücke, nämlich die Lage r^{I} des ersten Maximums und aus der Fläche unter demselben die Zahl nächster Nachbarn N^{I} entnommen werden.

Diskussion der Ergebnisse

1. Intensitätskurven

Das erste Maximum der in den Fig. 1 und 2 gezeigten Intensitätskurven verschiebt sich jeweils kontinuierlich mit steigendem Kohlenstoffgehalt von $s = 2{,}98\,\text{Å}$ beim reinen Eisen bis $2{,}90\,\text{Å}^{-1}$ bei einer Legierung aus Eisen mit 5 Gew.% C. Irgendwelche Nebenmaxima oder eine Aufspaltung des Hauptmaximums treten weder bei der Röntgen- noch bei der Neutronenbeugung auf. Die absolute Höhe des Hauptmaximums nimmt ebenfalls mit steigendem Kohlenstoffgehalt bei beiden Strahlenarten gleichmäßig ab, wobei sich die Halbwertsbreite erhöht.

Diese Befunde können damit erklärt werden, daß durch den Einbau der Kohlenstoffatome in das Eisengitter der Abstand der Eisenatome voneinander erstens vergrößert wird und zweitens die Schwankungsbreite dieser Abstände zunimmt.

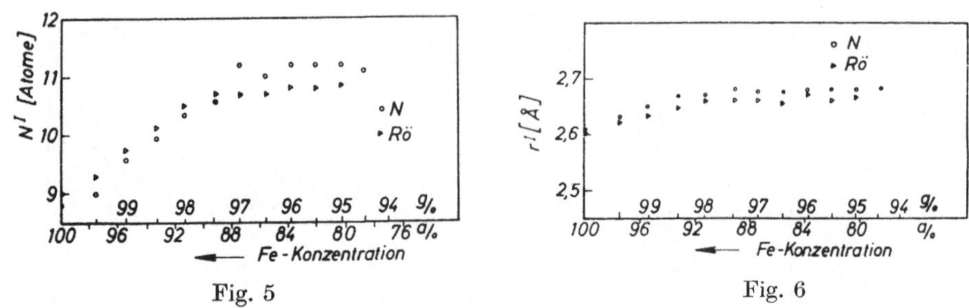

Fig. 5   Fig. 6

Fig. 5. System Fe-C: Zahl der Atome in der ersten Koordinationssphäre

Fig. 6. System Fe-C: Radius der ersten Koordinationssphäre

2. Atomverteilungskurven

Entsprechend den Integrationslängen $s_{max} = 13\ \text{Å}^{-1}$ bei der Röntgen- bzw. $9{,}1\ \text{Å}^{-1}$ bei der Neutronenbeugung weisen die Maxima der Atomverteilungskurven im Falle der Röntgenbeugung (vgl. Fig. 3) eine geringere Halbwertsbreite auf als diejenigen der Neutronenbeugung (vgl. Fig. 4). Die aus den Atomverteilungskurven berechneten Koordinationszahlen N^I sind in Fig. 5 in Abhängigkeit von der Konzentration für beide Strahlenarten aufgetragen. Für das reine Eisen ergibt sich aus den Röntgenmessungen ein Wert von 9,0 und aus den Neutronenmessungen ein solcher von 8,8 Atomen. In [2] wird für das reine Eisen aus Neutronenbeugungsuntersuchungen ein Wert von 9,5 angegeben. Dieser ist jedoch durch Integration über die Atomverteilungskurve von Null bis zum ersten Minimum nach dem Hauptmaximum berechnet worden. Nach dieser Methode würde sich in vorliegender Arbeit infolge des unsymmetrischen Maximums ein Wert von ungefähr 11 ergeben. Mit derselben Methode erhalten Schmitz-Pranghe und Kohlhaas [8] aus Röntgenmessungen eine Koordinationszahl von 11,6, während die Berechnung des symmetrischen Anteils jener Kurven zu einem Wert von 9,2 führt. 8,2 Atome in der ersten Koordinationssphäre wurden von Ruppersberg und Seemann [1] ebenfalls aus dem symmetrischen Anteil des Hauptmaximums berechnet. Nach dem Vergleich mit diesen Ergebnissen ist anzunehmen, daß die hier erhaltene Koordinationszahl N^I von etwa 9,0 Atomen den in der Eisen-Schmelze vorliegenden Verhältnissen am ehesten entspricht.

Mit steigendem Kohlenstoffgehalt ergibt sich für die Koordinationszahl, unabhängig von der verwendeten Strahlensorte, eine gleichmäßige Zunahme (vgl. Fig. 5), wobei die Röntgenwerte um ungefähr 0,15 Atome über den Neutronenwerten liegen. Ab dem Konzentrationsbereich von 2,5 bis 3 Gew.% nimmt die Koordinationszahl ebenfalls unabhängig von der zur Untersuchung verwendeten Strahlensorte einen konstanten Wert an, wobei diesmal die Neutronenwerte um ungefähr 0,4 Atome über den Röntgenwerten liegen. Aus den Röntgenmessungen ergibt sich so im Bereich von ungefähr 3 bis 5 Gew.% C eine Koordinationszahl von 10,8, während die aus den Neutronenmessungen erhaltene 11,2 beträgt.

In Fig. 6 sind die aus den Atomverteilungskurven der Fig. 3 und 4 ermittelten Werte der Abstände nächster Nachbarn r^I einmal für die Neutronenbeugungsuntersuchungen (o), zum anderen für die Röntgenbeugungsuntersuchungen ($\triangle$)

eingetragen. Aus beiden Versuchsreihen ergibt sich für das reine Eisen derselbe Wert von ungefähr 2,60 Å. Diese Werte stimmen recht gut mit dem von Waseda und Suzuki [2] mit Neutronen und dem von Schmitz-Pranghe und Kohlhaas [8] mit Röntgenstrahlen gefundenen Wert überein, der bei beiden 2,58 Å beträgt. Bei Ruppersberg und Seemann [6] beträgt der aus Röntgenuntersuchungen gewonnene Abstand der ersten Koordinationssphäre 2,52 Å.

Durch Zugabe von Kohlenstoff erhöht sich der r^{I}-Wert gleichmäßig, bis er bei ungefähr 1,8 Gew.% C einen konstanten Betrag annimmt, der sich mit steigendem Kohlenstoffgehalt nicht mehr erhöht. Der Wert, der sich aus Neutronenmessungen ergibt, liegt bei ungefähr 2,67 Å, während derjenige aus den Röntgenmessungen sich auf 2,66 Å beläuft. Diese Abweichung ist innerhalb der Fehlergrenze, die etwa $\pm$ 0,03 Å beträgt.

Insgesamt ergeben sich folgende Aussagen über die Struktur geschmolzener Fe-C-Legierungen: Der gesamte untersuchte Konzentrationsbereich von 0 bis 5,5 Gew.% kann in drei Teilbereiche unterteilt werden. Der erste erstreckt sich von 0 bis etwa 1,8 Gew.% C. In diesem Bereich nimmt der Radius der ersten Koordinationssphäre mit steigendem Kohlenstoffgehalt kontinuierlich zu. Diese Zunahme ist verbunden mit einer Zunahme der Koordinationszahl und zwar so, daß insgesamt die Packungsdichte der Schmelze größer wird. Im zweiten Bereich, der sich von ungefähr 1,8 bis etwa 3 Gew.% C erstreckt, ist zwar der Abstand nächster Nachbarn gleich, aber die Koordinationszahl nimmt noch zu. Bei Konzentrationen größer als 3 Gew.% C bleiben die beiden Größen r^{I} und N^{I} konstant, jedoch zeigt nach [4] die isotherme Kompressibilität noch einen gegenüber dem kontinuierlichen Abfall zusätzlichen Beitrag, der nach [4] auf Inhomogenitäten in den Schmelzen dieses Konzentrationsbereiches schließen läßt.

Abschließend sei bemerkt, daß das im vorhergehenden Abschnitt dargelegte Modell verträglich ist mit den an anderer Stelle durchgeführten Messungen der Viskosität [9], [3], der Dichte [10], des elektrischen Widerstandes [9], der Oberflächenspannung [9] und der von der Oberfläche ausgehenden, thermisch bedingten Emission elektromagnetischer Strahlung [11].

Dem Verein Deutscher Gießereifachleute, dem Rechenzentrum der Universität Stuttgart und dem Kernforschungszentrum Karlsruhe sei für die mannigfache Unterstützung dieser Arbeit gedankt.

Literatur

1. Ruppersberg, H., Seemann, H. J.: Z. Naturforsch. **21a**, 820 (1966)
2. Waseda, Y., Suzuki, K.: Phys. Status Solidi **39**, 669 (1970)
3. Krieger, W., Trenkler, H.: Arch. Eisenhüttenwesen **42**, Heft 3, 175 (1971)
4. Maier, U., Steeb, S.: Phys. kondens. Materie **15**, 1 (1973)
5. Knoll, W., Steeb, S.: J. phys. Chem. Liquids **4**, 39 (1973)
6. Bühner, H. F., Steeb, S.: Z. Metallkde **62**, 27 (1971)
7. Bacon, G. E.: Neutron Diffraction. Oxford: Clarendon Press 1962
8. Schmitz-Pranghe, N., Kohlhaas, R.: Z. Naturforsch. **25a**, 1752 (1970)
9. Samarin, A. M.: Structure and Properties of Liquid Metals. Report AEC-tr-4879 S. 157 (1960)

                            U. Maier und S. Steeb

10. Lucas, L. D.: Mem. Scient. Rev. Metallurg., LXI No. 2, 97 (1964)
11. Filippov, E. S., Grigorovich, V. K., Samarin, A. M.: Soviet Physics-Doklady 12, No. 3, 284 (1967)

Privat-Dozent Dr. Siegfried Steeb
Max-Planck-Institut für Metallforschung
Institut für Sondermetalle
D-7000 Stuttgart 1, Seestraße 92
Bundesrepublik Deutschland

Phys. cond. Matter 17, 17—36 (1973)

Displacement and Energy Density Correlations in Solids

P. F. Meier *

IBM Zürich Research Laboratory, CH-8803 Rüschlikon, Switzerland

Received February 21, 1973

A new approach to the microscopic treatment of a system of interacting phonons is developed based on techniques used by Mori to describe density fluctuations in liquids. Introduction of a matrix of retarded Greens functions allows a description of dynamic cooperative phenomena especially suited to investigate coupling of displacement and energy density correlations in the hydrodynamic domain. The general method is applied in conjunction with perturbation theory to the model of an anharmonic crystal. The usual expressions for frequency shift and damping of thermal phonons are reproduced. For long wavelengths and low frequencies a three-variable theory, which includes the fluctuations in the energy density, leads to expressions expected from phenomenological equations. This derivation is direct and avoids complicated treatment of hydrodynamic singularities of other microscopic theories. Finally, it is shown that a four-variable theory can account for second sound and its coupling to first sound.

1. Introduction

The theory of phonon interactions describing the frequency shift and line width of thermal phonons is well established. Using some Greens-function method one is able to derive by perturbation theory explicit expressions for the phonon self-energy in terms of the first few derivatives of the lattice potential energy [1]. Although the numerical evaluation of the resulting quantities is rather complicated one now gets results in fair agreement with neutron-scattering experiments, even in the case of quantum crystals where first a renormalization of the harmonic phonon energies has to be carried out in order to deal with the large anharmonicities [2, 3].

In the hydrodynamic limit where fluctuations of the system have wavelengths longer than the mean free paths of thermal phonons, theoretical treatment is more complicated. Usual perturbation theory breaks down since many collisions occur during a period of collective motions of the systems. In recent years a lot of theoretical work has been done in order to account for this collision-dominated situation. Following the general procedure developed by Kadanoff and Baym [4] non-equilibrium Greens functions have been used in Refs. [5]—[8] to derive transport equations for phonons from first principles. In this way one gets generalized forms of the Peierls equation [9] which are the starting points to a description of second sound and Poiseuille flow [10]. In a different approach, the self-energy of the equilibrium one-phonon Greens function has been examined in Refs. [11]—[14] where hydrodynamic singularities have to be treated in an appropriate manner. A microscopic foundation of phonon hydrodynamics was

---

* IBM Postdoctoral Fellow of the Institute for Theoretical Physics, University of Zürich, Switzerland.

obtained by Götze and Michel [12] who discussed the solutions of the coupled equations for the quasiparticle gas and the elastic waves. All these methods are rather complicated since finite-order perturbation theory does not lead to the required structure of the displacement correlation function.

In contrast to the situation in lattice dynamics, the low-frequency, hydrodynamic region for liquids is better understood than high-frequency domain where no theory is available. For density fluctuations in liquids Mori [15] has obtained microscopic expressions for intensities and widths of spectral lines.

In this work we apply the techniques developed by Mori to phonons in solids. The essential idea is the following. In all the work mentioned above on phonon hydrodynamics the one-phonon Greens function describing the displacement correlation has been subjected to a complicated perturbation analysis. This was necessary since in the hydrodynamic limit the phonon self-energy shows a pole due to fluctuations in heat density. Thus, starting from the displacement correlation function, one had to sum up infinite classes of diagrams or to solve integral equations for the vertex part in order to account correctly for coupling to fluctuations in the energy density. The idea worked out here is to consider a matrix of Greens functions for the displacement and the energy density response. In this way the effects of coupling of the fluctuations are taken into account at the beginning. The various poles corresponding to different possible excitations are then obtained in a direct way without the complicated treatment of hydrodynamic singularities of other microscopic theories.

The main advantage of this many-variable theory of Mori is that collective variables which couple strongly are treated on the same footing. The influence of coupling to other degrees of freedom leads to random forces which can be calculated in various approximations by perturbation theory. Furthermore, it is also possible by this method to reduce hydrodynamic transport coefficients to Kubo formulae without introducing a local equilibrium ensemble [15]. For the case of interacting phonons, the latter concept was thoroughly investigated by Enz [16] who developed a description of heat propagation using density matrices for local thermal equilibrium and local phonon drift. To this purpose an effective Liouville equation was derived. In contrast, the work of Mori and its application presented here is concerned with correlation functions evaluated in the canonical equilibrium ensemble. We can thus obtain second sound as a propagating collective mode of the energy density in a direct way.

In Sect. 2 we outline formal theoretical concepts according to various papers by Mori [15, 17, 18]. With the help of a projection-operator technique a generalized Langevin equation for a dynamic variable is derived. Corresponding equations for the correlation functions are put into relation with quantities used in Greens function methods. Introduction of a matrix of Greens functions then allows a description of dynamic cooperative phenomena. Further it is shown that the random forces can be represented as continued functions.

The Hamiltonian of our system and the necessary notations are given in Sect. 3. We assume that the harmonic approximation is a good starting point and consider the effects of cubic and quartic anharmonicities as small corrections. This allows the application of usual perturbation theory.

The fourth section is devoted to the derivation of expressions for the frequency shift and damping of phonons. The well-known results are reproduced in a simple way.

The main advantage of using a matrix of Greens functions is exhibited in Sect. 5 where fluctuations in the displacement and energy density and their coupling is investigated. The pole structure of correlation functions in the hydrodynamic limit is easily obtained by matrix inversion. The resonances of the displacement correlation function correspond to the Brillouin doublet and the Landau-Placzek peak, in agreement with previous works [13, 14, 19] and with the phenomenological equations of elasticity [20]. This aspect has been worked out very clearly in a recent review article by Götze and Michel [21]. Also using projection-operator techniques they started from general anharmonic interactions and achieved a self-consistent description of phonons, accounting in this way for quantum crystals. In particular, they related the various thermodynamic quantities occurring in the phenomenological theory to static correlation functions and calculated the corresponding microscopic expressions.

In Sect. 6 we discuss the second-sound mode which may be excited in perfect crystals at low temperatures where Umklapp processes are very rare. It has been detected in solid He [22] and, more recently, also in NaF [23] and in Bi [24]. Taking into account fluctuations also in the energy current we analyze a Greens-function matrix of four variables. The displacement response function then shows up two pairs of poles corresponding to first and second sound. The relative weight of the second pole is smaller by an order of magnitude in the anharmonicities. The retarded Greens function for the energy density exhibits the same behavior but with the second sound dominating. These results are in complete agreement with the structure of the equations obtained by Kwok and Martin [5] and Sham [25]. Our formulae, however, imply that the response functions contain a pole due to a mode propagating with the velocity of driftless second sound. This excitation has been considered by Griffin [26] and Enz [16, 27]. Although the value of the corresponding velocity is not very different from that of drifting second sound [10], the concepts are not the same. We study here, as in Refs. [26, 27], the behavior of energy-density and energy-current operators without considering the momentum density. In the derivation of drifting second sound, however, a Boltzmann equation for the phonon number density is used to obtain local conservation laws for energy and momentum [10]. Then the energy current is proportional to the momentum density with a factor given by the velocity of drifting second sound. In order to meet this situation in our formalism we need to investigate the coupling of fluctuations in energy and momentum densities. This does indeed lead to the drifting second-sound mode, but we have to assume a form of the momentum-density operator. The microscopic form of the latter is, however, not as well founded as for the energy density.

2. Equations of Motion for Collective Variables

In this section we first introduce some definitions to apply a projection-operator technique to the derivation of a generalized Langevin equation for a dynamical variable. This leads to an expression for the corresponding retarded

Greens function in terms of static correlation functions and random forces which are certain special dynamic correlation functions. Next the equations are extended to the case of several variables which couple strongly with one another. In this way we get a matrix of retarded Greens functions which is suited to describe dynamic cooperative phenomena. Finally, we comment on the evaluation of the random forces by means of a continued-fraction representation.

All main ideas presented here are due to Mori who gave in Ref. [17] the derivation of the generalized Langevin equation with the help of the projection-operator technique. The idea of considering a set of variables to describe the collective motion was outlined in Ref. [15], whereas the continued-fraction representation was given in Ref. [18].

Let us consider a conservative system described by the Hamiltonian H. The equation of motion of a dynamic variable X in the Heisenberg representation is

$$\frac{d}{dt} X(t) = \dot{X}(t) = i[H, X(t)] = iLX(t), \tag{2.1}$$

where L denotes the Liouville operator acting in the linear space of dynamic variables. In this space we can define a form by

$$(X, Y) = \int_0^\beta d\lambda \langle e^{\lambda H} X e^{-\lambda H} Y \rangle, \tag{2.2}$$

where an average over a canonical ensemble is introduced,

$$\langle \ldots \rangle = \mathrm{tr}\,(e^{-\beta H} \ldots)/\mathrm{tr}\, e^{-\beta H} \tag{2.3}$$

and $\beta = (k_\mathrm{B} T)^{-1}$. The following properties are easily derived,

$$(X, Y) = (Y, X) \tag{2.4}$$

and

$$(X, X^+) \geqq 0. \tag{2.5}$$

We can thus define a scalar product

$$X, Y \to (X, Y^+) = \int_0^\beta d\lambda \langle e^{\lambda H} X e^{-\lambda H} Y^+ \rangle, \tag{2.6}$$

which is linear in the first argument and satisfies

$$(X, Y^+)^* = (X^+, Y). \tag{2.7}$$

In this way a Hilbert space of dynamic variables can be constructed in which the Liouville operator is Hermitian,

$$(LX, Y^+) = (X, (LY)^+). \tag{2.8}$$

We now define an operator P_X by

$$P_X Z = (Z, X^+)(X, X^+)^{-1} X. \tag{2.9}$$

P_X is linear and Hermitian,

$$(P_X Z, Y^+) = (Z, (P_X Y)^+) \tag{2.10}$$

and idempotent ($P_X^2 = P_X$). These properties define a projection operator; the corresponding complement is denoted by $Q_X = 1 - P_X$.

We should now like to relate the correlation functions defined with the help of the scalar product (2.6) to the quantities introduced by Greens-function techniques. Following the notation of Schneider and Meier [28] we consider the definition of the retarded Greens function

$$G^{\mathrm{ret}}_{XY^+}(t) = -i\,\Theta(t)\langle[X(t),Y^+]\rangle \tag{2.11}$$

with Θ denoting the step function and $[\ ,\ ]$ the commutator. Evaluating the trace in the energy representation it can be shown that the scalar product (2.6) is connected with the spectral function $\chi(\omega)$ by

$$(X,Y^+) = P\int\frac{d\omega}{2\pi}\frac{1}{\omega}\chi_{X,Y^+}+(\omega) = P\int\frac{d\omega}{2\pi}\frac{1}{\omega}\int dt\,e^{i\omega t}\langle[X(t),Y^+]\rangle. \tag{2.12}$$

From that, the following relations are obtained,

$$(\dot{X},Y^+) = -(X,\dot{Y}^+) = -i\langle[X,Y^+]\rangle. \tag{2.13}$$

Further we define a dynamic correlation function Φ by

$$\Phi_{XY^+}(t) = (X(t),Y^+) \tag{2.14}$$

and its Laplace transform

$$\Phi_{XY^+}(z) = \int\limits_0^\infty dt\,e^{-zt}\,\Phi_{XY^+}(t) = \int\limits_0^\infty dt\,e^{-zt}(e^{iLt}X,Y^+). \tag{2.15}$$

This allows a representation of the retarded Greens function as

$$G^{\mathrm{ret}}_{XY^+}(t) = \Theta(t)(\dot{X}(t),Y^+) = \Theta(t)(d/dt)\Phi_{XY^+}(t). \tag{2.16}$$

The Laplace transform of G^{ret} which is related to the Fourier transform $G^{\mathrm{ret}}(\omega)$ by

$$G^{\mathrm{ret}}(\omega) = \lim_{\varepsilon\to 0^+} G^{\mathrm{ret}}(z=-i\omega+\varepsilon) \tag{2.17}$$

can therefore be written in the form

$$G^{\mathrm{ret}}_{XY^+}(z) = -\Phi_{XY^+}(t=0) + z\Phi_{XY^+}(z). \tag{2.18}$$

We now derive the equation of motion for the quantity $\Phi_{XX^+}(z)$. Definition (2.15) is equivalent to

$$\Phi_{XX^+}(z) = \left(\frac{1}{z-iL}X,X^+\right). \tag{2.19}$$

The resolvent is transformed according to the operator identity

$$\frac{1}{z-iL} = \frac{1}{z-iLQ-iLP} = \frac{1}{z-iLQ} - \frac{1}{z-iLQ}(-iLP)\frac{1}{z-iL}, \tag{2.20}$$

where we have dropped the subscript X on projection operators P_X and Q_X. Since $QX = 0$ we have

$$\frac{1}{z-iLQ}X = \frac{1}{z}X$$

and, using (2.9),

$$P\frac{1}{z-iL}X = \Phi_{XX^+}(z)(X,X^+)^{-1}X.$$

This yields

$$z\Phi_{XX^+}(z) = (X, X^+) + \Phi_{XX^+}(z)\,(X, X^+)^{-1}\left(\frac{z}{z - iLQ}\dot{X}, X^+\right). \qquad (2.21)$$

The last scalar product occurring in (2.21) can be written as

$$\left(\frac{z - iLQ + iLQ}{z - iLQ}\dot{X}, X^+\right) = (\dot{X}, X^+) - F_{XX^+}(z),$$

where we have defined the so-called random forces F by

$$F_{XX^+}(z) = \left(\frac{-iLQ}{z - iLQ}\dot{X}, X^+\right). \qquad (2.22)$$

The solution of (2.21) leads to the following expression for the correlation function Φ,

$$\Phi_{XX^+}(z) = (X, X^+)\,[z(X, X^+) - (\dot{X}, X^+) + F_{XX^+}(z)]^{-1}(X, X^+), \qquad (2.23)$$

and for the retarded Greens function we get from (2.18)

$$G^{\mathrm{ret}}_{XX^+}(z) = (X, X^+)\,[z(X, X^+) - (\dot{X}, X^+) + F_{XX^+}(z)]^{-1}\,[(\dot{X}, X^+) - F_{XX^+}(z)]. \quad (2.24)$$

These formulae are already written in a way which allows an extension to the case of several variables $X_1, X_2, \ldots, X_n$. X then denotes the column vector

$$X = (X_1, X_2, \ldots, X_n)^{\mathrm{T}} \qquad (2.25)$$

and (X, X^+) stands for the $n \times n$ matrix with elements

$$(X, X^+)_{ij} = (X_i, X_j^+). \qquad (2.26)$$

P_X is a projector on the subspace spanned by X_1 to X_n and operates according to

$$P_X Y = \sum_{i,j} (Y, X_i^+)\,(X, X^+)^{-1}{}_{ij}X_j. \qquad (2.27)$$

In this matrix formulation, (2.24) is an equation for the $n \times n$ matrix of Greens functions which can also be written in the form

$$G^{\mathrm{ret}}_{X,X^+}(z) = [z - (\dot{X}, X^+)(X, X^+)^{-1} + F_{XX^+}(z)\,(X, X^+)^{-1}]^{-1}$$
$$\times\,[(\dot{X}, X^+) - F_{XX^+}(z)], \qquad (2.28)$$

which for computational reasons is more convenient than (2.24).

Finally, we observe that using (2.8) definition (2.22) can be transformed into

$$F_{XX^+}(z) = \left(\frac{Q^2}{z - iLQ}\dot{X}, \dot{X}^+\right) = \left(\frac{1}{z - iQLQ}Q\dot{X}, (Q\dot{X})^+\right) \qquad (2.29)$$

which shows that

$$F_{XX^+}(t) = (e^{iQLQt}Q\dot{X}, (Q\dot{X})^+) \qquad (2.30)$$

is a correlation function which evolves under a dynamics from which fluctuations in the variable X are projected out. If we denote

$$QLQ = L_1 \quad \text{and} \quad Q\dot{X} = X_1 \qquad (2.31)$$

we can calculate

$$F_{XX^+}(z) = \left(\frac{1}{z - iL_1} X_1, \, X_1^+\right) \tag{2.32}$$

in the same way as the correlation function (2.19). The above projection method then again leads to an equation of the form (2.23) where all quantities now have a subscript *one* and where the dot denotes a time derivative according to L_1. This leads to a new random force $F_{X_1 X_1^+}$ which can again be treated in the same manner. This process can be continued and the original F_{XX^+} is thus expressed as a continued fraction.

These formal methods have been applied to various physical problems such as density fluctuations in fluids and spin correlations in magnetic systems. For a detailed discussion we refer to Mori's original papers [15, 17, 18]. We just mention that with the help of the projection-operator technique generalized (frequency and momentum-dependent) expressions for transport coefficients can be obtained without the use of a local equilibrium ensemble which in the static limit reduce to Kubo formulae. The many-variable theory which treats all quantities which couple strongly in a collective motion, on the same footing, is especially suited for a microscopic theory of hydrodynamic phenomena. The representation of memory functions by means of continued fractions has been applied successfully to a discussion of excitations in liquids [29, 30]. The matrix form (2.28) has also been derived in Refs. [21] and [31].

3. Phonon Operators

Following the notation of Refs. [7] and [32] we summarize briefly the basic definitions of phonons operators. The system to be considered is a perfect crystal with one atom per unit cell. The displacement $\boldsymbol{u}_n$ of the n-th atom with mass m from its equilibrium position $\boldsymbol{R}_n$ and the corresponding momentum $\boldsymbol{P}_n$ are transformed according to

$$\boldsymbol{u}_n = \frac{1}{\sqrt{mN}} \sum_k \boldsymbol{e}_k A_k e^{i\boldsymbol{k}\boldsymbol{R}_n} \tag{3.1}$$

and

$$\boldsymbol{P}_n = -i\sqrt{\frac{m}{N}} \sum_k \boldsymbol{e}_k B_k e^{i\boldsymbol{k}\boldsymbol{R}_n}. \tag{3.2}$$

We abbreviate $\pm k = (\pm \boldsymbol{k}, \lambda)$ where $\boldsymbol{k}$ represents the phonon wave vector and λ the polarization index. It is convenient to work with operators A_k and B_k which are related to the usual phonon creation and annihilation operators a_k and a_k^+ by

$$A_k = A_{-k}^+ = \frac{1}{\sqrt{2\,\omega_k}} (a_k + a_{-k}^+) \tag{3.3}$$

and

$$B_k = -B_{-k}^+ = \sqrt{\frac{\omega_k}{2}} (a_k - a_{-k}^+), \tag{3.4}$$

and which obey the commutation relation

$$[A_k, B_{k'}^+] = \delta_{k,k'}. \tag{3.5}$$

We assume here that a harmonic approximation with ω_k as eigenvalues and e_k as eigenvectors of the harmonic dynamic matrix makes sense, and consider corrections by anharmonic terms to be small. This procedure excludes a discussion of quantum crystals [3] which, however, following the line of approach of Ref. [32] could also be incorporated in the subsequent analysis. We therefore start from the Hamiltonian

$$H = \sum_k \omega_k a_k^+ a_k + V_3 + V_4 \tag{3.6}$$

with

$$V_\nu = \frac{1}{\nu} \sum_{k_1,\ldots,k_\nu} V^{(\nu)}(k_1, \ldots, k_\nu) A_{k_1}, \ldots, A_{k_\nu}. \tag{3.7}$$

The anharmonic coupling parameters $V^{(\nu)}$ are connected with the ν-th derivative $\Phi^{(\nu)}$ of the potential energy by

$$V^{(\nu)}(k_1, \ldots, k_\nu) = \frac{1}{(\nu - 1)!\,(mN)^{\nu/2}} \sum_{n_1 i_1, \ldots, n_\nu i_\nu} \Phi^{(\nu)}(n_1 i_1, \ldots, n_\nu i_\nu)$$
$$\times\, e_{i_1}(k_1) \ldots e_{i_\nu}(k_\nu)\, e^{i k_1 R(n_1) + \cdots + i k_\nu R(n_\nu)}. \tag{3.8}$$

They are completely symmetric in the indices k_i and obey

$$V^{(\nu)}(-k_1, \ldots, -k_\nu) = V^{(\nu)*}(k_1, \ldots, k_\nu). \tag{3.9}$$

As the eigenvectors may be chosen real, $V^{(3)}$ is purely imaginary and $V^{(4)}$ real. From (3.8) we see that inversion symmetry implies $V^{(\nu)}(\ldots, k_i = 0, \lambda_i, \ldots) = 0$.

From (3.5) to (3.7) we obtain the equations of motion

$$i\dot{A}_k = B_k, \tag{3.10}$$

$$i\dot{B}_k = \omega_k^2 A_k + \sum_{k_1 k_2} V^{(3)}(-k, k_1, k_2) A_{k_1} A_{k_2}$$

$$+ \sum_{k_1, k_2, k_3} V^{(4)}(-k, k_1, k_2, k_3) A_{k_1} A_{k_2} A_{k_3}. \tag{3.11}$$

For later use we also give here the expression for the energy density operator h_k,

$$h_k = \sum_{q, \lambda} \omega_{q\lambda} a_{q-k/2, \lambda}^+ a_{q+k/2, \lambda}$$
$$+ \tfrac{1}{3} \sum_{q_1, q_2, q_3} V^{(3)}(q_1, q_2, q_3) A_{q_1 + k, \lambda_1} A_{q_2} A_{q_3} \tag{3.12}$$
$$+ \tfrac{1}{4} \sum_{q_1, q_2, q_3, q_4} V^{(4)}(q_1, q_2, q_3, q_4) A_{q_1 + k, \lambda_1} A_{q_2} A_{q_3} A_{q_4}.$$

There is no unique definition of the energy density operator in the literature [16, 33, 34], but the differences are of minor importance for our work as they consist in small corrections due to polarization mixing. Eq. (3.12) corresponds to the energy density of Enz [16].

It is well known that phonon interactions described by V_3 and V_4 lead to a shift in frequency, Δ, and a damping, Γ, of the phonons which are usually calculated by means of some Greens-function technique. In the next section we briefly show that the formalism presented in Sect. 2 reproduces the same results in a straightforward manner.

4. Damping and Frequency Shift

A quantity of physical relevance is the retarded Greens function $G_{A_k A_k^+}^{\mathrm{ret}}(\omega)$. It describes the linear response of the system which is subjected to a mechanical force disturbing the mean displacement. Its singularities determine the elementary excitations, the phonons.

In the subsequent analysis we shall work out approximations for this Greens function by calculating the effects of the anharmonic terms V_3 and V_4 by means of perturbation theory, and confining ourselves to terms quadratic in V_3 and linear in V_4. We neglect polarization mixing, i.e., we assume $G_k^{\mathrm{ret}}(\omega)$ to be diagonal in the polarization indices (it has been shown elsewhere [32] that the corresponding effects are of higher order in the anharmonicities).

In this section we apply the techniques outlined in Sect. 2 to the case of variables A_k and B_k reasoning that displacement and momentum are the only variables strongly coupled. Thus we have a two-variable theory with the vector

$$X = (A_k, B_k)^T. \tag{4.1}$$

First we have to evaluate various correlation functions. Using (3.5), (3.10), and (2.13) we find

$$(A_k, B_k^+) = 0\,, \quad (B_k, B_k^+) = 1\,, \quad (\dot{A}_k, B_k^+) = (\dot{B}_k, A_k^+) = -i\,. \tag{4.2}$$

Abbreviating

$$(A_k, A_k^+) = s_k^{-2} \tag{4.3}$$

(s_k^{-2} will be calculated later) we get the following matrices

$$(X, X^+) = \begin{pmatrix} s_k^{-2} & 0 \\ 0 & 1 \end{pmatrix}; \quad (\dot{X}, X^+) = \begin{pmatrix} 0 & -i \\ -i & 0 \end{pmatrix}. \tag{4.4}$$

Let us first neglect the matrix F. Then (2.24) yields

$$G^{\mathrm{ret}}(k, z) \equiv G_{XX^+}^{\mathrm{ret}}(z) = \frac{-1}{z^2 + s_k^2} \begin{pmatrix} 1 & iz \\ iz & s_k^2 \end{pmatrix}. \tag{4.5}$$

Using (2.17) we obtain undamped phonons described by δ-function resonances of $G^{\mathrm{ret}}(k, \omega)$ with a renormalized frequency s_k. The latter quantity can be calculated most easily if we replace A_k in definition (4.3) by its values occurring in the equation of motion (3.11). We can then use (2.13) and obtain

$$s_k^{-2} = \omega_k^{-2} \left\{ 1 - \sum_{k_1, k_2} V^{(3)}(-k, k_1, k_2)(A_{k_1} A_{k_2}, A_k^+) \right.$$
$$\left. - \sum_{k_1 k_2 k_3} V^{(4)}(-k, k_1, k_2, k_3)(A_{k_1} A_{k_2} A_{k_3}, A_k^+) \right\}. \tag{4.6}$$

In these higher-order correlation functions we may again apply the same trick and replace A_k^+ according to (3.11). This leads to

$$s_k^{-2} = \omega_k^{-2} \left\{ 1 + \omega_k^{-2} \sum_{\substack{k_1, k_2 \\ q_1, q_2}} V^{(3)}(-k, k_1, k_2)\, V^{(3)}(k, q_1, q_2)(A_{k_1} A_{k_2}, A_{q_1} A_{q_2}) \right. \tag{4.7}$$
$$\left. - 3\,\omega_k^{-2} \sum_{k_1, k_2} V^{(4)}(-k, k, k_1, k_2)\langle A_{k_1} A_{k_2}\rangle \right\}.$$

The remaining correlation functions now occur in combination with anharmonic terms of order $V^{(3)^2}$ and $V^{(4)}$ and may, therefore, in the sense of our perturbation theory, be evaluated in the harmonic approximation [denoted by $(\ )^0$ or $\langle\ \rangle^0$, respectively]. Defining the Bose distribution function

$$N(\omega_k) = \frac{1}{e^{\beta\omega_k} - 1} \tag{4.8}$$

we obtain

$$\langle A_{k_1} A_{k_2}\rangle^0 = \delta_{k_1,\,-k_2} \frac{1}{\omega_{k_1}} [N(\omega_{k_1}) + 1/2]. \tag{4.9}$$

$(A_{k_1} A_{k_2}, A_{q_1} A_{q_2})^0$ is obtained in the Appendix as the static value of a time-dependent correlation. Abbreviating

$$N^+(\omega_1, \omega_2) = \frac{N(\omega_1) + N(\omega_2) + 1}{\omega_1 + \omega_2}, \tag{4.10}$$

$$N^-(\omega_1, \omega_2) = \frac{N(\omega_1) - N(\omega_2)}{\omega_2 - \omega_1} \tag{4.11}$$

we can write the resulting frequency shift in the form

$$s_k^2 - \omega_k^2 = 3 \sum_{k_1} V^{(4)}(k, -k, k_1, -k_1) \frac{N(\omega_{k_1}) + 1/2}{\omega_{k_1}}$$
$$- \sum_{k_1, k_2} |V^{(3)}(-k, k_1, k_2)|^2 \frac{1}{\omega_{k_1} \omega_{k_2}} [N^+(\omega_{k_1}, \omega_{k_2}) + N^-(\omega_{k_1}, \omega_{k_2})]. \tag{4.12}$$

This result equals the static contribution $\Delta(k, \omega = 0)$ (in the notation used in Ref. [7]) of the real part of the self-energy, which is usually calculated by Greens-function techniques. Considering Eq. (2.12) this is evident as

$$s_k^2 = P \int \frac{d\omega}{2\pi} \frac{\chi(k, \omega)}{\omega} = G(k, z = 0). \tag{4.13}$$

We should like to mention that the above procedure is also best suited to get the renormalized harmonic approximation (see [2]) for quantum crystals. This has been worked out in detail by Götze and Michel [21].

In a next step we take into account the damping and the dynamic effects of shift described by $F(z)$. To this end we first have to calculate the projection $Q\dot{X} = (1 - P_X)\dot{X}$. From (4.1), (4.4), and (2.27) we get

$$Q\dot{X} = Q \begin{pmatrix} \dot{A}_k \\ \dot{B}_k \end{pmatrix} = \begin{pmatrix} 0 \\ f_k \end{pmatrix} \tag{4.14}$$

with

$$f_k = \dot{B}_k + is_k^2 A_k. \tag{4.15}$$

Equation of motion (3.11) and Eq. (4.12) then yield up to orders $V^{(4)}$ and $V^{(3)^2}$,

$$f_k = -i \sum_{k_1, k_2} V^{(3)}(-k, k_1, k_2) A_{k_1} A_{k_2}. \tag{4.16}$$

Thus the matrix F reads

$$F(z) = \begin{pmatrix} 0 & 0 \\ 0 & F_{22}(z) \end{pmatrix} \tag{4.17}$$

with

$$F_{22}(z) = \left(\frac{1}{z - iQLQ}\, f_k,\, f_k^+ \right), \tag{4.18}$$

and from (2.24) we get

$$G_{XX^+}^{\text{ret}}(k, z) = \frac{-1}{z^2 + z F_{22}(z) + s_k^2} \begin{pmatrix} 1 & iz \\ iz & s_k^2 + z F_{22}(z) \end{pmatrix}. \tag{4.19}$$

Inspection of (4.16) shows that the quantity (4.18) is proportional to $V^{(3)^2}$. It may thus be calculated by using the harmonic density matrix and the harmonic time evolution. The corresponding time-dependent correlation function

$$F_{22}^0(t) = (e^{iL_0 t} f_k,\, f_k^+)^0 \tag{4.20}$$

is evaluated in the Appendix with the result

$$\begin{aligned} F_{22}^0(z) = \sum_{k_1, k_2} |V^{(3)}(-k, k_1, k_2)|^2\, \frac{1}{2\,\omega_1 \omega_2} \\ \times \left\{ N^+(\omega_1, \omega_2)\left(\frac{1}{z + i(\omega_1 + \omega_2)} + \frac{1}{z - i(\omega_1 + \omega_2)} \right) \right. \\ \left. + N^-(\omega_1, \omega_2)\left(\frac{1}{z - i(\omega_1 - \omega_2)} + \frac{1}{z + i(\omega_1 - \omega_2)} \right) \right\}, \end{aligned} \tag{4.21}$$

where $\omega_i = \omega_{k_i}$. Performing the limit (2.17),

$$F_{22}(z = -i\omega + \varepsilon) = i \left(\delta_k(\omega) - \frac{i}{2}\, \gamma_k(\omega) \right), \tag{4.22}$$

we get expressions for the shift δ and the damping γ which are well known from results of Greens-function techniques [35, 36].

5. Hydrodynamic Singularities

The results of the previous section are adequate to describe the line shape of phonon resonances measured by inelastic neutron scattering. They fail, however, to give an account of the behavior in the hydrodynamic regime, where ω and k tend to zero.

Mathematically, this failure is seen from the fact that the term proportional to N^- in expression (4.21) for $F_{22}(z)$ becomes singular for $k \to 0$ and $z \to 0$. Since

$$V^{(3)}(-k, k_1, k_2) \propto \Delta(k - k_1 - k_2), \tag{5.1}$$

we can write

$$k_1 = q + k/2; \quad k_2 = -q + k/2 \tag{5.2}$$

making use of the momentum conservation implied by the function Δ. For $\lambda_1 = \lambda_2$ (polarization indices) we get

$$\omega_1 - \omega_2 \approx k \cdot v_q, \tag{5.3}$$

where

$$v_q = \frac{\partial \omega_q}{\partial q} \tag{5.4}$$

is the group velocity. Thus in the hydrodynamic limit, (4.21) has singular terms

$$\frac{1}{z - i\,k\,v_q} + \frac{1}{z + i\,k\,v_q} \,. \tag{5.5}$$

(For a detailed discussion see Refs. [11, 12].)

Phenomenologically, the equations of elasticity [20] lead to a pole in the expression for the phonon self-energy which is due to heat conduction. This pole cannot, however, be obtained in finite order of perturbation theory which breaks down in the hydrodynamic regime.

In recent years there have been many theoretical efforts to obtain the correct hydrodynamic behavior in a microscopic way, starting from a Hamiltonian of the form (3.6). Equilibrium methods have been used by Götze and Michel [12], Sham [11], Klein and Wehner [13], and Beck [14], whereas the non-equilibrium methods of Kwok and Martin [5], Niklasson [6], and Meier [7] concentrated on the direct derivation of a transport equation. All these works are based on complicated techniques.

In this section we show that our method described here is able to account for the pole in the phonon self-energy, which is due to heat conduction, in a simple way. Reasoning that in the limit of long wavelengths the sound waves are coupled with heat conduction, we include the energy density h_k (3.12) in the set X of our collective variables. We have now to deal with a three-variable theory with the vector

$$X = (A_k, B_k, h_k)^{\mathrm{T}}. \tag{5.6}$$

Scalar products

$$(B_k, h_k^+) = (\dot{A}_k, h_k^+) = (\dot{h}_k, h_k^+) = 0 \tag{5.7}$$

vanish by symmetry. $(\dot{B}_k, h_k^+)$ can be evaluated with the help of (2.13) and (4.9) with the result

$$(\dot{B}_k, h_k^+) = (\dot{h}_k, B_k^+) = b_k, \tag{5.8}$$

where

$$b_k = \frac{2}{3} \sum_{k_1} |V^{(3)}(-k; k_1; -k_1 + k, \lambda_1)| \frac{N(\omega_1) + 1/2}{\omega_1} \tag{5.9}$$

[according to the remarks made in conjunction with Eq. (3.9) b_k is real].

We abbreviate

$$(A_k, h_k^+) = -(h_k, A_k^+) = i\,\frac{a_k}{\omega_k^2} \tag{5.10}$$

and evaluate this expression by replacing A_k with the equation of motion (3.11). We can then use (2.13) again. The resulting commutator of B_k with the anharmonic part of the energy density yields $-i b_k/\omega_k^2$. The remaining harmonic correlation function involving h_k^0 is easily calculated (see Appendix) and we obtain

$$a_k = b_k + \sum_{k_1} |V^{(3)}(-k; k_1; -k_1 + k, \lambda_1)| N^-(\omega_{k_1}, \omega_{k_1 - k}\lambda_1). \tag{5.11}$$

Whereas the correlation functions (5.8) and (5.10) are anharmonic quantities proportional to $V^{(3)}$, the correlation function

$$(h_k, h_k^+) \equiv \varepsilon_k \tag{5.12}$$

is non-vanishing in the harmonic approximation where it takes the value

$$\varepsilon_k^0 = \sum_{k_1} \omega_1^2 \, N^- \left[\omega\left(k_1 + \frac{k}{2}, \lambda_1\right), \, \omega\left(k_1 - \frac{k}{2}, \lambda_1\right) \right]. \tag{5.13}$$

As shown in Refs. [21, 25] $\varepsilon_{k=0}$ is connected with the specific heat C:

$$\varepsilon_{k=0} = TC. \tag{5.14}$$

From (4.4), (5.7), (5.8), (5.10), and (5.12) we get the following matrices:

$$(X, X^+) = \begin{pmatrix} s_k^{-2} & 0 & i\,a_k\,\omega_k^{-2} \\ 0 & 1 & 0 \\ -i\,a_k\,\omega_k^{-2} & 0 & \varepsilon_0 \end{pmatrix} \tag{5.15}$$

and

$$(\dot{X}, X^+) = \begin{pmatrix} 0 & -i & 0 \\ -i & 0 & b_k \\ 0 & -b_k & 0 \end{pmatrix}. \tag{5.16}$$

Anticipating the fact that the only non-vanishing elements of matrix $F(z)$ are F_{22} and F_{33} [see (5.26)] we can now calculate the retarded Greens functions using (2.24) or (2.28). After some matrix algebra[1] we obtain

$$G_{AA^+}^{\mathrm{ret}}(k, z) = \cfrac{-1}{z^2 + z F_{22}(z) + s_k^2 + z \cfrac{(a_k - b_k)^2}{\varepsilon_k[z + F_{33}(z)/\varepsilon_k]}}, \tag{5.17}$$

and for the energy density Greens function

$$G_{hh^+}^{\mathrm{ret}}(k, z) = - \frac{F_{33}(z)}{z + F_{33}(z)/\varepsilon_k} + i z \sum_{\lambda} b_k \, G_{AA^+}^{\mathrm{ret}}(k, z). \tag{5.18}$$

These equations show the expected structure as discussed at the beginning of this section. In the hydrodynamic limit, F_{33} is directly connected with the thermal conductivity, and F_{22} with the viscosity (see [15, 21, 31]).

From (5.17) we see that the density response has three resonances. Besides the two phonon peaks (Brillouin doublet) there is also a central (Landau-Placzek) peak corresponding to a purely diffusive process. Whereas the intensity of the latter is usually much smaller than that of the phonon resonances it will dominate near a phase transition. Recent neutron-scattering experiments from Nb_3Sn, $SrTiO_3$, and $KMnF_3$ have indeed revealed the existence of this central peak at

---

1 Actually one has to deal with 7 by 7 matrices as the variables A_k and B_k both comprise three polarization indices. By matrix inversion, non-diagonal contributions also occur, which are, however, smaller than the diagonal ones by a factor of $V^{(3)\,2}$, consistent with the remarks made in Sect. 3. Therefore we may continue to consider only terms diagonal in the polarizations and we are again left with 3 by 3 matrices.

temperatures closely above the structural phase transition which occurs in these crystals. For a further discussion of these aspects we refer to Beck and Meier [37].

The main contribution to the energy density correlation stems from the first term in (5.18) and corresponds to a resonance of a purely diffusive process (heat conduction). The second term describes coupling to the displacement correlations through b_k and vanishes if only the harmonic part of the energy density h_k (3.12) is retained, as seen from (5.8). In this context we also note that in the anharmonic term $(a_k - b_k)^2$ occurring in (5.17) the effects of the anharmonic part of h_k cancel.

The hydrodynamic limit of (5.17) and (5.18), and relation of the various forms with thermodynamic quantities have been discussed thoroughly by Götze and Michel [19, 21].

We have already made use of the form of matrix F. To calculate the matrix elements we first have to evaluate the projected variables $Q\dot{X}$. From (2.27), (5.6), (5.15), and (5.16) we obtain

$$Q\dot{X} = (0, f_k, g_k)^T \tag{5.19}$$

with

$$f_k \equiv \dot{B}_k + i\omega_k^2 A_k - \frac{b_k - a_k}{\varepsilon_k} h_k \tag{5.20}$$

and

$$g_k \equiv \dot{h}_k - \sum_\lambda b_{k,\lambda} B_{k,\lambda} . \tag{5.21}$$

Inspection of (4.15) and (4.16) shows that in the three-variable theory at hand we have an additional term in f_k. If $F_{22}(z)$ is calculated in the same way as in Sect. 4 the effect of this term is to cancel the singular contributions (5.5) in the limit $k \to 0$ and $z \to 0$. A treatment more appropriate in the hydrodynamic limit is to use the continued-fraction method mentioned in Sect. 2.

In the hydrodynamic limit the energy density satisfies a local conservation law of the form

$$\dot{h}_k = i\,\boldsymbol{k} \cdot \boldsymbol{S} \tag{5.22}$$

with $\boldsymbol{S}$ being the energy current operator (see Hardy [34]). We now neglect in (5.21) coupling to the sound waves described by the term bB which is due to the anharmonic part of h_k (consequently $\dot{Q}h_k = \dot{h}_k$). Inspection of (2.30) then shows that the corresponding expression for F_{33} is

$$F_{33}(z = \eta \to 0) = k_i\,k_j \int_0^\infty dt\,e^{-\eta t}(S^i(t), S^j) . \tag{5.23}$$

Regarding the definition (2.2) of the scalar product, the integral in (5.23) corresponds to the Kubo formula [38] for the thermal conductivity tensor K^{ij} (up to a factor T). Using (5.14) we get therefore for the term $F_{33}(z)/\varepsilon_0$ entering (5.17) in the corresponding limits,

$$F_{33}/\varepsilon = (1/C)\,k^i\,k^j\,K^{ij} . \tag{5.24}$$

This expression is to be expected from the phenomenological equations of elasticity which also relate F_{22} to the viscosity. As regards the additional term in (5.21), which we have neglected, we refer to Kadanoff and Martin [39] who discuss equivalent expressions for transport coefficients and show that in the limits considered the term in question does not contribute.

Deo and Behera [40] have evaluated (5.23) assuming a harmonic expression for S and using a simple Greens-function technique to calculate the correlation function which corresponds to a solution of Peierls' equation for thermal conductivity by a relaxation time. Hardy [41] used a master equation and discussed also the influences of various correction terms to S. In our formalism we can follow Mori [18] and apply the continued-fraction method. According to (2.30) to (2.32) one has to consider the two variables

$$X_1 = (f_k, g_k)_T \tag{5.25}$$

and evaluate, as before, the corresponding matrices of correlation functions. From (5.20), (5.21), (5.7), and (5.8) it follows that $(f_k, g_k^+) = 0$. Hence we now see from (5.19) that the matrix F_{XX^+} has indeed the structure assumed previously, i.e.,

$$F_{XX^+}(z) = \begin{pmatrix} 0 & 0 & 0 \\ 0 & F_{22}(z) & 0 \\ 0 & 0 & F_{33}(z) \end{pmatrix}. \tag{5.26}$$

Eq. (2.23) for

$$\Phi_{X_1 X_1}(z) = \left(\frac{1}{z - iL_1} X_1, X_1^+ \right) \tag{5.27}$$

then allows the writing down of explicit expressions for F_{22} and F_{33}, since the new random forces $F_{X_1 X_1^+}$ can, in the sense of perturbation theory, be expressed in terms of the anharmonic coefficients $V^{(3)}$. The corresponding expressions for the thermal conductivity and the viscosity have been calculated by Götze and Michel [21].

Finally, we should like to mention that in the high-frequency domain the three-variable theory considered in this section reduces to the expressions of the previous section. The denominator of (5.17) then simplifies to

$$z^2 + z F_{22}(z) + s_k^2 + (a_k - b_k)^2 / \varepsilon_k.$$

The last term just compensates the difference between F_{22} obtained from (5.20) and F_{22} in (4.19) resulting from (4.15).

6. Second Sound

Eq. (5.18) exhibits that the response of the system to a variation in temperature will result in a diffusive process. The heat is transported by thermal conduction. It is known that for perfect crystals at low temperatures, where Umklapp processes are very rare, a propagating heat mode, called second sound, may be excited.

In this section it is shown that we can account for this propagating temperature mode by including the operator $\dot{h}_k$ into the set of our collective variables. We shall thus investigate the retarded Greens functions for the displacement and the energy density correlations by the use of a four-variable theory according to the vector

$$X = (A_k, B_k, h_k, \dot{h}_k)^T. \tag{6.1}$$

In addition to the correlation functions already known we have to evaluate

$$(\dot{A}_k, \dot{h}_k^+) = (\ddot{h}_k, A_k^+) = i b_k \tag{6.2}$$

which results from (3.10) and (5.8), with the help of (2.7) and (2.13). The quantities

$$(\dot{B}_k, \dot{h}_k^+) = (\ddot{h}_k, B_k^+) = (\ddot{h}_k, \dot{h}_k^+) = 0 \tag{6.3}$$

vanish by symmetry. If we abbreviate

$$(\dot{h}_k, \dot{h}_k^+) = -(\ddot{h}_k, h_k^+) \equiv \varepsilon_k t_k^2 \tag{6.4}$$

we can write down the matrices

$$(X, X^+) = \begin{pmatrix} s_k^{-2} & 0 & i\,a_k\,\omega_k^{-2} & 0 \\ 0 & 1 & 0 & -b_k \\ -i\,a_k\,\omega_k^{-2} & 0 & \varepsilon_k & 0 \\ 0 & -b_k & 0 & \varepsilon_k t_k^2 \end{pmatrix} \tag{6.5}$$

and

$$(\dot{X}, X^+) = \begin{pmatrix} 0 & -i & 0 & i\,b_k \\ -i & 0 & b_k & 0 \\ 0 & -b_k & 0 & \varepsilon_k t_k^2 \\ i\,b_k & 0 & -\varepsilon_k t_k^2 & 0 \end{pmatrix}. \tag{6.6}$$

For a first discussion we again neglect random forces F_{XX^+}. Inserting (6.5) and (6.6) into (2.28) the relevant retarded Greens functions can be calculated by matrix algebra. This leads to the following expressions,

$$G_{AA^+}^{\text{ret}}(k, z) = -\frac{z^2 + \bar{t}_k^2 + (a_k - b_k)\,b_k/\varepsilon_k}{(z^2 + \bar{s}_k^2)(z^2 + \bar{t}_k^2) + (a_k - b_k)(b_k \bar{s}_k^2 - a_k \bar{t}_k^2)/\varepsilon_k}, \tag{6.7}$$

$$G_{hh^+}^{\text{ret}}(k, z) = -\frac{\varepsilon_k t_k^2(z^2 + \bar{s}_k^2) + z^2 b_k(a_k - b_k)/\varepsilon_k}{(z^2 + \bar{s}_k^2)(z^2 + \bar{t}_k^2) + (a_k - b_k)(b_k \bar{s}_k^2 - a_k \bar{t}_k^2)/\varepsilon_k}, \tag{6.8}$$

where we abbreviated the renormalized quantities

$$\bar{s}_k^2 = s_k^2 + a_k(a_k - b_k)/\varepsilon_k \tag{6.9}$$

and

$$\bar{t}_k^2 = t_k^2 + \sum_\lambda a_k(a_k t_k/s_k - b_k)/\varepsilon_k. \tag{6.10}$$

The Greens functions now have two pairs of poles, one corresponding to sound propagating with frequency $\bar{s}_k$ and a new mode propagating with frequency $\bar{t}_k$. The weights of these poles, however, are different. For the displacement response function (6.7) the residue of the first-sound pole is one, whereas the second-sound can only be excited with a weight proportional to cubic anharmonicities squared. This behavior is reversed in the energy density response (6.8) where the poles at $z = \pm i\bar{t}_k$ dominate and the first-sound poles have much smaller residues.

These results have already been discussed by Kwok and Martin [5] and by Sham [25] who obtained the same equations by means of an approximate solution of the Boltzman equation. The fact that both first and second sound appear as resonances of the same correlation function is due to coupling of the displacement and the phonon fields. These aspects have been discussed by Meier [7] and, extensively, by Beck [14].

There is, however, an important difference in our results (6.7) to (6.10) compared to those of Refs. [5, 7, 14, 25]. We can calculate (6.4) by virtue of (2.13) and the time derivative of h_k^+. The result has a non vanishing harmonic part given by

$$\varepsilon_k t_k^2 = \sum_{k_1} \omega_{k_1}^2 [\omega(k_1 + k/2, \lambda_1) - \omega(k_1 - k/2, \lambda_1)]^2$$
$$\times N^-[\omega(k_1 + k/2, \lambda_1), \omega(k_1 - k/2, \lambda_1)]. \tag{6.11}$$

The expansion for small wave vectors yields

$$\varepsilon_0 t_k^2 = k_i k_j \sum_{k_1} \omega_{k_1}^2 v_{k_1}^i v_{k_1}^j N^-(\omega_{k_1}, \omega_{k_1}). \tag{6.12}$$

For a Debye model where $v_k^i = C_\lambda k_i/k$ this expression can be easily evaluated. Upon comparison with (5.13) we see that the sum over the momenta gives $1/3\,\delta_{ij}\,\varepsilon_0$ up to the summation over polarization indices. Thus we obtain

$$t_k^2 = C_{II}'^2 k^2 \tag{6.13}$$

with

$$C_{II}'^2 = \tfrac{1}{3} \sum_\lambda C_\lambda^{-1} / \sum_\lambda C_\lambda^{-3} \tag{6.14}$$

being the velocity of driftless second sound [16] which is different from

$$C_{II}^2 = \tfrac{1}{3} \sum_\lambda C_\lambda^{-3} / \sum_\lambda C_\lambda^{-5} \tag{6.15}$$

the corresponding velocity of drifting second sound [10]. Before we comment further on this difference we first discuss the damping terms described by F. To this end we have to calculate the projected values $Q_X \dot{X} = (1 - P_X)\dot{X}$. From (2.27), (6.5), and (6.6) we get

$$Q\dot{X} = (0, f_k, 0, l_k)^T \tag{6.16}$$

with f_k given by (5.20) and with

$$l_k = Q\ddot{h}_k = \ddot{h}_k + t_k^2 h_k - i \sum_\lambda (b_k s_k^2 - a_k t_k^2) A_k. \tag{6.17}$$

As $(f_k, l_k^+) = 0$ the only non-vanishing matrix elements of F_{XX^+} are F_{22} and F_{44}. If these are taken into account in the evaluation of (2.28) we have to modify expressions (6.7) and 6.8) according to

$$z^2 + \bar{s}_k^2 \rightarrow z^2 + z F_{22}(z) + \bar{s}_k^2 \tag{6.18}$$

and

$$z^2 + \bar{t}_k^2 \rightarrow z^2 + z F_{44}(z)/\varepsilon_k t_k^2 + \bar{t}_k^2. \tag{6.19}$$

F_{22} gives the damping of first sound as in the previous sections, whereas F_{44} is responsible for the damping of second sound. These expressions again agree with those of other authors [5, 25].

The fact that our treatment with the four variables (6.1) yields driftless second sound is understandable with regard to the derivations of Griffin [26] and Enz [16, 27], who also obtain a mode with velocity (6.14) by considering the energy density correlation without recourse to a momentum conservation law. In the derivation of drifting second sound, however, one uses the local conservation laws for energy and momentum density, respectively. In a Debye-model approximation the heat current S is then proportional to the momentum density P:

$$S = 3 \cdot C_{\mathrm{II}}^2 \, P \tag{6.20}$$

and the velocity (6.15) enters through this relation. If we assume an expression of the (harmonic) momentum density operator p to be given by

$$\boldsymbol{p_k} = \sum_{k_1} \boldsymbol{k_1} \, a^+_{\boldsymbol{k_1} - k/2, \lambda_1} a_{\boldsymbol{k_1} + k/2, \lambda_1} \tag{6.21}$$

we can investigate this coupling of energy density h_k and momentum density p_k simply by considering

$$X = (A_k, B_k, h_k, \boldsymbol{p_k})^{\mathrm{T}}. \tag{6.22}$$

We neglect, for simplicity of the argument, coupling to first sound as well as damping. In a Debye model we get in the limit of long wave-lengths

$$(p_{\boldsymbol{k}}^\alpha, p_{\boldsymbol{k}}^{\beta^+}) = \delta_{\alpha, \beta} \, \varepsilon_0 \, \tfrac{1}{3} \sum_\lambda C_\lambda^{-5} \Big/ \sum_\lambda C_\lambda^{-3} \,, \tag{6.23}$$

and

$$(\dot{h}_{\boldsymbol{k}}, p_{\boldsymbol{k}}^{\beta^+}) = - \, i \, \tfrac{1}{3} \, \varepsilon_0 \, k_\beta. \tag{6.24}$$

Using the same technique as above one can calculate the retarded Greens function for the energy density with the result

$$G_{hh^+}^{\mathrm{ret}}(\boldsymbol{k}, z) = - \, \frac{\varepsilon_0 \, k^2 \, C_{\mathrm{II}}^2}{z^2 + C_{\mathrm{II}}^2 \, k^2} \,. \tag{6.25}$$

Thus by coupling energy and momentum densities we get an excitation with the velocity of drifting second sound. The weak point in this discussion is, however, the form (6.21) for the pseudo-momentum density operator which is not at all obvious. It corresponds to a summation over the pseudo-momenta times a Wigner-density operator. In contrast to the energy density (3.12) there exists no microscopic foundation of a pseudo-momentum density and (6.21) cannot be extended to include anharmonic corrections.

It seems possible, however, to investigate these two second-sound modes further on the basis of the theory developed here. A detailed calculation of the damping of driftless second sound and its relation to the drifting mode will be discussed in a separate publication.

Acknowledgements. The author would like to express his gratitude to Dr. T. Schneider for introducing him to the Mori theory of cooperative phenomena. He also thanks Prof. Dr. A. Thellung for valuable discussions and Prof. Dr. G. Scharf for critical reading of the manuscript.

Appendix

Evaluation of Harmonic Correlation Functions

In the framework of our perturbation theory, correlation functions can be calculated by transformation to a form involving anharmonic coupling parameters

$V^{(3)}$ and $V^{(4)}$. This is achieved by successive use of equations of motion (3.10) and (3.11). All correlation functions occurring in combination with $V^{(3)^2}$ and $V^{(4)}$ may be evaluated in the harmonic approximation. This can be done by some Greens-function technique in conjunction with (2.12) or by direct calculation using definition (2.2). Since the operators evolve according to

$$A_k(t-i\lambda) = \frac{1}{\sqrt{2\,\omega_k}}\,(e^{-i\omega_k(t-i\lambda)}a_k + e^{i\omega_k(t-i\lambda)}a^+_{-k})\,, \qquad (A.1)$$

the trace can be easily evaluated.

For convenience we write down the results for the functions used in Sects. 4 and 5. From

$$\langle A_{k_1} A_{k_2}\rangle^0 = \delta_{k_1,-k_2}\,\frac{1}{\omega_{k_1}}\,[N(\omega_{k_1}) + 1/2] \qquad (A.2)$$

we get

$$(a_1(t)\,a^+_{-2}(t), A^+_3 A^+_4)^0 = \frac{e^{-i(\omega_1-\omega_2)t}}{2\sqrt{\omega_1\omega_2}}\,(\delta_{1,3}\delta_{2,4} + \delta_{1,4}\delta_{2,3})\,N^-(\omega_1,\omega_2) \qquad (A.3)$$

and

$$(a_1(t)\,a_2(t), A^+_3 A^+_4)^0 = \frac{e^{-i(\omega_1+\omega_2)t}}{2\sqrt{\omega_1\omega_2}}\,(\delta_{1,3}\delta_{2,4} + \delta_{1,4}\delta_{2,3})\,N^+(\omega_1,\omega_2)\,, \qquad (A.4)$$

where the definitions (4.10) and (4.11) have been used. Combining (A.3), (A.4), and the corresponding equations for the complex conjugate quantities we obtain

$$\begin{aligned}
(A_1(t)\,A_2(t), A^+_3 A^+_4)^0 = \frac{1}{4\,\omega_1\omega_2}\,&(\delta_{1,3}\delta_{2,4} + \delta_{1,4}\delta_{2,3})\\
\times\,\{&N^+(\omega_1,\omega_2)\,(e^{+i(\omega_1+\omega_2)t} + e^{-i(\omega_1+\omega_2)t})\\
+\,&N^-(\omega_1,\omega_2)\,(e^{+i(\omega_1-\omega_2)t} + e^{-i(\omega_1-\omega_2)t})\}
\end{aligned} \qquad (A.5)$$

which leads to (4.21). The static limit of (A.5) was used in (4.12) to obtain the renormalized frequency s_k.

References

1. See, for example, the review article by Kwok, P. C.: Solid State Physics **20**, 214 (1967)
2. Choquard, Ph.: The anharmonic crystal. New York: Benjamin 1967
3. Guyer, R. A.: Solid State Physics **23**, 413 (1969)
4. Kadanoff, L. P., Baym, G.: Quantum statistical mechanics. New York: Benjamin 1962
5. Kwok, P. C., Martin, P. C.: Phys. Rev. **142**, 495 (1966)
6. Niklasson, G.: Fortschr. Phys. **17**, 235 (1969)
7. Meier, P. F.: Phys. kondens. Materie **8**, 241 (1969)
8. Beck, H., Meier, P. F.: Z. Phys. **247**, 189 (1971)
9. Peierls, R. E.: Quantum theory of solids. London: Oxford Univ. Press 1955
10. Sussmann, J. A., Thellung, A.: Proc. Phys. Soc. **81**, 1122 (1963)
11. Sham, L. J.: Phys. Rev. **156**, 494 (1967)
12. Götze, W., Michel, K. H.: Phys. Rev. **156**, 963 (1967); **157**, 738 (1967)
13. Klein, R., Wehner, R. K.: Phys. kondens. Materie **8**, 141 (1968); **10**, 1 (1969)
14. Beck, H.: Phys. kondens. Materie **12**, 330 (1971)
15. Mori, H.: Progr. theor. Phys. (Kyoto) **28**, 763 (1962)
16. Enz, C. P.: Ann. Phys. (N.Y.) **46**, 114 (1968)
17. Mori, H.: Progr. theor. Phys. (Kyoto) **33**, 423 (1965)

18. Mori, H.: Progr. theor. Phys. (Kyoto) **34**, 399 (1965)
19. Götze, W., Michel, K. H.: Z. Phys. **223**, 199 (1969)
20. Landau, L. D., Lifshitz, E. M.: Theory of elasticity. London: Pergamon Press 1959
21. Götze, W., Michel, K. H.: In: Lattice dynamics. Eds.: Maradudin, A. A., Horton, G. K., North-Holland (to appear)
22. Ackerman, C. C., Bertman, B., Fairbank, H. A., Guyer, R. A.: Phys. Rev. Letters **16**, 789 (1966)
23. Jackson, H. E., Walker, C. T., McNelly, T. F.: Phys. Rev. Letters **25**, 26 (1970)
24. Narayanamurti, V., Dynes, R. C.: Phys. Rev. Letters **28**, 1461 (1972)
25. Sham, L. J.: Phys. Rev. **163**, 401 (1969)
26. Griffin, A.: Phys. Letters **17**, 208 (1965)
27. Enz, C. P.: Phys. Letters **20**, 442 (1966)
28. Schneider, T., Meier, P. F.: To be published
29. Sears, V. F.: Can. J. Phys. **47**, 199 (1969); **48**, 616 (1970)
30. Lovesey, S. W.: J. Phys. C **4**, 3057 (1971)
31. Murase, C.: J. Phys. Soc. Japan **29**, 549 (1970)
32. Beck, H., Meier, P. F.: Phys. kondens. Materie **12**, 16 (1970)
33. Choquard, Ph.: Helv. Phys. Acta **36**, 415 (1963)
34. Hardy, R. J.: Phys. Rev. **132**, 168 (1963)
35. Kashcheev, V. N., Krivoglaz, M. A.: Fiz. Tverd. Tela **3**, 1528 (1961) [Soviet Physics Solid State **3**, 1107 (1961)]
36. Maradudin, A. A., Fein, A. E.: Phys. Rev. **128**, 2589 (1962)
37. Beck, H., Meier, P. F.: To be published
38. Kubo, R.: J. Phys. Soc. Japan **12**, 570 (1957)
39. Kadanoff, L. P., Martin, P. C.: Ann. Phys. (N.Y.) **24**, 419 (1963) (in particular Sec. G)
40. Deo, B., Behera, S. N.: Phys. Rev. **141**, 738 (1966)
41. Hardy, R. J.: J. Math. Phys. **6**, 1749 (1965)

Dr. P. F. Meier
IBM Zürich
Research Laboratory
CH-8803 Rüschlikon, Switzerland

Phys. cond. Matter 17, 37—53 (1973)
© by Springer-Verlag 1973

Specific Heat of Nickel-Iron and Nickel-Cobalt Alloys between 600° K and 1500° K

John Orehotsky* and Klaus Schröder
Department of Chemical Engineering and Materials Science
Syracuse University

Received November 27, 1972 / In Revised Form March 5, 1973

The λ-specific heat anomalies of ferromagnetic fcc Ni-Co and Ni-Fe alloys were investigated near the Curie temperatures. The magnetic contribution to the specific heat was found to be logarithmically divergent $[c_M = \log |T/T_c - 1|^A]$ above the Curie temperature. The exponent A was composition dependent. Entropy and energy values associated with magnetic transitons were determined experimentally and compared with theoretical predictions of the Heisenberg and Ising localized electron models obtained from series expansion calculations.

Introduction

The specific heat of a large number of ferromagnetic elements and compounds has been measured over extended temperature ranges. Of particular interest is the region near the critical temperature, T_c, where the magnetic contribution $c_M(T)$ to the specific heat has been determined and its temperature dependence has been compared with existing theories. However, an extensive investigation of this type has not been made on the disordered substitional alloy systems involving iron, nickel and cobalt as the components. This is due to both experimental difficulties since changes of atomic ordering by diffusion may effect the magnetic specific heat contribution and due to theoretical difficulties since disordered alloys lack the symmetry of pure elements.

The molecular field theory was the first theory which allowed the calculation of magnetic specific heats, c_M. One assumes in this model that the magnetic interaction of all the other atoms of the sample on one individual atom is given by a molecular field which is proportional to the magnetization of the sample. This theory predicts a sharp drop of c_M at T_c, and zero c_M values above T_c. The drop of c_M at T_c is equal to

$$\Delta c_M = \frac{5j(j+1)}{j^2 + (j+1)^2}\, n_0\, R,$$

where n_0 is the number of Bohr magnetons per atom, and j the total angular momentum quantum number [1]. This theory can be applied to alloys. The results of the Bethe-Peierls-Weiss theory, a modification of this approach, predicts also a sharp λ-peak at T_c. c_M drops continuously at T_c, but non-zero c_M values exist above T_c. Unfortunately neither theory agrees well with experimental findings.

---

* Present address: Wilkes College, Department of Engineering, Wilkes Barre, Pennsylvania.

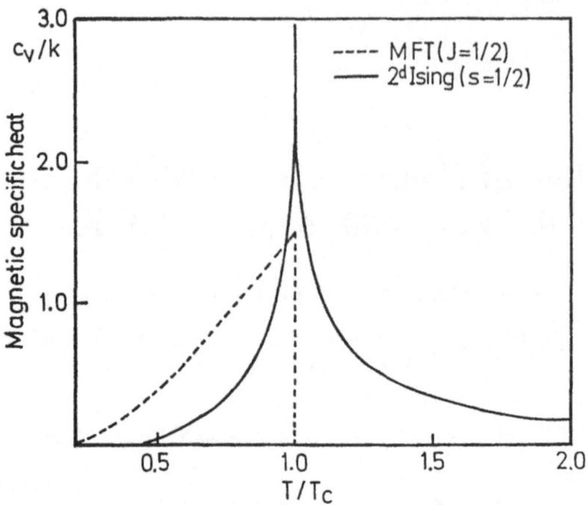

Fig. 1. Magnetic specific heat as derived from the Weiss molecular field theory (MFT) for $J = 1/2$ and for the 2-dimensional Ising model with $s = 1/2$

More recent calculations of the magnetic specific heat are based on the Ising or Heisenberg localized electron model. The Onsager solution of the two-dimensional Ising model gives a divergency in c_M at T_c, but c_M will not change discontinuously (see Fig. 1). Analytical solutions to the three-dimensional Ising or Heisenberg model are not available, but series expansion techniques make it possible to calculate the temperature dependence of c_M near T_c for the Ising and Heisenberg models. These two theories are most appropriate for describing magnetic specific heat behavior in electrically insulating materials and the behavior of several materials closely approximates the predictions of the theory.

Attempts were made by Stoner [2] and Wohlfahrt [3] to determine magnetic properties of metals with a collective (itinerant) electron model. Recently Katsuki [4] calculated with this model the specific heat as a function of temperature not only for pure elements but also for alloys using density of states curves determined from liquid He specific heat measurements. However, the model contains adjustable parameters which are not always known. Therefore, one frequently calculates both the magnetic energy and the temperature dependence of $c_M(T)$ from the localized electron model of ferromagnetic insulators and applies the results to metals.

These models might give a reasonable description of magnetism in transition metals and alloys due to the nearly localized character of the magnetic d-electrons. However, the conduction electrons will be partly polarized by the field of the d-electrons and this should yield experimental results that differ somewhat from the predictions of the insulator model. As a result, the magnetic properties of many pure elements appear to have both localized and itinerant electron characteristics.

Nickel is usually classified as an itinerant electron ferromagnet. This means that non-localized d-electrons, and partly polarized s-electrons are responsible for ferromagnetic ordering below a critical temperature. The Stoner itinerant electron model would explain the slight curvature in the $1/\chi$ versus temperature curve in this element. However, the susceptibility near the Curie temperature (λ-anomaly) is also in agreement with the predictions of the localized (an highly unrealistic) Ising model with the quantum number of $S = 1/2$ [5]. An experimental confirmation of the scaling laws has been demonstrated from experimental data on nickel [6] which again suggests localized electron characteristics. Iron displays a sizeable specific heat λ-peak. This transition is a second order transformation. The magnetic energy and entropy values associated with the peak along with the temperature dependence of the susceptibility [7] suggests that iron can be described with a Heisenberg model. It is therefore assumed that most of the magnetization in iron is due to localized (non-conducting) d-electrons of Heisenberg model character with the conduction electrons partially polarized. Cobalt again shows both itinerant and localized electron features. It may be closer to nickel in its properties, where the itinerant character gives usually a better description, than iron, where only a small fraction of the ferromagnetic ordering should be due to itinerant conduction electrons. The election of the Heisenberg or Ising model is to a certain degree arbitrary, since each model is frequently able to duplicate features of the other model [8].

Since a unified model of magnetism in the iron group metals is not available, a specific heat study of magnetic alloys involving iron, nickel and cobalt might yield valuable additional information on magnetism in the $3d$ transition metals. Only a few high temperature specific heat measurements on alloys are presently available. Measurements in the neighborhood of the Curie temperature demonstrated that small additions of copper impurities to nickel broaden the λ-peak considerably [9]. The peak disappears entirely when the alloy contains more than 25% copper even though alloys having up to 50% copper are ferromagnetic and should display anomalous behavior at their Curie temperatures. The broadening and disappearance of the peak could be attributed to the presence of non-magnetic impurities in a magnetic host lattice which decreases the effective long range correlation necessary for critical behavior. Another possible explanation is that compositional inhomogenities exist in the alloy resulting in a distribution of Curie temperatures and a broadened λ-peak. This argument [10] has been advanced to explain magnetic specific heats in PdCo and PdFe alloys. The rather rounded λ-peak in these alloys could be quantitatively accounted for by assuming a Gaussian distribution of the chemical composition of small volume elements even if each volume element has a sharp λ-peak at its Curie temperature. This model does not appear to be applicable to the nickel rich Ni-Cu alloys. There it is difficult to envision small magnetic volume elements acting independently of each other since in e.g. the $Ni_{.75}Cu_{.25}$ alloy each Ni-atom has in the average 8 nickel atoms as nearest neighbors. Large non-magnetic or only weakly magnetic boundary sections in between nickel rich volume elements are unlikely. A sharp λ-function has been found for FeCr alloys with up to 10 at % chromium [11], indicating that compositional fluctuations in an alloy with non-ferromagnetic impurities in a ferromagnetic matrix with localized electrons would exhibit a similar specific heat as the pure element.

To determine whether compositional fluctuations in microscopic volume elements or a break-up of spin alignments between a nickel atom and a non-magnetic impurity atom are responsible for the rounding of the λ-peak in the specific heat of NiCu alloys, the specific heat of nickel alloys with iron or cobalt elements as a second component, was investigated. Both elements are ferromagnetic in their pure state and most likely have some localized moments, which should magnetically couple with the nickel matrix. Therefore, if the λ-peak in the specific heat of disordered nickel base Ni-Co and Ni-Fe is sharp then the rounded specific heat curves of NiCu alloys cannot be due to fluctuations of the chemical composition in small volume elements.

Experimental Technique

Samples were prepared in an arc-melter similar in design to that of Nielson and Johns [12]. The suppliers and the purity of the metallic elements used for making the alloys were: Nickel 99.9% Whitehead Metals, Inc., Iron 99.9% Westinghouse "Purion", and Cobalt 99.9% Electronics Space Products.

The appropriate amounts of the elements were mixed and placed into a water cooled copper hearth under a vacuum jar. The system was evacuated to a pressure of about 100 microns and then flushed with argon, re-evacuated to 100 microns, and reflushed. This procedure was repeated several times before sealing off the pump and admitting the 3 to 5 psi pressure of argon necessary to maintain a suitable arc for melting. Once the charge was completely molten, the arc was passed back and forth over the length of the liquid alloy for about one minute. This provided a considerable amount of agitation in the liquid for obtaining compositional homogeneity in the ingot.

After cooling, the sample was turned over and melted from the backside without opening the system to air. To insure homogeneity, this procedure was repeated several times. The material loss due to melting was about 0.02% of the original weight. These samples should be disordered, since the cooling rate of the sample in the water cooled copper hearth after melting is very high. The samples were not reannealed and quenched, since they had to be reheated for several minutes to $1500-1800\,°\mathrm{K}$ before they cooled in the calorimeter for the c_p measurements.

The essential features of the "heat exchange" calorimeter and the analytic technique to evaluate the measurements has been described by Orehotsky and Schröder [13]. In this calorimeter, the hot sample was inserted into a copper can. The heat loss of the sample, $\dot{Q}_s = C_s \dot{T}_s$ was equal to the heat absorption of the copper can, $C_c \cdot T_c$ corrected for heat losses. Since C_c, T_c and T_s are known, C_s can be calculated.

Orehotsky and Schröder [13] demonstrated the accuracy ($\pm 2\%$) and reliability of this calorimeter by determining specific heats of iron in the temperature range $600\,°\mathrm{K}$ to $1300\,°\mathrm{K}$. They found excellent agreement with published results to Hultgren [14], Braun, Kohlhaas and Vollmer [15] and Pallister [16] but lower values than the results of Lyman [17] and Schröder and MacInnes [18]. Experiments on an $\mathrm{Fe_{90}Al_{10}}$ alloy showed that changes in short range order during heating, cooling, or short time annealing should not effect these measurements [19]. Five down quenching runs were made on this alloy specimen where each run was quenched from a different temperature: $1425\,°\mathrm{K}$, $1225\,°\mathrm{K}$, $1100\,°\mathrm{K}$, $995\,°\mathrm{K}$ and $875\,°\mathrm{K}$. The measured $c_p(T)$ values for each run coincided with the overlapping experimental points of the other runs showing that no changes in short range ordering influenced the measurements.

Results and Discussion

1. Specific Heat Values

The experimentally determined specific heat values of our samples are shown in Figs. 2 to 4. Fig. 2 summarizes the specific heat data for the nickel-cobalt alloys. It shows that the general features of the c_p-temperature plots are similar for all samples. A λ-peak is found for all alloys.

Figs. 3 and 4 give the specific heat of the NiCo and NiFe alloys in more detail. The curves show rounding near the Curie temperature. This rounding extends over

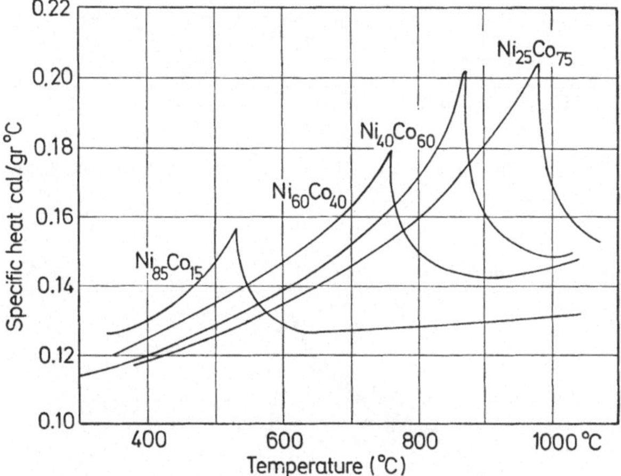

Fig. 2. Total specific heat of cobalt-nickel alloys

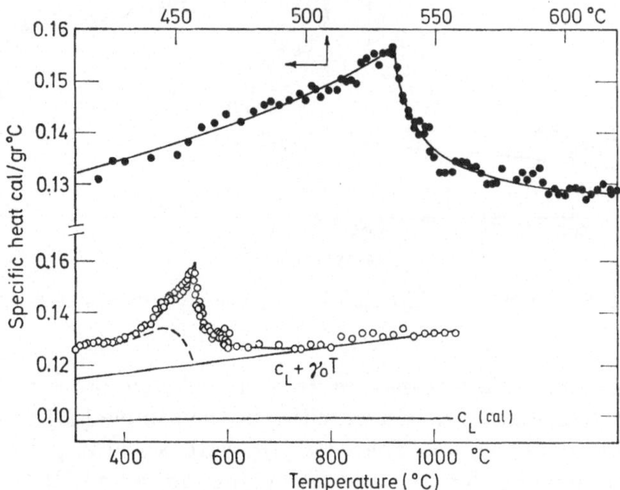

Fig. 3a. Specific heat of $Co_{15}Ni_{85}$ (dashed line gives $c_L + c_E$, if $A^+ = A^-$). Top line gives details near T_c

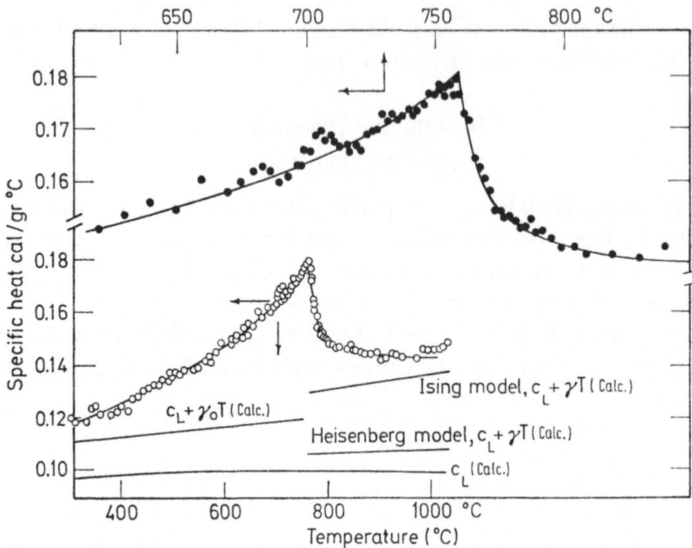

Fig. 3b. Specific heat of $Co_{40}Ni_{60}$. Top line gives details near T_c

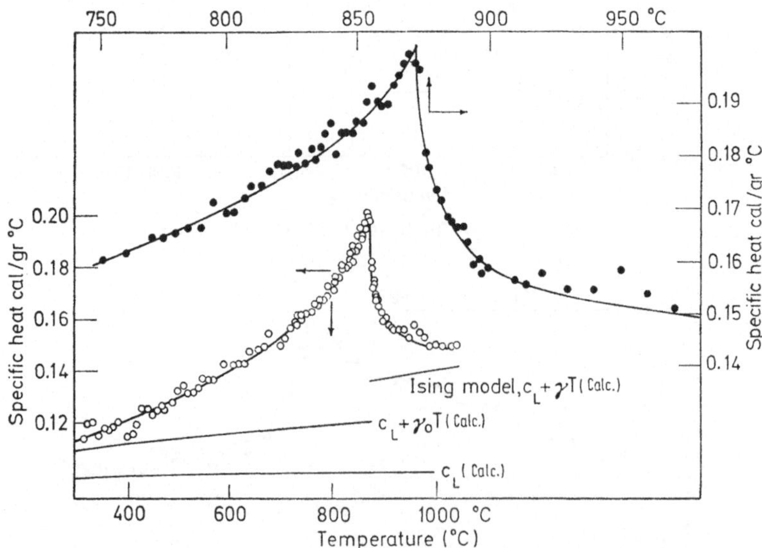

Fig. 3c. Specific heat of $Co_{60}Ni_{40}$. Top line gives details near T_c

a temperature range of a few degrees. It is of the same order of magnitude as expected from temperature inhomogeneities in the sample [13]. High resolution experiments by MacInnes and Schröder [18] with a "heat pulse" calorimeter showed similar rounding for iron and iron-chromium alloys. It seems therefore likely that higher resolution measurements would not lead to noticeably different specific heat data in these alloys.

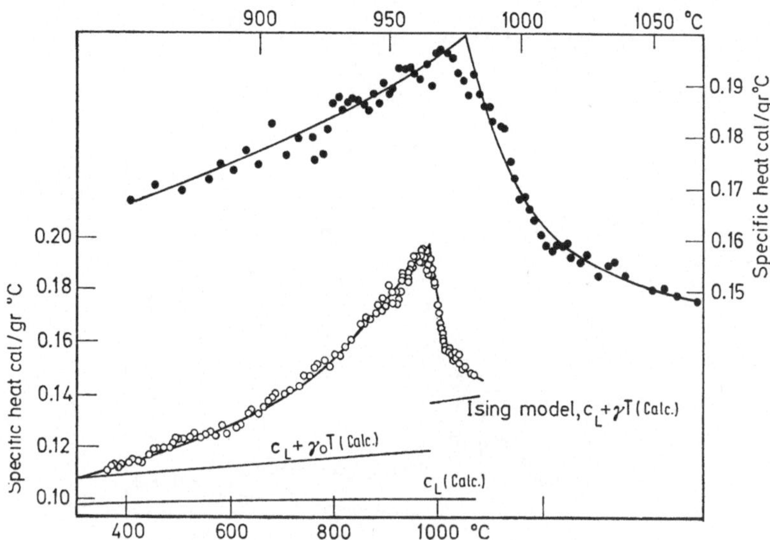

Fig. 3d. Specific heat of $Co_{75}Ni_{25}$. Top line gives details near T_c

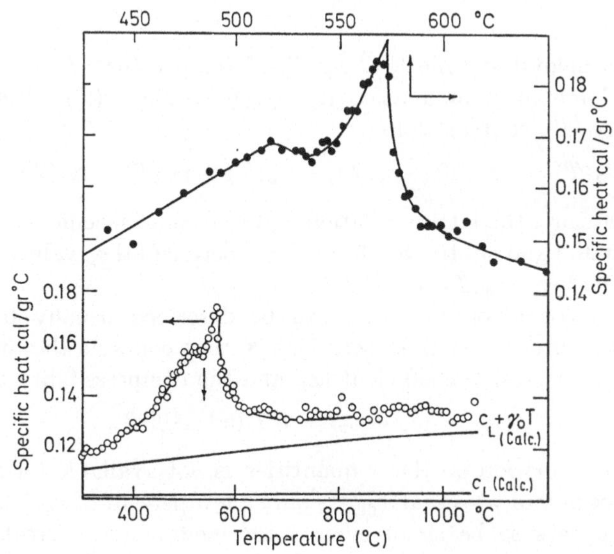

Fig. 4a. Specific heat of $Fe_{22}Ni_{78}$. Top line gives details near T_c

The data can be analyzed in the following way. If the magnetic ordering contribution to the specific heat $c_M(T)$ can be separated from the other contributions then both the temperature dependence of $c_M(T)$ and the energies and entropies involved in the magnetic transformation can be determined and compared with the predictions of the localized electron models. At high temperatures,

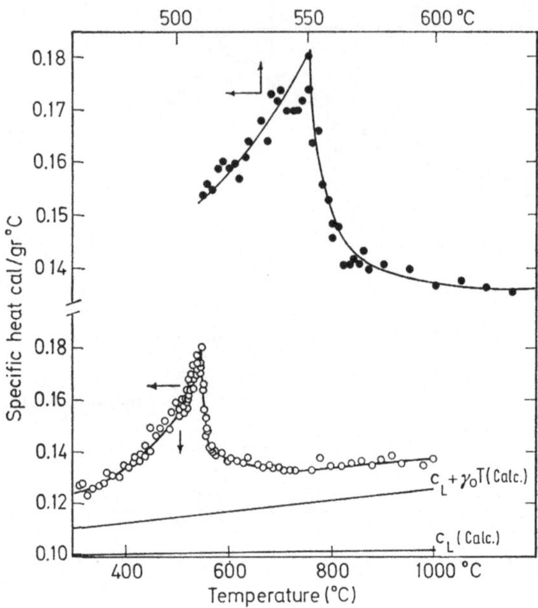

Fig. 4 b. Specific heat of $Fe_{44}Ni_{56}$. Top line gives details near T_c

the total experimental specific heat $c_p(T)$ at temperature T is assumed to be composed predominantly of dilation $(c_p - c_v)$, lattice $c_L(T)$, electronic $c_E(T)$, and magnetic $c_M(T)$, contributions:

$$c_p(T) = [(c_p(T) - c_v(T)] + c_L(T) + c_E(T) + c_M(T) . \tag{1}$$

At each temperature, the lattice, dilation and electronic specific heat background has to be estimated and subtracted from the experimental c_p values to obtain the magnetic specific heat, $c_M(T)$.

The dilation correction $(c_p - c_v)$ can be described usually by a function linear in temperature, through a coefficient K that contains the atomic volume (V), the thermal expansion coefficient (α), and the compressibility (β).

$$c_{\text{dila.}} = (c_p - c_v) = KT \, (9 \, V \alpha^2/\beta) \, T . \tag{2}$$

Experimental information on these quantities is not available for alloys but an empirical approach indicates that $c_{\text{dila.}}$ is only a few percent of $c_p(T)$. The dilation contribution appears to be small even if thermodynamic derivatives like the thermal expansion and compressibility do exhibit anomalous behavior in the neighborhood of the critical point [20, 21]. In nickel $c_p - c_v$ is about $0.03 \, c_v$. Therefore, $c_p - c_v$ was neglected in this alloy investigation.

The lattice contribution, excluding for the moment anharmonic effects, can be determined quite easily since the Debye specific heat is expected to give an adequate representation of the lattice specific heat $c_L(T)$ and since the characteristic Debye temperature, T_D, is known. Even variation of T_D by 50 °K will not noticeably effect the evaluation of the data, since our measurements are performed usually at $T > 1.5 \, T_D$.

Kittel [22] suggests that the anharmonic contribution to c_L is a linear function of temperature. Calculations [23] on b.c.c. iron show that the anharmonic specific heat of iron range from $+2.82 \times 10^{-4}\,T$ to $7.62 \times 10^{-4}\,T$ calories/(mole degree) depending on the particular model chosen for the calculation. c_{anh} is about 1/4 to 1/2 of c_E. The anharmonic specific heat term shows the same temperature dependence as the electronic specific heat. Therefore anharmonic effects could not be separated from c_E. Experimental c_E values obtained from the specific heat term linear in the absolute temperature will therefore always include this anharmonic lattice specific heat contribution.

The electronic contribution at temperatures below T_c, was calculated from the relationship:

$$c_E(T) = \gamma_0\,T \tag{3}$$

where the γ_0 coefficient (Table 1) was taken from low temperature specific heat studies [24—28]. More recent measurements by Ho and Viswanathan [29] give higher γ_0 values. Using their data would not modify our results markedly.

Table 1. Debye temperature θ and low temperature electronic specific heat coefficients γ_0 (Values for alloys obtained by interpolation from values given in references)

	Alloy	θ (K)	γ_0 (cal/mole°K^2)	Reference
f.c.c.	Ni	472	16.8 (10^{-4})	[24]
	$Co_{15}Ni_{85}$	480	15.0	[25]
	$Co_{40}Ni_{60}$	480	11.0	[25]
	$Co_{60}Ni_{40}$	490	10.0	[25]
	$Co_{75}Ni_{25}$	490	9.0	[25]
	$Fe_{44}Ni_{56}$	420	10.3	[27]
	$Fe_{22}Ni_{78}$	410	10.7	[28]

Deviations from Eq. (3) at higher temperature due to shifting of the degeneracy of the electron gas are probably unimportant for our investigation. $c_E(T) = \gamma_0\,T$ was added to the previously calculated lattice contribution. This $c_L + c_E$ value was taken as the background to be subtracted from the experimental $c_p(T)$ curves yielding the temperature dependent magnetic contribution $c_M(T)$. The temperature dependence of $c_M(T)$ and the associated magnetic energies and entropies can be determined at these temperatures and compared with the theoretical predictions for temperature regions below T_c.

The density of states changes due to the shifting energy band structure [4] when the samples pass from the ferromagnetic to the paramagnetic state. In the paramagnetic region above the Curie temperature, the low temperature γ_0 coefficients are no longer appropriate for determing the electronic specific heat contribution. Very little is known about the electronic specific heat coefficients (i.e. density of states) in the temperature interval above the Curie point except for some recent calculations by Katsuki. There appears to be no theoretical or experimental work on the alloys of this investigation.

The magnetic properties of the nickel alloys will be analyzed with localized electron Ising and Heisenberg models since such models have been used to

                    J. Orehotsky and K. Schröder

describe properties of the pure elements with more success than models based on the collective electron model. The addition of iron or cobalt to nickel should strengthen the localized electron character of our samples.

Onsager's [30] group theoretical treatment of a two-dimensional Ising magnet leads to logarithmically divergent temperature dependencies of the magnetic heat capacity around critical points [34]

$$c_{\mathrm{M}}(T) = A^{\pm} \ln \frac{T - T_{\mathrm{c}}}{T_{\mathrm{c}}} + B^{\pm} \tag{4}$$

Table 2. Magnetic entropy (S) and energy (E) values below T_{c}

			$(S_{\mathrm{c}} - S_0)/k_{\mathrm{B}}$	$(E_{\mathrm{c}} - E_0)/k_{\mathrm{B}} T_{\mathrm{c}}$
	Theory			
Ising	f.c.c. ($q = 12$)	$s = 1/2$	0.591	0.463
(Ref. 34)		$s = 1$	0.983	0.721
		$s = 2$	1.486	0.990
		$s = \infty$	∞	1.541
Heisenberg	f.c.c. ($q = 12$)	$s = 1/2$	0.473	0.297
(Ref. 34)		$s = 1$	0.810	0.555
		$s = 2$	1.305	0.833
		$s = \infty$	∞	1.406
	Experimental			
	f.c.c.	$Co_{75}Ni_{25}$	0.497	0.416
		$Co_{60}Ni_{40}$	0.502	0.432
		$Co_{40}Ni_{60}$	0.440	0.379
		$Co_{15}Ni_{85}$	0.172	0.183

Table 2 (Continuation). Magnetic entropy (S) and energy (E) values above T_{c}

			$(S_{\infty} - S_{\mathrm{c}})/k_{\mathrm{B}}$	$(E_{\infty} - E_{\mathrm{c}})/k_{\mathrm{B}} T_{\mathrm{c}}$	Electronic Specific Heat Coefficient
Theory					
Ising	f.c.c.	$s = 1/2$	0.102	0.150	
(Ref. 34)	($q = 12$)	$s = 1$	0.116	0.160	
		$s = 2$	0.123	0.167	
		$s = \infty$	0.131	0.175	
Heisenberg	f.c.c.	$s = 1/2$	0.265	0.439	
(Ref. 36)	($q = 12$)	$s = 1$	0.289	0.449	
		$s = 2$	0.304	0.459	
		$s = \infty$	0.322	0.474	
Experimental					
Alloy	Background				$\gamma \left[\dfrac{10^{-4}\ \mathrm{cal}}{\mathrm{mole}\,{}^{\circ}\mathrm{K}^2}\right]$
f.c.c.					
$Co_{75}Ni_{25}$	Ising $s = 1/2$		0.133	0.150	17.7
$Co_{60}Ni_{40}$	Ising $s = 1/2$		0.124	0.157	17.7
$Co_{40}Ni_{60}$	Heis. $s = 1/2$			0.440	3.9
	Ising $s = 1/2$		0.125	0.147	17.1

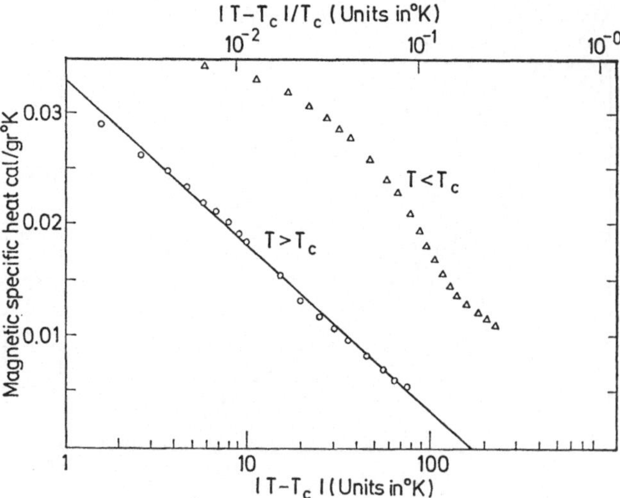

Fig. 5a. Magnetic specific heat of $Co_{15}Ni_{85}$

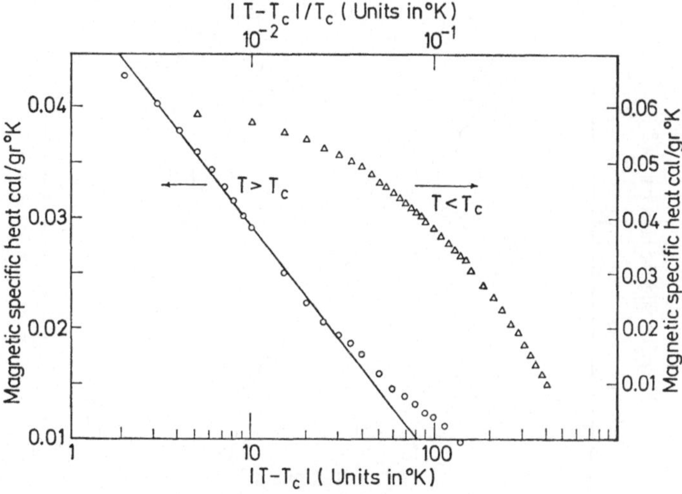

Fig. 5b. Magnetic specific heat of $Co_{40}Ni_{60}$

where $A^\pm$ and $B^\pm$ are the critical values characterizing the temperature dependence of the magnetic specific heat above ($+$) and below ($-$) the critical temperature. An additional result of the theoretical treatments are the values of the energies and entropies associated with the magnetic transformation as a function of temperature. Table 2 gives theoretical values of

$$(E_\infty - E_c)/k_B T_c, \quad (S_\infty - S_c)/k_B, \quad (E_c - E_0)/k_B T_c \quad \text{and} \quad (S_c - S_0)/k_B,$$

obtained from series expansion methods and used in the evaluation of our c_p data. $E_\infty - E_c$ would be the magnetic ordering energy above the critical temperature, and $E_c - E_0$ would be the magnetic ordering energy below T_c. S_∞, S_c and S_0

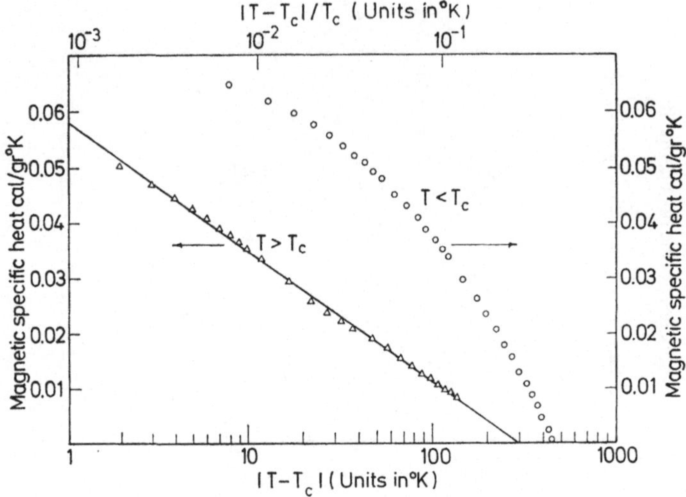

Fig. 5c. Magnetic specific heat of $Co_{60}Ni_{40}$

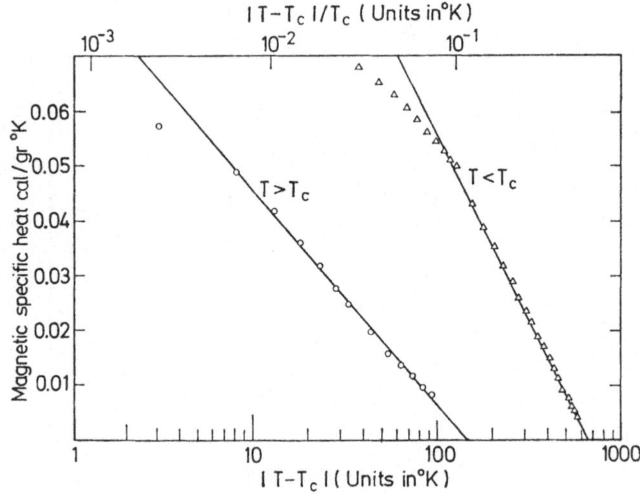

Fig. 5d. Magnetic specific heat of $Co_{75}Ni_{25}$

would be the entropy at $T = \infty$, $T = T_c$ and $T = 0\,°K$ respectively. Table 2 shows that the $(E_\infty - E_c)/kT_c$ values are nearly independent of the spin quantum number, but differ markedly for the Ising and Heisenberg model, whereas the $(E_c - E_0)/kT_c$ values depend mostly on the spin quantum number, but are nearly the same for the Heisenberg and Ising model with the same spin.

The analysis for $T < T_c$ proceeded in the following way: knowing the background below T_c from $c_L(T) + c_E(T)$, the temperature dependence of $c_M(T)$ and the experimental magnetic energy and entropy associated with the transition below T_c can be determined and compared with the predicted values for the Heisenberg and Ising models. Figs. 5 and 6 give plots of $c_M(T)$ as a function of

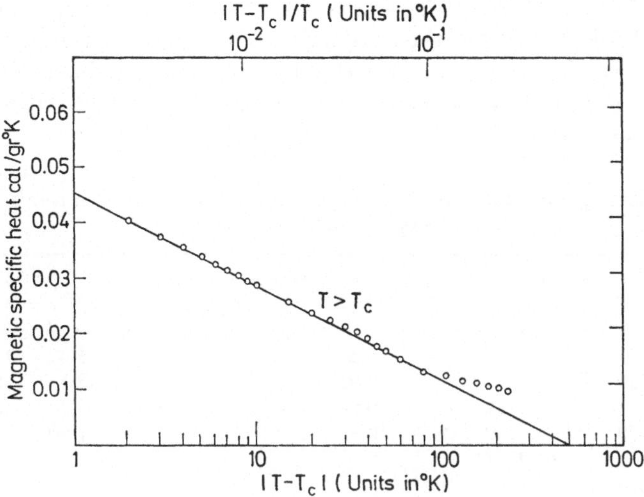

Fig. 6a. Magnetic specific heat of $Fe_{22}Ni_{78}$

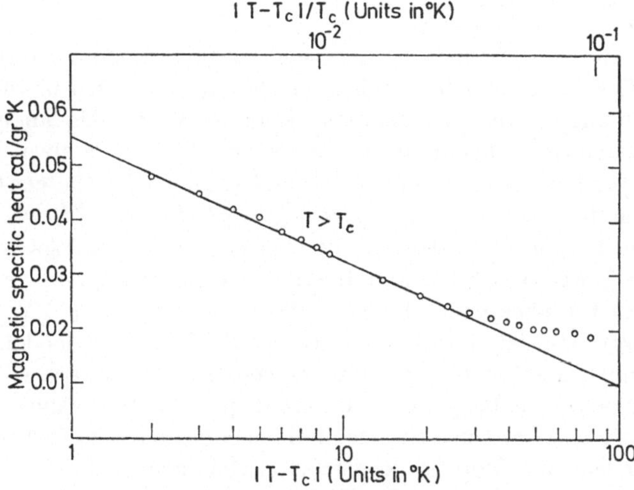

Fig. 6b. Magnetic specific heat of $Fe_{44}Ni_{56}$

$\log |T - T_c| \cdot c_M(T)$ does not follow a power law temperature dependency below T_c for any of the Fe-Ni alloys and only the alloy $Co_{75}Ni_{25}$ displayed a logarithmic dependency (Fig. 5d) from which the A^- value can be termined. This value along with the A^- values determined from literature data on pure Ni and Co are presented in Table 3. A possible reason for the non-logarithmic behavior of the $c_M(T)$ data and the sizeable transitional rounding just below T_c are changes in the magnetization of the alloys as their temperature approaches T_c. Band shifts and changes in the density of states occur. As a result, the electronic contribution to the specific heat neat T could rapidly diverge from the assumed linear $\gamma_0 T$ behavior used in the analysis to obtain $c_M(T)$. It can be shown that c_E may even be divergent for $T \to T_c$ [19]. The results of the energy and entropy determinations are given in Table 2. Additional contributions to the specific heat at temperatures below T_c due to crystallographic ordering [31] prevented a similar determination of $c_M(T)$ below T_c and the associated energies, entropies and critical values A^- or α^- for the f.c.c. alloys $Fe_{22}Ni_{78}$ and $Fe_{44}Ni_{56}$.

Table 3. Experimental critical values A^+ and A^-, and Curie temperatures T_c
Experimental results

	A^+	Range $(T - T_c)/T_c$	A^-	Range $(T_c - T)/T_c$	A^-/A^+	e/a	T_c (°K)
Ni	0.149		0.234		1.57	10.00	628
Co	0.420		0.910		2.16	9.00	1388
Fe	0.875	$2 \times 10^{-3} - 2 \times 10^{-2}$	0.988	$1 \times 10^{-2} - 3 \times 10^{-1}$	1.13	8.00	1038
$Ni_{25}Co_{75}$	0.479	$8 \times 10^{-3} - 9 \times 10^{-2}$	0.862	$1 \times 10^{-1} - 5 \times 10^{-1}$	1.80	9.26	1255
$Ni_{40}Co_{60}$	0.305	$3 \times 10^{-3} - 1 \times 10^{-1}$				9.40	1145
$Ni_{60}Co_{40}$	0.288	$3 \times 10^{-3} - 3 \times 10^{-2}$				9.60	1033
$Ni_{85}Co_{15}$	0.186	$3 \times 10^{-3} - 1 \times 10^{-1}$				9.82	803
$Ni_{56}Fe_{44}$	0.282	$2 \times 10^{-3} - 3 \times 10^{-2}$				9.12	825
$Ni_{78}Fe_{22}$	0.215	$2 \times 10^{-3} - 1 \times 10^{-1}$				9.56	843

The data above T_c cannot be discussed in this fashion. However, the c_p data above T_c can be used to obtain a crude approximation of the density of states for each alloy in the paramagnetic region above T_c. The analysis requires two major assumptions: first that the lattice specific heat is given by the temperature independent Dulong-Petit value and second that the electron specific heat is a linear function of temperature: $c_E = \gamma T$ for $T > T_c$ where γ is proportional to the paramagnetic density of states. $c_L + c_E$ above T_c was constructed in such a way (γ was selected as an adjustable parameter) that the enclosed area between it and the experimental specific heat curve $c_p - (c_L + c_E)$ equals the magnetic energy associated with the theoretical Heisenberg model. In exactly the same manner, another background was constructed for each alloy from the energy predictions of the Ising model. Different spin quantum values did not have to be taken into account since the magnetic energy above T_c in both models is nearly independent of s. Both the constructed Heisenberg and Ising background are presented in Fig. 3b for the $Co_{40}Ni_{60}$ alloy. For the other alloys only the Heisenberg or the Ising model matches up reasonably well with $c_p - (c_L + c_E)$. γ obtained with this approach is presented in Table 2. It shows that the density of states above T_c is higher than below T_c for the Ising model (see Table 1).

$c_M(T)$ values above T_c for all the alloys were found to be logarithmically temperature dependent (Figs. 5 and 6) for all the alloys. A^+ values are presented in Table 3. This analysis above T_c may yield an incorrect A^+ value because of an improper background choice. However, the logarithmic dependency and the associated A^+ value were found to be independent of the background choice for the $Co_{40}Ni_{60}$ alloy where both the Ising and Heisenberg models gave a reasonable match with the high temperature experimental data. This indicates that A^+ is most likely independent of the background calculation for all alloys in this investigation.

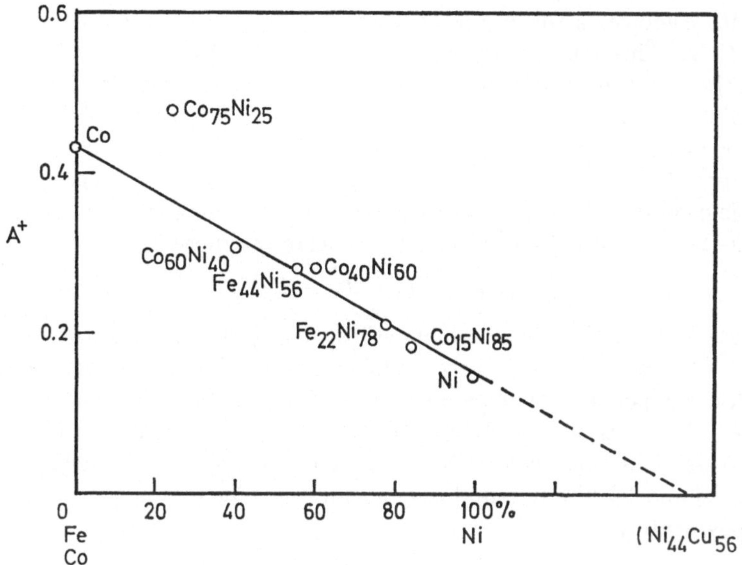

Fig. 7. Composition dependence of the critical value A^+ for f.c.c. crystals

The feature that the $c_M(T)$ data of all the alloys displays a logarithmic temperature dependence above T_c and, in general, not below T_c is similar to the behavior found in the insulating materials EuO and RbMnF$_3$ [32, 33]. The non-logarithmic behavior below T_c suggests that transitional rounding is severe in the CoNi alloys and can be attributed to intrinsic effects such as magnetic domains or magnetostriction. This severe rounding means that the calculated entropy $S_c - S_0$ and energy $E_c - E_0$ values for the CoNi alloys was anomalously small.

Another feature of the data (see Table 2), is that the constants A^+ and A^- determined for each alloy system from the logarithmic dependencies do not agree with either theoretical predictions of the Ising ($s = 1/2$) model [2] or with previous results on other magnetically ordered system [33, 35] that are considered usually as localized electron ferromagnets. Stoner excitation might account for this discrepancy.

The compositional dependency of A^+ is presented in Fig. 7. The A^+ values for the four f.c.c. materials Co, Ni, CoNi alloys and FeNi alloys are a linear function of nickel concentration. The linear plot extrapolated to A^+ (f.c.c.) $= 0$ intercepts the abscissa at approximately the composition Ni$_{40}$Cu$_{60}$ where the d band is filled and the spontaneous magnetization is zero. It is presently not possible to explain why A^+ should be a linear function of composition. A^+ may be also a linear function of T_c, since T_c is in first approximation also a linear function of composition.

Summary and Conclusion

The specific heat of ferromagnetic f.c.c. alloys with iron, cobalt, and nickel elements were measured with a "heat exchange" calorimeter in the temperature range from 600 to 1500 °K. It was found that the specific heat near the Curie temperature showed a sharp γ-peak. Rounding occurred in a temperature interval of 5 to 10 °K. This was slightly larger than expected from experimental rounding and nearly the same as that found for pure elements in our experimental system. This indicates that marked rounding of the peak in the magnetic specific heat, as found for nickel copper alloys, is not typical for all disordered alloys.

The specific heat of the alloys was separated into the lattice, electronic and magnetic contributions. In practically all alloys, the magnetic specific heat above the Curie temperature followed the theoretically predicted relation:

$$c_M = \log | T/T_c - 1 |^{A^+},$$

independent of whether the non-magnetic specific heat contribution was estimated with the concept of a Heisenberg or Ising model for the magnetic specific heat. The critical exponent A^+ depended linearly on the composition. This equation described the data accurately for $10^{-3} < | T/T_c - 1 | < 10^{-1}$ above the Curie temperature. However, below the Curie temperature, such a linear relationship between c_M and $\log T$ was found over a much smaller temperature range only for the $Co_{40}Ni_{60}$ alloy.

Acknowledgment. The authors acknowledge gratefully the financial support by the National Science Foundation and frequent discussions with Dr. W. MacInnes and Dr. N. Baum.

References

1. Morrish, A. H.: The Physical Principles of Magnetism. New York: J. Wiley & Sons 1965, p. 271
2. Stoner, E. C.: Proc. roy. Soc. A **165**, 372 (1938); A **169**, 339 (1939)
3. Wohlfahrt, E. P.: Rev. mod. Phys. **25**, 211 (1953)
4. Katsuki, A.: J. of the Faculty of Science, Shinshu University 2, No. 1, 19 (1967)
5. Domb, C., Sykes, M. F.: Proc. roy. Soc. **240**, 214 (1957)
6. Kouvel, J. S., Robell, D. S.: Phys. Rev. Letters **18**, 215 (1967)
7. Domb, C.: Magnetism, Vol. II A. Eds: Rado, G. T., Suhl, H. New York: Academic Press 1965, p. 35
8. Herring, C.: Magnetism, Vol. IV. Eds: Rado, G. T., Suhl, H. New York: Academic Press 1966, p. 136
9. Pawel, R. E., Stansbury, E. E.: J. Phys. Chem. Solids **26**, 607 (1965)
10. Takahashi, T., Shimizu, M.: J. Phys. Soc. Japan **23**, 945 (1967)
11. MacInnes, W. M., Schröder, K.: Dynamical Aspects of Critical Phenomena. New York: Gordon and Breach 1972, p. 305
12. Nielson, M. L., Johns, I. B.: Rev. Sci. Instrum. **25**, 596 (1954)
13. Orehotsky, J. L., Schröder, K.: J. phys. E. **3**, 889 (1970)
14. Hultgren, R. R.: Thermodynamic Properties of Metals and Alloys. New York: John Wiley & Sons 1963
15. Braun, M., Kohlhaas, R., Vollmer, D.: Z. angew. Phys. **25**, 365 (1968)
16. Pallister, P. L.: J. Iron and Steel Inst. **161**, 87 (1949)
17. Lyman, T., Ed.: Metals Handbook, American Society for Metals 1948
18. Schröder, K., MacInnes, W. M.: J. phys. E. **2**, 959 (1969)
19. Orehotsky, J.: Magnetic Critical Point Phenomena in 3-d Transition Metal Alloys. Dissertation, Syracuse University, Syracuse, N.Y. (1971)
20. Philp, J. W., Gonano, R., Adams, E. D.: Phys. Rev. **188**, 973 (1969)
21. Bozorth, R. M.: Ferromagnetism, New York: Van Norstrand 1951, p. 736
22. Kittel, C.: Introduction to Solid State Physics, Third Edition. New York: John Wiley & Sons 1967, p. 195
23. Foreman, A. J. E.: Proc. physical. Soc. **79**, 1124 (1962)
24. Phillips, N. E.: CRC Critical Reviews in Solid State Sciences, Dec. 1971, p. 467

25. Walling, J. C., Bunn, P. B.: Proc. physical. Soc. **74**, 417 (1959)
26. Gupta, K. P., Cheng, C. H., Beck, P. A.: J. Phys. Radium (Paris) **23**, 721 (1962)
27. Gupta, K. P., Cheng, C. H., Beck, P. A.: J. Phys. Chem. Solids **25**, 73 (1964)
28. Keeson, W. H., Kurrelmeyer, B.: Physica **7**, 1003 (1940)
29. Ho, J. C., Viswanathan, R.: Phys. Rev. **172**, 705 (1968)
30. Onsager, L.: Phys. Rev. **65**, 117 (1944)
31. Hansen, M.: Constitution of Binary Alloys. New York: McGraw-Hill 1958
32. Teaney, D. T., Moruzzi, V. L., Argyle, B.: J. appl. Physics. **37**, 1122 (1967)
33. Teaney, D. T.: Phys. Rev. Letters **14**, 898 (1965)
34. Fisher, M. E.: Reports on Prog. in Phys. XXX, Part II, 615 (1967)
35. Fisher, M. E.: Phys. Rev. A **136**, 1599 (1964)
36. Domb, C., Miedema, A. R.: Progress in Low Temp. Physics. Amsterdam: North-Holland Publishing Co. 1964, p. 296

Dr. John Orehotsky
Wilkes College
Department of Engineering
Wilkes Barre, Pennsilvania
USA

Phys. cond. Matter 17, 55—61 (1973)

Determination of the Conduction Band Width of Li from the K Emission Spectrum

Helmut Bross

Sektion Physik der Universität München

Received July 9, 1973

The shape of the K emission band of Lithium is investigated by considering electron-electron interaction in the conduction band. The width of the conduction band turns out to be 3.4 eV. The fitting procedure applied may also be used for interpreting the K spectra of other metals.

1. An Approximation for the Emission Spectrum in the Many-Body Scheme

Recent calculations of the band structure of Lithium [1] showed that the width of the conduction band depends on the potential used in the one-particle Schrödinger equation. Comparison of such calculations with experimental results then allows to determine the most appropriate potential. In the one-electron scheme the breadth of the soft X-ray emission spectrum gives directly the width of the conduction band. However, due to many-body effects the uncertainty in determining this magnitude from measurements is considerable. In the case of Lithium not only the bottom of the conduction band is obscured by the low energy tail but also the X-ray emission spectrum near the Fermi edge bends over in a very round way producing a premature peak 0.4 eV below the edge [2, 3]. In spite of the different extensive investigations [4—8] the origin of this Fermi edge behaviour is not completely explained up to date. For that reason we shall restrict the consideration to the influence of many-body effects on the low energy side. Nevertheless we get a good result near the Fermi edge by a reasonable assumption. Neglecting the relaxation of the core hole the intensity of X-rays emitted with frequency ω is given by [9]

$$I(\omega) = \omega^2 \sum_k |\mathbf{p}_{cv}(k)|^2 A(k, E_c(k) + \omega) \Theta(E_F - \omega - E_c(k)). \tag{1}$$

$\mathbf{p}_{cv}(k)$ is the matrix element of the momentum operator between a core state ck and a valence state vk, $E_c(k)$ and $E_v(k)$ are the corresponding energies. In the case of Lithium the k-dependence of the core energies turns out to be negligible. Θ is the Heaviside step-function and $A(k, \omega)$ the spectral density function for one-electron excitations. It is related to the one-electron Green function through

$$A(k, \omega) = \frac{1}{\pi} \operatorname{Im} G(k, \omega). \tag{2}$$

For further details of the many-body aspects the reader is referred to relevant review articles [10].

The calculation of $I(\omega)$ using Eq. (1) is impossible because $A(\mathbf{k}, \omega)$ depending both on the vector $\mathbf{k}$ and the frequency ω, is not known for a real crystal at the moment. All the present investigations restrict themselves to the case of the homogeneous electron gas where the $\mathbf{k}$-dependence of $A(\mathbf{k}, \omega)$ comes only from the energy $E(\mathbf{k})$, which, in turn, is a function of the absolute value of $\mathbf{k}$. Then the spectral function $A(\mathbf{k}, \omega)$ depends on two scalar variables: the energy $E(\mathbf{k})$ and the frequency ω. Therefore we may first integrate in Eq. (1) over a surface of constant energy yielding as one factor the unbroadened spectrum I_0 defined by

$$I_0(\omega) = \omega^2 \sum_{\mathbf{k}} |\mathbf{p}_{\mathrm{cv}}(\mathbf{k})|^2 \, \delta(\omega + E_{\mathrm{c}} - E_{\mathrm{v}}(\mathbf{k})). \tag{3}$$

The emitted intensity is given by the following integral

$$I(\omega) = \omega^2 \int_{-\infty}^{\infty} \frac{I_0(\omega')}{\omega'^2} \, A[k(\omega' + E_{\mathrm{c}}), \omega + E_{\mathrm{c}}] \, d\omega' \, \Theta(E_{\mathrm{F}} - \omega - E_{\mathrm{c}}). \tag{4}$$

From the investigations done for the homogeneous electron gas [11, 12, 13] we know that the spectral density function exhibits two different kinds of structures. One strong peak in the ω-dependence may be attributed to the excitation of a quasi-particle characterized by the wave-vector $\mathbf{k}$ and the energy $E(\mathbf{k})$. Besides this structure the spectral density function shows two pronounced continuous side bands at energies satisfying $|E - E_{\mathrm{F}}| > \hbar \omega_{\mathrm{p}}$ (ω_{p}: plasma frequency) due to the excitation of plasmons. As we are interested in the main emission spectrum and not in the plasmon satellite it suffices to consider the quasi-particle peak in the spectral density function only. This restriction is desirable since for performing the integration in Eq. (4) we need the spectral function $A(\mathbf{k}, \omega)$ as a function of $\mathbf{k}$ for fixed values of ω. Even in the case of the homogeneous electron gas this information is at our disposal only in the neighbourhood of the energy-shell $\omega = E(\mathbf{k})$. Elsewhere the ω-dependence of $A[k(E), \omega]$ is known for some values of $|\mathbf{k}|$ only. For $\omega \sim E(\mathbf{k})$ we approximate the Green function by its quasi-particle pole

$$G(\mathbf{k}, \omega) \sim \frac{Z[E(\mathbf{k})]}{\omega - E(\mathbf{k})} \tag{5}$$

where the residue $Z(E(\mathbf{k}))$ of the pole may be determined from the self-energy function. Both $Z(\mathbf{k})$ and $E(\mathbf{k})$ are complex magnitudes. The corresponding real and imaginary values will be denoted by subscripts 1 and 2, respectively. The renormalized energy $E_1(\mathbf{k})$ will differ little from the energy eigenvalue obtained by solving a Schrödinger equation where many-body effects are taken into account. $E_2(\mathbf{k})$ may be interpreted as the inverse life-time of the quasi-particle state. Assuming (5) to be valid in the whole integration region of Eq. (4) we get

$$I(\omega) = \frac{1}{\pi} \, \omega^2 \int_{-\infty}^{\infty} \frac{d\omega'}{(\omega' - E_{\mathrm{c}})^2} \, \frac{Z_1(\omega') \, E_2(\omega') + Z_2(\omega') \, (\omega - \omega' + E_{\mathrm{c}})}{(\omega - \omega' + E_{\mathrm{c}})^2 + E_2(\omega')^2} \, I_0(\omega' - E_{\mathrm{c}}). \tag{6}$$

Note that this result of the many-body theory is different from a previous procedure [14] which takes into account a finite line width of the one-particle state by folding I_0 with an appropriate Lorentzian. The two procedures differ from one another by the term $Z_2(\omega')\,(\omega - \omega' + E_c)$ in the numerator. Since in general the imaginary part of the energy $E_2(\omega)$ is small in comparison with the band width, the integrand will be peaked at $\omega' = \omega + E_c$ with the result that this additional term will contribute only little. This conclusion does not hold below the bottom of the conduction band and thus we expect that both procedures will differ in the low energy tail.

2. Details of the Calculation and Results

Using Bohn's MAPW-results the unbroadened emission and absorption spectra of Lithium were thoroughly investigated by Faust [15]. He found that the matrix elements $p_{\mathrm{cv}}(k)$ determining the K emission spectrum are weakly orientational dependent and that they may be approximated as a linear function of the difference of the energy E and the energy of the bottom of the conduction band $-E_{\mathrm{B}}$. Further, up to E_{F}, the quasi-particle energy $E(k)$ was found to be a quadratic function of $|k|$. The unbroadened K emission spectrum then has the well-known form

$$I_0(\omega) \sim (\omega + E_c - E_{\mathrm{B}})^{3/2} \quad \text{for} \quad E_{\mathrm{B}} - E_c < \omega < E_{\mathrm{F}} - E_c . \tag{7}$$

Detailed calculations of Faust [15] had the result that the contributions to the integral (6) from $\omega > E_{\mathrm{F}} - E_c$ are negligible.

Numerical values of $Z_1(E)$, $Z_2(E)$ and $E_2(E)$ were calculated by Hedin [16] for the jellium model, for different values of $|k|$ and electronic densities characterized by r_{s}. We have parametrized the k-dependence in the following way: The imaginary part of the energy is approximated by

$$E_2(E) = \gamma\,(E_{\mathrm{F}} - E_{\mathrm{B}})\left(\frac{E - E_{\mathrm{F}}}{E_{\mathrm{B}} - E_{\mathrm{F}}}\right)^2 \tag{8}$$

in agreement with the general results of the many-body theory [17]. Similiar approximations hold for the real and imaginary part of $Z(k)$, too.

$$Z_1(E) = \zeta_{1,1} + \zeta_{1,2}\left(\frac{E - E_{\mathrm{F}}}{E_{\mathrm{B}} - E_{\mathrm{F}}}\right)^2 \tag{9}$$

$$Z_2(E) = \zeta_2\left(\frac{E - E_{\mathrm{F}}}{E_{\mathrm{B}} + E_{\mathrm{F}}}\right)^2 . \tag{10}$$

In the jellium model the numerical values of γ, $\zeta_{1,1}$, $\zeta_{1,2}$ and ζ_2 depend on r_{s} only. For all values of r_{s} encountered in metals, $E_2(E)$ turns out to be smaller than the band width $(E_{\mathrm{F}} - E_{\mathrm{B}})$. Hence we expect that both procedures of folding i.e. the one based on Eq. (4) and the heuristic one due to Landsberg yield similar results. Due to the factor $\dfrac{E - E_{\mathrm{F}}}{E_{\mathrm{B}} - E_{\mathrm{F}}}$ the agreement becomes the better the nearer the X-ray transition is to the Fermi edge.

As no other information about Lithium is known, the results deduced for the homogeneous electron gas will be used to study the influence of the electron-electron interactions in the conduction band on the K emission spectrum. Even better agreement with the measured spectrum may be achieved by an appropriate choice of the parameters in Eqs. (8), (9) and (10). Since the influence of the term

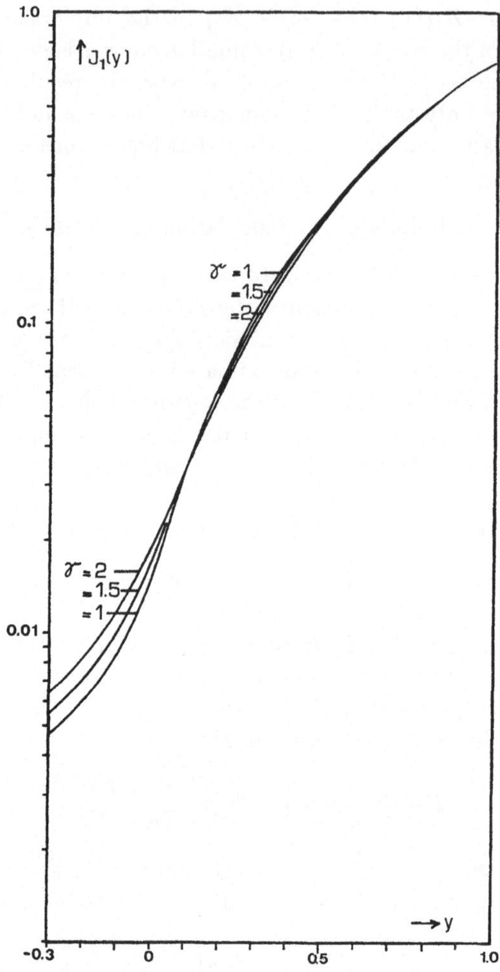

Fig. 1. The frequency dependence of the integral J_1

$(\omega - \omega' + E_c) Z_2(\omega')$ in the numerator of Eq. (6) is small in the main spectrum and as further the real part of Z varies slowly with the energy, only the parameter γ was adjusted, whereas we took $\zeta_{1,1} = 0.70$, $\zeta_{1,2} = -0.069$ and $\zeta_2 = -0.138$ from Hedin's paper. With these approximations the intensity turns out to be proportional to the integral

$$J_1(y) = \frac{1}{\pi} \int_0^1 \frac{(y - x) Z_2(x) + \gamma (1 - x)^2 Z_1(x)}{(y - x)^2 + \gamma^2 (1 - x)^4} x^{3/2} \, dx \tag{11}$$

with

$$\omega = -E_c + E_B (1 - y) + y E_F \tag{12}$$

and

$$x = \frac{E - F_B}{E_F - E_B} \tag{13}$$

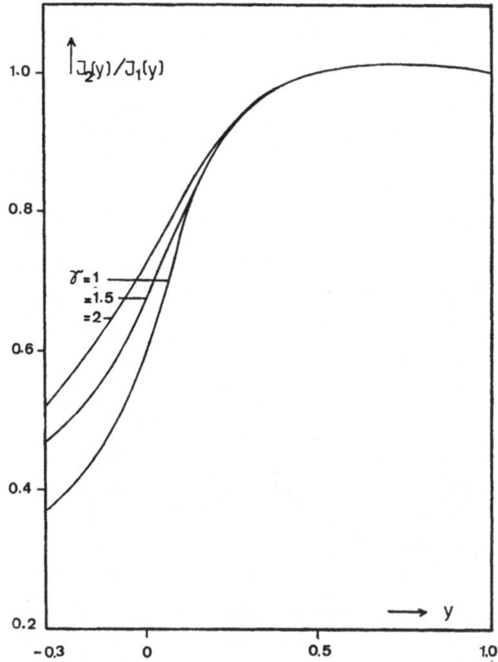

Fig. 2. The frequency dependence of the ratio $J_2(y)/J_1(y)$

as dimensionless variables. In Fig. 1, $J_1(y)$ is shown for different values of γ. In addition, an integral $J_2(y)$ was calculated for the heuristic broadening procedure which is formally obtained from J_1 by setting $Z_2(x) = 0$. Fig. 2 showing the ratio $J_2(y)/J_1(y)$ confirms the· above suggestion that both procedures yield results differing only in the low energy tail of the main spectrum. As Z_1 varies slowly with energy, the integral $J_2(y)$ is almost independent of the solid considered and hence may be used for the K spectra of other materials, too.

Finally, with an appropriate choice of E_B and E_F the expression $J_1(y)$ was mapped on the K emission spectrum measured by Crisp and Williams [2]. This was achieved by approximating the measured intensity through the expression

$$I(\omega) = a + b J_1(\omega). \tag{14}$$

The coefficients a and b refer to the background and the unknown scaling ratio of intensities respectively. They and the energy E_B were determined by minimizing the quadratic deviation in the energy range $1.3\,E_B - 0.3\,E_F < E < 0.4\,E_B + 0.6 E_F$. The value of the Fermi energy was determined from the intersection of the straight line $a + \frac{1}{2} b$ with the measured spectrum. That means that the Fermi edge is at the 50% point of the spectrum without background.

This procedure allows to locate the bottom of the conduction band at $E_B = 51.27$ eV and the Fermi edge at $E_F = 54.63$ eV. The resulting width of the conduction band $E_F - E_B = 3.36$ eV differs slightly from the value evaluated by Ham [18]. It can be shown [19] that the band structure calculations approximating many-body effects through a non-local screened exchange potential yield the same value of the band width, when the screening parameter is appropriately chosen.

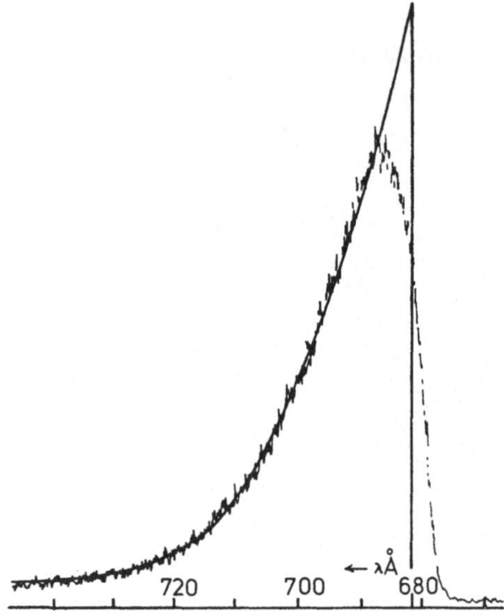

Fig. 3. Comparison of the optimized theoretical emission spectrum with the recorder trace of LiK measured by Crisp and Williams [2]

The broadening of the line width at the bottom of the conduction band turns out to be $\gamma(E_F - E_B) = 0.41$ eV, a little smaller than McMullen's value [20]. The spectrum corresponding to the best choice of the parameters is shown in Fig. 3 together with the experimental result. The good agreement of both curves up to 0.4 eV below the Fermi edge is remarkable although the fitting procedure was only applied in the region 50.3 eV $< E < 53.3$ eV. The difference of both spectra in the vicinity of the Fermi edge is obvious. At the moment it is not possible to decide which one of the mechanisms proposed to explain the edge singularity [4, 5, 7, 21—24] describes this discrepancy in the case of Li. It is remarkable that Haensel *et al.* [25] have found the same value of the K edge by photo-absorption measurements.

Acknowledgement. The author is grateful to Dipl.-Phys. A. Faust for making available the results of his work and for many illuminating discussions.

References

1. Bohn, G.: Dissertation München 1970
2. Crisp, R. S., Williams, S. E.: Phil. Mag. (8) 5, 525 (1960)
3. Sagawa, T.: In: Soft X-Ray Band Spectra, (Ed.: D. J. Fabian). London and New York: Academic Press 1968
4. Mahan, G. D.: Phys. Rev. **153**, 882 (1967); **163**, 612 (1967)
5. Nozières, P., de Dominicis, C. T.: Phys. Rev. **178**, 1097 (1969)
6. Langreth, D. C.: Phys. Rev. **182**, 973 (1969)
7. Ausman, G. A., Glick, A. J.: Phys. Rev. **183**, 687 (1969)
8. Schotte, K. D., Schotte, U.: Phys. Rev. **185**, 509 (1969)
9. Hedin, L.: Solid State Commun. **5**, 541 (1967)
10. See for Example: Hedin, L., Lundqvist, S. In: Solid State Physics. (Ed.: F. Seitz, T. Turnbull, and H. Ehrenreich), Vol. 23, p. 1. New York and London: Academic Press 1969

11. Bose, S. M., Bardasis, A., Glick, A. J., Hone, D., Longe, P.: Phys. Rev. 155, 379 (1967)
12. Lundqvist, B. I.: Phys. kondens. Materie 6, 193 and 205 (1967)
13. Lundqvist, B. I.: Phys. kondens. Materie 7, 117 (1968)
14. Landsberg, P. T.: Proc. Phys. Soc. A 62, 806 (1949)
15. Faust, A.: Diploma thesis, Munich 1971
16. Hedin, L.: Phys. Rev. 139, A 796 (1965)
17. Luttinger, J. M.: Phys. Rev. 121, 942 (1961)
18. Ham, F. S.: Phys. Rev. 128, 2524 (1962)
19. Bross, H., Bohn, G.: to be published
20. McMullen, T.: J. Phys. C.: Solid State Phys. 3, 2178 (1970)
21. Mizuno, Y., Ishikawa, K.: J. Phys. Soc. Japan 25, 627 (1960)
22. Shuey, R. T.: Phys. kondens. Materie 5, 192 (1966)
23. Allotey, F. K.: Phys. Rev. 157, 467 (1967)
24. Goodings, D. A.: Proc. Phys. Soc. (Lond.) 86, 75 (1965)
25. Haensel, R., Keitel, G., Sonntag, B., Kunz, C., Schreiber, P.: Phys. status solidi (a) 2, 85 (1970)

Prof. Dr. Helmut Bross
Sektion Physik
der Universität München
D-8000 München
Federal Republic of Germany

Phys. cond. Matter 17, 63—70 (1973)
© by Springer-Verlag 1973

Further Comments on Thermodynamic Response
of Mie-Gruneisen Materials

R. O. Davis

Department of Civil Engineering, University of Canterbury, Christchurch, New Zealand

Received June 13, 1973

Results obtained in a previous paper [1] are modified to consider materials which exhibit a linear shock velocity — particle velocity Hugoniot, and for which the specific heat at constant volume is a function of temperature. Also, a method for synthesizing functional dependence of the Gruneisen parameter on strain is illustrated. A Mie-Gruneisen equation of state for water is obtained using this method.

1. Introduction

In a recent paper [1], we presented expressions for entropy and temperature in Mie-Gruneisen materials, for which both the Hugoniot pressure and the Gruneisen parameter were given by infinite power series in the volumetric strain. It was also assumed in [1] that the specific heat at constant volume was a constant. In this work we further discuss and amplify the results obtained in [1]. Our purpose is threefold. First, we show how a material which exhibits a linear shock velocity — particle velocity Hugoniot may be represented by the analysis previously presented. Second, we discuss a general approach toward representation of materials for which the specific heat varies as a function of temperature. Third, we describe a method for determination of the Gruneisen parameter from isothermal compression data. In connection with this, we present data concerning the Gruneisen parameter for water.

2. Linear Shock Velocity-Particle
Velocity Hugoniots

We consider a plane shock wave propagating into a material initially at rest. In a spatial co-ordinate system, conservation of mass and linear momentum are expressed by the well known jump equations:

$$\varrho_0 U = \varrho (U - u), \tag{1}$$

$$p_\mathrm{H} - p_0 = \varrho_0 U u \tag{2}$$

where ϱ_0 and ϱ denote mass densities leading and following the shock, U denotes the shock velocity, u denotes the particle velocity of the material following the shock, and p_0 and p_H are the pressures leading and following the shock. For a large number of materials, it has been empirically found that shock response may be adequately represented by a linear relationship between U and u.

$$U = c + \sigma u. \tag{3}$$

McQueen *et al.* [2] give numerical values for the constants c and σ for more than 40 materials. Combining Eqs. (1), (2) and (3), we find that the Hugoniot pressure p_H is given by

$$p_H = \frac{\varrho_0\, c^2\, \eta}{(1 - \sigma\eta)^2} \tag{4}$$

where the natural volumetric strain η is defined by

$$\eta = 1 - \varrho_0/\varrho \tag{5}$$

and the initial pressure p_0 has been ignored. To insure that p_H remain finite, we restrict η so that $\sigma\eta < 1$.

In [1], the Hugoniot pressure was represented in the following way

$$p_H = \sum_{n=0}^{\infty} a_n\, \eta^n. \tag{6}$$

A similar expression can be obtained from Eq. (4) by noting that

$$\frac{1}{(1 - \sigma\eta)^2} = \sum_{n=0}^{\infty} (n + 1)\,(\sigma\eta)^n. \tag{7}$$

Convergence of this series is assured by our earlier restriction on η. Using Eq. (7) in Eq. (4) gives the following expression for the Hugoniot pressure.

$$p_H = \frac{\varrho_0\, c^2}{\sigma} \sum_{n=1}^{\infty} n\,(\sigma\eta)^n. \tag{8}$$

Replacing Eq. (6) by Eq. (8), we can now follow the analysis given in [1]. We assume that the pressure p is given by the Mie-Gruneisen equation:

$$p = p_H(1 - \tfrac{1}{2}\, v_0\, \eta\, g) + g(\varepsilon - \varepsilon_0) \tag{9}$$

where $v_0 = 1/\varrho_0$ denotes the unstrained specific volume, ε denotes specific internal energy, and g is the product.

$$g = \varrho\,\gamma, \tag{10}$$

γ being the Gruneisen parameter. As before, the function g is assumed to be given by a power expansion in η.

$$g = \sum_{n=0}^{\infty} b_n\, \eta^n. \tag{11}$$

If we then assume that the specific heat at constant volume, c_v, is a constant, direct substitution into the analysis of [1] gives the following expressions for entropy, s, and temperature, θ, at any point (η, ε).

$$s(\eta, \varepsilon) = s_0 + c_v \ln\left[1 + \frac{\varepsilon - \varepsilon_0 - \sum_{n=1}^{\infty} \beta_n\, \eta^n}{\theta_0\, c_v (1 + \sum_{n=1}^{\infty} \alpha_n\, \eta^n)} \right], \tag{12}$$

$$\theta(\eta, \varepsilon) = \theta_0\left(1 + \sum_{n=1}^{\infty} \alpha_n\, \eta^n\right) + \frac{1}{c_v}\left(\varepsilon - \varepsilon_0 - \sum_{n=1}^{\infty} \beta_n\, \eta^n\right).$$

In these equations, $s_0 = s(0, \varepsilon_0)$, $\theta_0 = \theta(0, \varepsilon_0)$, and

$$\alpha_{n+1} = \frac{v_0}{n+1} \sum_{k=0}^{n} \varphi(n, k)\, b_{n-k},$$

$$\beta_{n+1} = \frac{v_0}{n+1} \sum_{k=0}^{n} \varphi(n, k)\, P_{n-k}, \tag{13}$$

$$\varphi(0, n) = 1 \quad \text{for all} \quad n$$

$$\varphi(k + 1, n) = \frac{v_0}{n - k} \sum_{m=0}^{k} \varphi(m, n)\, b_{k-m}, \quad k = 0, 1, 2, \ldots$$

and

$$P_{n+1} = \varrho_0\, c^2 \left[(n + 1)\, \sigma^n - \frac{v_0}{2} \sum_{k=1}^{n} n\, \sigma^{k-1} b_{n-k} \right], \quad n = 0, 1, 2, \ldots.$$

Expressions representing isothermal compression and isentropic release from a Hugoniot state may be easily obtained from these results by the methods given in [1].

3. Variable c_v Materials

It was shown in [1] that the characteristic curves comprising the entropic equation of state surface, $s = s(\eta, \varepsilon)$, are themselves isentropes. To determine the value of s on any particular characteristic, we must prescribe initial data for s along some curve which nowhere has a characteristic direction. The most likely place to prescribe this initial data is in zero-strain, constant volume plane. In [1] we assumed that the specific heat at constant volume was a constant. A straight forward calculation then yielded the following initial entropy data.

$$s(0, \varepsilon) = s_0 + c_v \ln\left(1 + \frac{\varepsilon - \varepsilon_0}{\theta_0\, c_v}\right). \tag{14}$$

Although the constant c_v description is adequate for many materials, a variable c_v offers wider application. In this section we consider materials for which c_v is a function of temperature

$$c_v = c_v(\theta). \tag{15}$$

The development in [1] employed η and ε as independent variables. In order to exploit a relationship such as Eq. (15), it is convenient to interchange ε with θ as the second independent variable. Thus we transform the functions

$$\theta = \theta(0, \varepsilon) \quad \text{and} \quad s = s(0, \varepsilon) \tag{16}$$

into new functions

$$\varepsilon = \hat{\varepsilon}(0, \theta) \quad \text{and} \quad s = \hat{s}(0, \theta). \tag{17}$$

In these equations, we have shown the functional dependence with $\eta = 0$ to emphasize that our calculations are confined to the zero strain plane. Using Eqs. (17) we can now employ the thermodynamic relationships

$$\frac{\partial \hat{\varepsilon}}{\partial \theta} = c_v \quad \text{and} \quad \frac{\partial \hat{s}}{\partial \theta} = \frac{c_v}{\theta}. \tag{18}$$

For any given function $c_v(\theta)$, we must solve both Eqs. (18) to obtain the functional forms expressed by Eqs. (17). Eq. $(17)_1$ is then inverted to obtain Eq. $(16)_1$ and this result is finally employed in Eq. $(17)_2$ to give the final initial data, Eq. $(16)_2$. In general, numerical techniques may be needed to perform these manipulations. However, for certain functions $c_v(\theta)$, exact solutions may be obtained. We illustrate the method with a simple linear model for c_v.

Let c_v be given by

$$c_v = \lambda + \mu(\theta - \theta_0) \tag{19}$$

where λ and μ are constants. Substituting Eq. (19) into Eqs. (18) and integrating the resulting differential equations gives

$$\hat{\varepsilon}(0, \theta) = \hat{\varepsilon}(0, \theta_0) + \lambda(\theta - \theta_0) + \tfrac{1}{2}\mu(\theta - \theta_0)^2, \tag{20}$$

$$\hat{s}(0, \theta) = \hat{s}(0, \theta_0) + \mu(\theta - \theta_0) + (\lambda - \mu\theta_0)\ln(\theta/\theta_0). \tag{21}$$

Inverting Eq. (20) we find (provided $\mu \neq 0$)

$$\theta(0, \varepsilon) = \theta_0 - \frac{1}{\mu}\left(\lambda - \sqrt{\lambda^2 + 2\mu(\varepsilon - \varepsilon_0)}\right). \tag{22}$$

This result is then used in Eq. (21) to give the following initial entropy data

$$\begin{aligned}
s(0, \varepsilon) = s_0 - \lambda + \sqrt{\lambda^2 + 2\mu(\varepsilon - \varepsilon_0)} \\
+ (\lambda - \mu\theta_0)\ln\left[1 - \frac{\lambda - \sqrt{\lambda^2 + 2\mu(\varepsilon - \varepsilon_0)}}{\mu\theta_0}\right].
\end{aligned} \tag{23}$$

We emphasize that this equation applies only in the $\eta = 0$ plane. The general expression for entropy as a function of both η and ε is obtained by combining Eq. (23) with the equation of the characteristic ground curves given in [1].

Other expressions for c_v will also lead to exact equations for $s(0, \varepsilon)$. For example, if

$$c_v = [\lambda + \mu(\theta - \theta_0)]^{-1}$$

we find that

$$s(0, \varepsilon) = s_0 - \frac{1}{\lambda - \theta_0\mu}\left\{\mu(\varepsilon - \varepsilon_0) + \ln\left[1 - \frac{\lambda}{\theta_0\mu}(1 - e^{\mu(\varepsilon - \varepsilon_0)})\right]\right\}.$$

More complicated functions linking c_v to θ may be used, but numerical procedures will probably be involved. For many $c_v(\theta)$ models, Eqs. (18) may be readily integrated, leaving only the inversion of Eq. $(17)_1$ to be done numerically.

4. Synthesis of the Gruneisen Parameter

An important problem in shock mechanics and related fields of high pressure physics is determination, for a given material, of the functional dependence of γ on η. Although direct measurements have been made [3, 4, 5], they are confined to elastic stresses and evidently have little effect on high pressure research at present. Other approaches centre on the equations of Slater [6] or Dugdale and McDonald [7], or on shock wave experiments with porous samples (for example, see the work of Ahrens et al. [8]). Both of these approaches also contain drawbacks. The developments in [6] and [7] have been criticized by Knopoff and Shapiro [9]. Porous sample experiments are restricted to materials which are available, or may be synthesized, in distended form. In this section, we discuss a new method, suggested in [1], for synthesizing the Gruneisen parameter from isothermal compression data. Our aim is to determine the coefficients b_n in the expansion [11] for g. In what follows, we employ the development of [1], where the Hugoniot pressure is given by Eq. (6). In general, we assume that both Eqs. (6) and (11) are truncated after an arbitrary, but finite number of terms.

If c_v is assumed to be constant, the pressure p_R on an isotherm at temperature θ_R will be

$$p_R = p_H\left(1 - \tfrac{1}{2}v_0\,\eta\,g\right) + g\left\{\sum_{n=1}^{\infty}\beta_n\,\eta^n + c_v\left[\theta_R - \theta_0\left(1 + \sum_{n=1}^{\infty}\alpha_n\,\eta^n\right)\right]\right\} \quad (24)$$

where α_n and β_n are given by Eqs. (13), with the exception that

$$P_{n+1} = a_{n+1} - \tfrac{1}{2}v_0\sum_{k=0}^{n}a_k\,b_{n-k}, \qquad n = 0, 1, 2, \ldots. \quad (25)$$

We assume that the Hugoniot pressure p_H is centred at zero, so that $a_0 = 0$ in Eq. (6). Eqs. (24) and (25) are derived in detail in [1].

We can employ Eq. (24) to synthesize the function g from measured isothermal compression data in the following way. Initially, numerical values for the coefficients a_n in Eq. (6) are determined by directly fitting measured Hugoniot data. We assume that an appropriate value for c_v is known. The coefficients b_n in Eq. (11) can then be varied to achieve a best fit between Eq. (24) and measured isothermal data. Of course, the coefficients b_n do not appear in Eq. (24) linearly. Hence classical least square fitting techniques are not especially helpful. Numerical techniques which obtain rapid approximate solutions to nonlinear fitting problems are in wide use, however. In actual application, we have utilized the steepest descent method to minimize the squared error in pressure. Computational time depends mainly on how many coefficients b_n are employed in the expansion for g, and on how good the initial guesses for the coefficients are.

4 a. Gruneisen Parameter for Water

We illustrate the ideas given above with a specific example. A fourth order representation for the Gruneisen parameter for water is determined below. Water was selected as an example material since it evidently possesses an unusually complex $\gamma - \eta$ relationship, and because ample experimental data is readily available.

Present knowledge of the dynamic behaviour of water at high pressures stems from the work of Walsh and Rice [10]. They experimentally determined sixteen points on the water Hugoniot. Subsequently, Rice and Walsh [11] proposed a water equation of state for the pressure range 25 to 250 kbar. Papetti and Fujisaki [12] extended the Rice and Walsh description to 450 kbar and extrapolated these results to very high pressures. The Walsh and Rice data have also been discussed by Cowperthwaite and Shaw [13], and by Gurtman, Kirsh, and Hastings [14].

We have used the Hugoniot data of Walsh and Rice, together with the data given by Lysne [15], to obtain a least square fit for the Hugoniot pressure p_H. Centering the Hugoniot at zero pressure, we set $a_0 = 0$. The next six coefficients a_n in the expansion (6) for p_H were determined, and these are given in Table 1.

Table 1. Coefficients in Water Equation of State

n	a_n (Mbar)	b_n
0	0	0.346
1	0.0220	1.27
2	0.0768	27.54
3	0.450	$-$ 52.90
4	$-$ 0.941	$-$ 25.84
5	7.930	—

A constant value of 0.82 cal/gm/°K was selected for c_v, based on the data of Vedam and Holton [16]. This value is slightly larger than the 0.78 cal/gm/°K used by both Cowperthwaite and Shaw [13] and by Gurtman *et al.* [14], but trial calculations employing both values for c_v have shown that the high pressure response is only marginally effected by the different values.

Eq. (24) can now be fitted to isothermal compression data to determine the coefficients b_n. The most useful data for this purpose are the 125 °C and 175 °C isotherms of Bridgman [17], primarily because the greatest range of specific volumes is considered. Unfortunately, for these two isotherms, Bridgman reports only the volume change as a function of pressure, starting from 5000 kg/cm². The initial volumes at 5000 kg/cm² are not given. Hence, in order to employ this data, we must first estimate the specific volume of water at 5000 kg/cm² and 125° and 175 °C.

To estimate Bridgman's missing initial volumes, we will establish a provisional expression for the function g, then invert Eq. (24) to give η for $p_\mathrm{R} = 5000$ kg/cm² and $\theta_\mathrm{R} = 125°$ and 175 °C. This provisional version of g will be based on the first two coefficients of Eq. (11). That is, we temporarily assume that

$$g = b_0 + b_1 \eta. \tag{26}$$

Referring to Eq. (24), the isothermal pressure at zero strain is given by

$$p_\mathrm{R} = b_0 c_v (\theta_\mathrm{R} - \theta_0). \tag{27}$$

We now interpolate the 80 °C isotherm reported by Vedam and Holton [16] to obtain a value of 0.630 kbar for the zero-strain pressure. Using this value in Eq. (27), and setting $\theta_\mathrm{R} = 80$ °C, $\theta_0 = 20$ °C, and $c_v = 0.82$ cal/gm/°C, we obtain a provisional value for b_0 of 0.305 gm/cm³.

Next, differentiating Eq. (24) with respect to η, and then setting $\eta = 0$, we find the following expression for the isothermal bulk modulus K_R at zero-strain.

$$K_\mathrm{R} = a_1 - v_0 c_v \theta_0 b_0{}^2 + b_1 c_v (\theta_\mathrm{R} - \theta_0). \tag{28}$$

A value for K_R at $\eta = 0$ may be interpolated from Vedam and Holton's data. For $\theta = 80$ °C, we find that $K_\mathrm{R} = 25.3$ kbar. Using this value in Eq. (28), to-

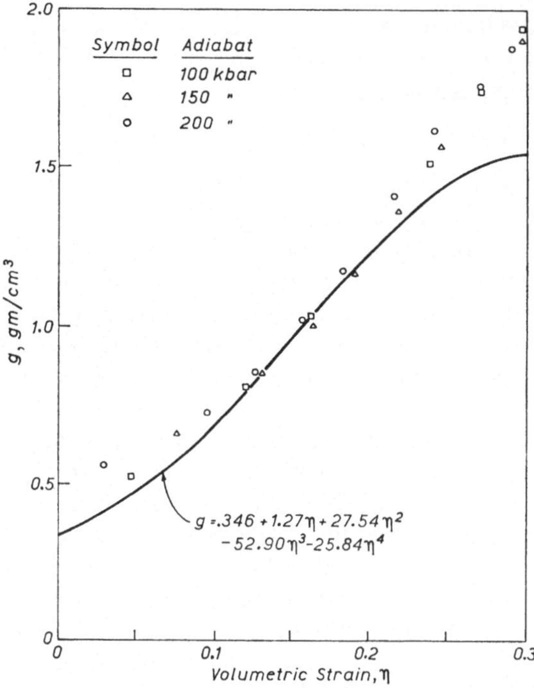

Fig. 1. Variation in g with strain, compared with data from Rice and Walsh [11]

gether with the above value for b_0, we may solve for the provisional value of b_1. This gives $b_1 = 1.63$ gm/cm^3.

Finally, setting $p_R = 5000$ kg/cm^2, and $\theta_R = 125\,°C$ and $175\,°C$, Eq. (24) can be inverted to give η at these two temperatures. The resulting estimated specific volumes are 0.909 cm^3/gm and 0.933 cm^3/gm respectively for $\theta_R = 125°$ and $175\,°C$. These same specific volumes were estimated by Rice and Walsh [11]. Their estimates were 0.914 cm^3/gm and 0.936 cm^3/gm.

Bridgman's data together with the initial volumes we have estimated, now completely specify the 125° and 175 °C water isotherms. Therefore, we discard the provisional representation for g used above, and simultaneously fit Eq. (24) to both isotherms, varying the first five coefficients in Eq. (11). The numerical values for $b_0, \ldots, b_4$ determined in this way are given in Table 1. A graph of g versus η is shown in Fig. 1. The range of validity for our representation is restricted by the isothermal data from which it is obtained. The largest value of η reported by Bridgman was 0.29, obtained from the 175 °C isotherm. Extrapolation of our polynomial representation for g above this value is unwarrented and could lead to gross errors.

A comparison between the polynomial form for g given above and the water equation of state of Rice and Walsh [11] can be made. Gurtman et al. [14], have shown that the isentropes synthesized by Rice and Walsh may be differenced to obtain values of γ (and hence g) at specific strains. This calculation was performed and the resulting data plotted on Fig. 1. Data corresponding to release isentropes from 100, 150, and 200 kbar, calculated by Rice and Walsh, are shown. It must be emphasized that this comparison is drawn between two models for water, not between our model and actual measurements.

4b. Discussion

The above method for determining γ appears to offer certain advantages over other determinations. Primary among these is the relative ease with which isothermal compression data may be obtained. Published isothermal data exist for a wide range of materials; hence, in many instances, no experimentation need be performed. With only a minor computational effort, one can obtain adequate representations for g, and hence γ. Moreover, we are in no way restricted to using the constant c_v model. The ideas described above can be applied equally well to the $c_v(\theta)$ case.

Most nonlinear fitting methods require starting guesses for the unknown coefficients b_n. Relatively good guesses for b_0 and b_1 can be easily obtained in the way our provisional values of b_0 and b_1 were found above. Care must be exercised when using these numerical procedures, however. It is always possible (yet difficult to detect) that one finds a local rather than absolute minimum in squared error. This "lumpy mattress" effect can be a source of error.

The major drawback to our method lies in its range of applicability. As was pointed out above, the resulting description for g is valid only over the range of strains corresponding to the isothermal data from which it was obtained. Often, values for g at strains in excess of those attainable in isothermal tests may be required. In these cases, one or another of the other determinations for γ must be employed.

References

1. Davis, R. O.: Phys. kondens. Materie 15, 230 (1972)
2. McQueen, R. G., Marsh, S. P., Taylor, J. W., Fritz, J. N., Carter, W. J.: In High Velocity Impact Phenomena, ed. R. Kinslow. New York: Academic Press 1970
3. Gauster, W. B.: Phys. Rev. B, 4, 1288 (1971)
4. Perry, F. C.: J. appl. Phys. 41, 5017 (1970)
5. Oswald, R. B., McLean, F. B., Schallhorn, D. R., Buxton, L. D.: Appl. Phys. Letters 16, 24 (1970)
6. Slater, J. C.: Introduction to Chemical Physics. New York: McGraw Hill 1939
7. Dugdale, J. S., McDonald, D. K. C.: Phys. Rev. 89, 832 (1953)
8. Ahrens, T. J., Takahashi, T., Davies, G. F.: J. Geophys. Res. 75, 310 (1970)
9. Knopoff, L., Shapiro, J. N.: J. Geophys. Res. 74, 1439 (1969)
10. Walsh, J. M., Rice, M. H.: J. Chem. Phys. 26, 815 (1957)
11. Rice, M. H., Walsh, J. M.: J. Chem. Phys. 26, 824 (1957)
12. Papetti, R. A., Fujisaki, M.: J. appl. Phys. 39, 5412 (1968)
13. Cowperthwaite, M., Shaw, R.: J. chem. Phys. 53, 555 (1970)
14. Gurtman, G. A., Kirsh, J. W., Hastings, C. R.: J. appl. Phys. 42, 851 (1971)
15. Lysne, P. C.: J. Geophys. Res. 75, 4375 (1970)
16. Vedam, R., Holton, G.: J. Acoust. Soc. Amer. 43, 108 (1967)
17. Bridgman, P. W.: Amer. Acad. Arts. Sci. 74, 399 (1942)

Dr. R. O. Davis
Department of Civil Engineering
University of Canterbury
Christchurch
New Zealand

Phys. cond. Matter 17, 71—97 (1974)
© by Springer-Verlag 1974

Non-Radiative Electronic Capture Rates by a Halide Vacancy in Alkali Halides

S. Fraser

Institut für Theoretische Physik, Universität Tübingen
and
Department of Engineering Science, Parks Road, Oxford

Received July 2, 1973

We treat non-radiative electronic capture rates induced by the non-adiabatic perturbation operator with corresponding transition probabilities. Therefore a dynamical electronic wavefunction is evaluated. By coupling the electron to "dressed phonon" modes, which connect the "local" optical mode to the acoustic phonons, exact energy balance in the transition probabilities is possible. The Franck-Condon-Integrals arising there are thermal averaged analytically. The method is applicable to transitions between two discrete electronic levels as well as band to discrete level transitions. Numerical results are given.

Introduction

An important physical problem is the treatment of electronic transitions between discrete electronic states in semiconducting substances. If energy is to be strictly conserved, as is physically required, then the discrete electronic states must be coupled to some form of continuous spectra. For radiative transitions the radiation field itself provides such a spectrum but non-radiative transitions, where the electronic energy is transformed only into phonon energy, are not so easy to treat.

Large molecules and crystals contain a quasi-continuous spectrum, that of the acoustic phonons, and if this is directly coupled to the electron then again the problem can be solved. Unfortunately there are many instances where the electron is coupled to phonon modes that do not lie within the acoustic branches. The F-centre in alkali halides is a well documented example.

Study of the experimental data for F-luminescence enables one to calculate an effective frequency for the coupled modes [1] which lies close to or inside the optical phonon band. In this case, calculation of non-radiative transition rates generally requires some manipulation of the Dirac delta function that expresses absolute energy conservation.

Various methods have been employed, the simplest being an approximation to a Kronecker delta on the coupled mode quantum numbers, implying an exact energy fit with harmonic phonon states. The popular alternative is to replace the delta function by a less stringent distribution function. Such a smearing of the delta function can be obtained by direct replacement by a Lorentz [2] or Gauss [3] distribution or by fourier transforming the delta function and restricting the resultant summations [4] or integrations [5]. While these methods might be motivated by physical argument, they are mathematical non-sequitors whereby the essential quality of the delta function is arbitrarily altered.

In 1957, to overcome this difficulty, Stumpf suggested [6, 7] taking into account the resonance character of the combined electron-phonon states leading thus to well defined Lorentz factors instead of delta-distributions. He performed calculations that agreed quite well with experiment. Later Stumpf showed [8, 9, 10] that the resonance character of these states can be obtained by introducing "dressed phonons" which lead to an indirect coupling of the electron to the acoustic phonons of the crystal. The "dressed phonons" couple the local mode to the acoustic phonons by the anharmonic phonon-phonon interaction. The use of dressed phonons instead of pure harmonic phonon states enables us to satisfy energy conservation exactly. It is interesting to note that the result is similar, though not exactly the same, to using the Lorentz approximation.

Stumpf's methods are essential for the sound mathematical treatment of multiphonon non-radiative transitions between discrete electronic levels. They are also applicable to transitions between a band and a discrete level since the coupling of an additional continuous spectrum greatly increases the range of electronic energies from which an energy conserving transition is possible.

In this paper we aim to show that numerical evaluations of Stumpf's model are tractable. As an example we use the F-centre in alkali halides and calculate non-radiative transitions from the bottom of the conduction band to the F-centre ground state. We do not explicitly use the continuous electronic spectrum so that we can demonstrate the applicability of our calculations to discrete levels. We are restricted to this transition at the moment solely by lack of suitable electronic wavefunctions.

1. The Stochastic Reaction Equations

We consider the non-radiative trapping of an electron by a negative-ion vacancy in ionic crystals to produce an F-centre in its ground state. After Stumpf [1, 2] we describe the defect electron system by the hamiltonian

$$\mathscr{H} = \mathscr{H}_0 + {}^t\mathscr{H} \tag{1.1a}$$

which we have separated into an adiabatic part $\mathscr{H}_0$ and a non-adiabatic perturbation ${}^t\mathscr{H}$. $\mathscr{H}_0$ satisfies the equation

$$\mathscr{H}_0 \,|n, m\rangle = E_m^n \,|n, m\rangle \tag{1.1b}$$

where n is the electronic quantum number and $[m] \equiv m_1, \ldots, m_t, \ldots, m_{3N}$ is the complete set of phonon quantum numbers.

Non-radiative electronic transitions induced by ${}^t\mathscr{H}$ have differential transition probabilities given by

$$^tW_{n'm',nm} = \frac{2\pi}{\hbar} \,|\langle n', m'|\,{}^t\mathscr{H}\,|n, m\rangle|^2 \,\delta(E_m^n - E_{m'}^{n'}). \tag{1.2}$$

We obtain the total electronic transition probability by summing over all final phonon states and thermal weighting with respect to the initial states

$$^tW_{n',n}(\beta) = \mathscr{N} \sum_{m',m} {}^tW_{n'm',nm}\, e^{-\beta \sum m_t \hbar \omega_t}, \tag{1.3a}$$

$$\mathscr{N} = \prod_t (1 - e^{-\beta \hbar \omega_t}); \quad \beta^{-1} = \varkappa T. \tag{1.3b}$$

In [2] Stumpf derives the stochastic equations in the case of non-radiative transitions. For only two electronic levels these simplify to

$$\dot{P}_n(t) = P_{n'}(t)\,{}^tW_{nn'}(\beta) - P_n(t)\,{}^tW_{n'n}(\beta)\,, \tag{1.4}$$
$$\dot{P}_{n'}(t) = -\,\dot{P}_n(t)$$

where $P_n(t)$ is the temporal occupation probability of the electronic state n. Using the initial conditions $P_n(0) = 1$, $P_{n'}(0) = 0$ we solve (1.4) to obtain:

$$P_n(t) = \Gamma^{-1}({}^tW_{nn'}(\beta) + {}^tW_{n'n}(\beta)\,e^{-\Gamma t})\,,$$
$$P_{n'}(t) = \Gamma^{-1}\,{}^tW_{n'n}(\beta)\,(1 - e^{-\Gamma t})\,, \tag{1.5}$$
$$\Gamma = {}^tW_{nn'}(\beta) + {}^tW_{n'n}(\beta)\,.$$

We assume that only one local phonon mode is coupled to the electron and so after a transition this mode is left highly excited. There is only a small probability, however, for re-excitation of the electron if the phonon modes decay in a time shorter than that characteristic of an electronic transition. This indeed seems to be the case. For temperatures up 300 K we neglect the transition into the conduction band as being highly improbable and describe the system solely with the rate ${}^tW_{\text{FC-CB}}(\beta)$.

2. The Wave Functions

The F-centre ground state wave function has been calculated by Schmid [11] and Renn [12] in the static case. From this wave function we derive a dynamic one by the procedure of Stumpf [10] which is amplified by Schwarz [13] and Heinzel [14]. We take the "dressed phonons" wave functions from Stumpf [8] which couple the "local mode" to the continuous spectrum of the acoustic modes of the crystal via the phonon-phonon interaction.

2.1 The Dynamical Electronic Wave Functions

This section is taken directly from [10] using Stumpf's simplified crystal model where the polarizability of the crystal is considered small. The longitudinal optical phonons are degenerated and the "local mode" is assumed to have the same frequency. We expand the electronic wave function into a set of h ortho-normal functions:

$$\Psi_n(r, Q) = \left(\sum_{k=1}^{h}\alpha_k^2\right)^{-1/2}\sum_{k=1}^{h}\alpha_k\,\Psi_n^k(r) \tag{2.1}$$

where

$$\alpha_k = \delta_{1,k} + \Delta_k \tag{2.2}$$

and [2] gives us:

$$\Delta_l = -\sum_{k\varrho}\sum_{r}\mathbf{A}_{lr}^{-1}\left(\frac{\partial}{\partial\alpha_r}\frac{\partial}{\partial R_{k\varrho}}\,U_n\right)_s X_{k\varrho}^n \tag{2.3}$$

and

$$\mathbf{A}_{lj} = \left(\frac{\partial}{\partial\alpha_l}\frac{\partial}{\partial\alpha_j}\,U_n\right)_s. \tag{2.4}$$

U_n is the total energy of the crystal and the subscript "s" means that the expression is to be evaluated in the static limit. The $X_{k\varrho}^n$ are the displacements of the $k\varrho$-th ion from its static equilibrium position when the electron is in the n-th state.

From [2] we have

$$\left(\frac{\partial}{\partial \alpha_r}\,\frac{\partial}{\partial \boldsymbol{R}_{k\varrho}}\,U_n\right)_s = 2\,(1+\alpha^e)^{-1}\,e\,a_{k\varrho}\,\boldsymbol{\nabla}_{k\varrho}\int \Psi_n^1(\boldsymbol{r})\,\Psi_n^r(\boldsymbol{r})\,(1-\delta_{1r})\,C\,(\boldsymbol{r},\boldsymbol{R}_{k\varrho})\,d\tau$$

$$(2.5)$$

in the notation of Renn. Substituting in (2.3) yields

$$\Delta_l = -\sum_{k\varrho}\sum_r 2\,\mathbf{A}_{lr}^{-1}(1-\delta_{1r})\,e\,a_{k\varrho}\,\boldsymbol{\nabla}_{k\varrho}\int\Psi_n^1(\boldsymbol{r})\,\Psi_n^r(\boldsymbol{r})\,C\,(\boldsymbol{r},\boldsymbol{R}_{k\varrho})\,d\tau\,X_{k\varrho}^n. \quad (2.6)$$

We have the following equality for the ion position vectors

$$\boldsymbol{R}_{k\varrho} = \boldsymbol{R}_{k\varrho}^n + \boldsymbol{X}_{k\varrho}^n \equiv \boldsymbol{R}_{k\varrho}^{n'} + \boldsymbol{X}_{k\varrho}^{n'} \qquad (2.7)$$

from which follows

$$\boldsymbol{X}_{k\varrho}^{n'} = \boldsymbol{X}_{k\varrho}^n + \boldsymbol{R}_{k\varrho}^n - \boldsymbol{R}_{k\varrho}^{n'}. \qquad (2.8)$$

We define a transformation into normal coordinates

$$\boldsymbol{X}_{k\varrho}^n = \sum_t M_{k\varrho}^{-1/2}\,\mathbf{B}_{k\varrho,t}^n\,\boldsymbol{q}_t^n. \qquad (2.9)$$

Under this transformation (2.8) becomes:

$$\sum_t \mathbf{B}_{k\varrho,t}^{n'}\,\boldsymbol{q}_t^{n'} = \sum_t \mathbf{B}_{k\varrho,t}^n\,\boldsymbol{q}_t^n + M_{k\varrho}^{1/2}\,(\boldsymbol{R}_{k\varrho}^n - \boldsymbol{R}_{k\varrho}^{n'}). \qquad (2.10)$$

We assume now that the transformation (2.9) is unaltered when the defect electronic configuration changes. This implies that the eigenfrequencies of the crystal do not change. Hence we have:

$$\mathbf{B}^n = \mathbf{B}^{n'} = \mathbf{B} \qquad (2.11)$$

and (2.10) becomes

$$\boldsymbol{q}_t^{n'} = \boldsymbol{q}_t^n + \boldsymbol{a}_t^{nn'} \qquad (2.12)$$

with

$$\boldsymbol{a}_t^{nn'} = \sum_{k\varrho}\mathbf{B}_{t,k\varrho}^{-1}\,M_{k\varrho}^{1/2}\,(\boldsymbol{R}_{k\varrho}^n - \boldsymbol{R}_{k\varrho}^{n'}). \qquad (2.13)$$

Renn [11] gives

$$\boldsymbol{R}_{k\varrho}^n - \boldsymbol{R}_{k\varrho}^{n'} = -\frac{\alpha^g\,d^3}{4\,\pi(1+\alpha^e+\alpha^g)}\,\frac{a_{k\varrho}}{e}\,\boldsymbol{\nabla}_{k\varrho}\int[\,|\Psi_n^1(\boldsymbol{r})|^2 - |\Psi_n^1(\boldsymbol{r})|^2]\,C\,(\boldsymbol{r},\boldsymbol{R}_{k\varrho})\,d\tau.$$

$$(2.14)$$

We substitute in (2.13) to obtain:

$$\boldsymbol{a}_t^{nn'} = \sum_{k\varrho}\mathbf{B}_{t,k\varrho}^{-1}\,M_{k\varrho}^{1/2}\,\frac{\alpha^g\,d^3}{4\,\pi(1+\alpha^e+\alpha^g)}\,\frac{a_{k\varrho}}{e}$$
$$\cdot\,\boldsymbol{\nabla}_{k\varrho}\int[\,|\Psi_{n'}^1(\boldsymbol{r})|^2 - |\Psi_n^1(\boldsymbol{r})|^2]\,C\,(\boldsymbol{r},\boldsymbol{R}_{k\varrho})\,d\tau \qquad (2.15)$$

and using (2.9) in (2.6) we get

$$\Delta_l = -\sum_{k\varrho}\sum_{r,t} 2\,\mathbf{A}_{l,r}^{-1}(1-\delta_{1r})\,\frac{e\,a_{k\varrho}}{(1+\alpha^e)}\,\boldsymbol{\nabla}_{k\varrho}\int\Psi_n^1(\boldsymbol{r})\,\Psi_n^r(\boldsymbol{r})\,C\,(\boldsymbol{r},\boldsymbol{R}_{k\varrho})\,d\tau\,M_{k\varrho}^{-1/2}\,\mathbf{B}_{k\varrho,t}\,\boldsymbol{q}_t^n$$

$$(2.16)$$

Because the L.O.P. branch is degenerate we can make a unitary transformation of the normal coordinates to produce an equivalent set.

$$Q_r^n = \sum_t \mathbf{U}_{r,t}\,\boldsymbol{q}_t^n \qquad (2.17)$$

and (2.12) becomes

$$Q_r^{n'} = Q_r^n + \sum_t \mathbf{U}_{r,t}\, a_t^{nn'} = Q_r^n + A_r^{nn'} \,. \tag{2.18}$$

We define the matrix $\mathbf{S}$ by

$$\mathbf{S}_{r,k\varrho} = \sum_t \mathbf{U}_{r,t}\, \mathbf{B}_{t,k\varrho}^{-1}\, M_{k\varrho}^{1/2} \tag{2.19}$$

and operating with $\mathbf{U}$ on (2.15) and (2.16) we arrive at

$$A_t^{nn'} = \sum_{k\varrho} \mathbf{S}_{t,k\varrho} \frac{\alpha^g d^3}{4\pi(1 + \alpha^e + \alpha^g)} \frac{a_{k\varrho}}{e}$$
$$\cdot \nabla_{k\varrho} \int [|\Psi_{n'}^1(r)|^2 - |\Psi_n^1(r)|^2]\, C(r, R_{k\varrho})\, d\tau \,, \tag{2.20}$$

$$\Delta_l = - \sum_{k\varrho} \sum_{r,t} 2\, A_{l,r}^{-1}(1 - \delta_{1r}) \frac{e\, a_{k\varrho}}{(1 + \alpha^e)}$$
$$\cdot \nabla_{k\varrho} \int \Psi_n^1(r)\, \Psi_n^r(r)\, C(r, R_{k\varrho})\, d\tau\, \mathbf{S}_{k\varrho,t}^{-1}\, Q_l^n \,. \tag{2.21}$$

Since the transformation $\mathbf{U}$ is arbitrary and independent of the electronic quantum number and the phonon k vector we can construct $\mathbf{S}$, which must also be from the subspace of the L.O.P.'s, out of an arbitrary but orthonormal set of base vectors $\boldsymbol{\xi}\!\left(m\,\middle|\,\begin{matrix}k\\\varrho\end{matrix}\right)$, $m = 1 \dots N$

$$\mathbf{S}_{r,k\varrho} = \boldsymbol{\xi}\!\left(r\,\middle|\,\begin{matrix}k\\\varrho\end{matrix}\right) M_{k\varrho}^{1/2} \tag{2.22}$$

so that

$$A_t^{nn'} = \sum_{k\varrho} \boldsymbol{\xi}\!\left(t\,\middle|\,\begin{matrix}k\\\varrho\end{matrix}\right) M_{k\varrho}^{1/2} \frac{\alpha^g d^3}{4\pi(1 + \alpha^e + \alpha^g)} \frac{a_{k\varrho}}{e}$$
$$\cdot \nabla_{k\varrho} \int [|\Psi_{n'}^1(r)|^2 - |\Psi_n^1(r)|^2]\, C(r, R_{k\varrho})\, d\tau \,, \tag{2.23}$$

$$\Delta_l = - \sum_{k\varrho} \sum_{r,t} 2\, A_{l,r}^{-1}(1 - \delta_{1r}) \frac{e^2}{(1 + \alpha^e)} \boldsymbol{\xi}\!\left(t\,\middle|\,\begin{matrix}k\\\varrho\end{matrix}\right) M_{k\varrho}^{-1/2} \frac{a_{k\varrho}}{e}$$
$$\cdot \nabla_{k\varrho} \int \Psi_n^1(r)\, \Psi_n^r(r)\, C(r, R_{k\varrho})\, d\tau\, Q_l^n \,. \tag{2.24}$$

We now make the expansion

$$\frac{a_{k\varrho}}{e} \nabla_{k\varrho} \int \Psi_n^1(r)\, \Psi_n^r(r)\, C(r, R_{k\varrho})\, d\tau = \sum_{m=1}^{h} {}^n C_m^r\, \boldsymbol{\xi}\,(m \mid k\varrho) \tag{2.25}$$

and a particular choice of eigenvectors.

$$\boldsymbol{\xi}\!\left(1\,\middle|\,\begin{matrix}k\\\varrho\end{matrix}\right) = C_0^{-1} \frac{a_{k\varrho}}{e} \nabla_{k\varrho} \int [|\Psi_{n'}^1(r)|^2 - |\Psi_n^1(r)|^2]\, C(r, R_{k\varrho})\, d\tau \,, \tag{2.26}$$

where normalization gives

$$C_0^2 = \sum_{k\varrho} |\nabla_{k\varrho} \int [|\Psi_{n'}^1(r)|^2 - |\Psi_n^1(r)|^2]\, C(r, R_{k\varrho})\, d\tau\,|^2 \,. \tag{2.27}$$

Using these representations in (2.33) and (2.24) along with the approximation $M_1 = M_2 = 2\, M_0$ we get:

$$A_t^{nn'} = \frac{\alpha^g d^3 \sqrt{2 M_0}}{4\pi(1 + \alpha^e + \alpha^g)} \sum_{k\varrho} \xi\left(t \,\middle|\, \begin{matrix} k \\ \varrho \end{matrix}\right) C_0 \,\xi\left(1 \,\middle|\, \begin{matrix} k \\ \varrho \end{matrix}\right) = \frac{\alpha^g d^3 \sqrt{2 M_0}}{4\pi(1 + \alpha^e + \alpha^g)} \, C_0 \,\delta_{1t} ,$$

$$\tag{2.28}$$

$$\Delta_l = - \sum_{k\varrho} \sum_{r,t} 2 \, \mathbf{A}_{lr}^{-1} \, (1 - \delta_{1r}) \, \frac{e^2}{(1 + \alpha^e)} \, \xi\left(t \,\middle|\, \begin{matrix} k \\ \varrho \end{matrix}\right) \sum_{m=1}^{h} {}^nC_m^r \, \xi\left(m \,\middle|\, \begin{matrix} k \\ \varrho \end{matrix}\right) Q_t^n \, (2 M_0)^{-1/2}$$

$$= - \sum_r 2^{1/2} \, \mathbf{A}_{l,r}^{-1} (1 - \delta_{1r}) \frac{e^2}{\sqrt{M_0}(1 + \alpha^e)} \sum_{m=1}^{h} {}^n C_m^r \, Q_m^r . \tag{2.29}$$

Our special choice of lattice eigenvectors results in only one mode suffering a coordinate shift. This we call the "local mode". Of the h modes coupled to the electron in (2.29) this is the most important since the operator $^t\mathcal{H}$ can induce large changes in its quantum number. Hence we neglect the coupling of the electron to all modes but the local mode and finally obtain:

$$A_t^{nn'} = \frac{\alpha^g d^3 \sqrt{2 M_0}}{4\pi(1 + \alpha^e + \alpha^g)} \, C_0 \,\delta_{1t} , \tag{2.30a}$$

$$\Delta_l = - \sqrt{\frac{2}{M_0}} \, \frac{e^2}{(1 + a^e)} \sum_{r=2}^{h} \mathbf{A}_{l,r}^{-1} \, {}^nC_1^r \, Q_1^n := C_{l'}^n \, Q_1^n . \tag{2.30b}$$

2.2. Choice of Wave Functions

We need only consider a dynamic wave function for one of our two electronic states so we develope the F-centre ground state wave function. The symmetric matrix $\mathbf{A}$ defined in (2.4) has $\mathbf{A}_{11} = 0$ as is shown in appendix A, so that for a non-trivial dynamical expansion we are forced to use a minimum of three functions in (2.1).

Both Schmid [11] and Renn [12] fitted the ground state with the static wave function

$$\Psi_F^1(r) = \sqrt{\frac{\beta^3}{7\pi}} \, (1 + \beta r) \, e^{-\beta r} \tag{2.31}$$

but since we need two move functions that are orthogonal to this we find it simpler to use a hydrogenic wave function — also fitted in [11] — to describe the static wave function. The other wave functions in the expansion must be s-type since the operator $^t\mathcal{H}$ transforms as Γ_1 under operations of the crystal point group. Thus we choose

$$\Psi_F^1(r) = \sqrt{\frac{\beta^3}{\pi}} \, e^{-\beta r} \tag{2.32a}$$

$$\Psi_F^2(r) = \sqrt{\frac{\beta^2}{8\pi}} \left(1 - \frac{1}{2}\beta r\right) e^{-\frac{1}{2}\beta r} , \tag{2.32b}$$

$$\Psi_F^3(r) = \sqrt{\frac{\beta^3}{27\pi}} \left(1 - \frac{2}{3}\beta r + \frac{2}{27}\beta^2 r^2\right) e^{-\frac{1}{3}\beta r} . \tag{2.32c}$$

Then using (2.30b) and (2.32) we expand (2.1) as a function of Q_1^F, taking terms in the power series to first order to obtain

$$\Psi_F(r, Q) = \Psi_F^1(r) + (C_{21}^F \Psi_F^2(r) + C_{31}^F \Psi_F^3(r)) \, Q_1^F . \tag{2.33}$$

For the conduction band wave function we take a plane wave normalized to the volume of the micro-block. The micro-block contains one defect

$$\Psi^1_{\mathrm{CB}}(r) = \Omega^{-1/2}\, e^{ik\cdot r}\ .$$

We further assume that transitions occur only from points near the bottom of the band, $|k| \sim 0$ and so have:

$$\Psi^1_{\mathrm{CB}}(r) = \Omega^{-1/2}\ . \tag{2.34}$$

2.3. Phonon Wave Functions — the Dressed Phonons

The "dressed phonons" are obtained from the lattice wave functions in the harmonic approximation by solving the anharmonic phonon-phonon interaction problem. A derivation based on the resolvent techniques of Rampacher [15] is given by Stumpf [8]. The problem has been considered fully in [14]. We restrict ourselves to those terms diagonal in the resolvent and consider phonon-phonon interaction arising solely from the third order term of the crystal energy expansion. A further simplification is obtained by performing the calculation in the perfect crystal. Schwarz [13] gives:

$$\Psi^n_m(R) = |\,m\rangle + \sum_{m'} \frac{\langle m'|\,{}^a\mathscr{H}\,|\,m\rangle}{E^n_m - E^n_{m'} + \dfrac{i}{2}\,\Gamma_{m'}}\,|\,m'\rangle \tag{2.35}$$

where

$$\Gamma_m = 2\pi \sum_{m'} |\langle m|\,{}^a\mathscr{H}\,|\,m'\rangle|^2\,\delta(E^n_m - E^n_{m'})\ . \tag{2.36}$$

We use the notation

$$|\,m\rangle = \prod_{t=1}^{3N+1} |\,m_t\rangle \tag{2.37a}$$

and in particular

$$|\,m\rangle = |\,l\rangle \prod_{t=2}^{3N+1} |\,m_t\rangle \tag{2.37b}$$

where $|\,l\rangle$ is the local mode wave function.

3. The Anharmonic Interaction

We express the crystal lattice energy by the phenomenological potential

$$U_0 = \frac{1}{2} \sum_{i,j}{}^{+} \left\{ \frac{-e^2}{|R_i - R_j|} + \frac{b}{|R_i - R_j|^n} \right\} \tag{3.1}$$

and consider small displacements from the static equilibrium positions

$$R_{k\varrho} = R^0_{k\varrho} + X^n_{k\varrho}\ . \tag{3.2}$$

In a Taylor expansion of (3.1) about the static positions the first order term is considered zero as a requisit for equilibrium. Then the zero order and second order terms yield the crystal hamiltonian in the harmonic approximation. Phonon-phonon coupling is provided by the higher order terms in the expansion. We

78   S. Fraser

restrict ourselves to the third order term

$$
{}^{a}\mathscr{H} = \frac{1}{3!} \sum_{\substack{ll'l'' \\ \mu\mu'\mu'' \\ \alpha\alpha'\alpha''}} G_{\alpha\alpha'\alpha''}\begin{pmatrix} l\ l'\ l'' \\ \mu\mu'\mu'' \end{pmatrix} X_\alpha\begin{pmatrix} l \\ \mu \end{pmatrix} X_{\alpha'}\begin{pmatrix} l' \\ \mu' \end{pmatrix} X_{\alpha''}\begin{pmatrix} l'' \\ \mu'' \end{pmatrix} \tag{3.3}
$$

where

$$
G_{\alpha\alpha'\alpha''}\begin{pmatrix} l\ l'\ l'' \\ \mu\mu'\mu'' \end{pmatrix} = \left(\frac{\partial}{\partial X_\alpha}\begin{pmatrix} l \\ \mu \end{pmatrix} \frac{\partial}{\partial X_{\alpha'}}\begin{pmatrix} l' \\ \mu' \end{pmatrix} \frac{\partial}{\partial X_{\alpha''}}\begin{pmatrix} l'' \\ \mu'' \end{pmatrix} U_0 \right)_s . \tag{3.4}
$$

Heinzel [14] has considered a more general approach involving terms of higher order.

We transform the displacements $X_\alpha\begin{pmatrix} l \\ \mu \end{pmatrix}$ into reduced coordinates

$$
Y_\alpha\begin{pmatrix} l \\ \mu \end{pmatrix} = M_\mu^{1/2}\, X_\alpha\begin{pmatrix} l \\ \mu \end{pmatrix} \tag{3.5}
$$

and expand these new displacements into the normal modes of the lattice. The transformation matrix is found from the eigenvectors of the harmonic problem and these satisfy ortho-normality and completeness relations

$$
\sum_{l\mu\alpha} \xi_\alpha^*\begin{pmatrix} l & k \\ \mu & \sigma \end{pmatrix} \xi_\alpha\begin{pmatrix} l & k' \\ \mu & \sigma' \end{pmatrix} = \delta_{kk'}\,\delta_{\sigma\sigma'} , \tag{3.6}
$$

$$
\sum_{k\sigma} \xi_\alpha^*\begin{pmatrix} l & k \\ \mu & \sigma \end{pmatrix} \xi_{\alpha'}\begin{pmatrix} l' & k \\ \mu' & \sigma \end{pmatrix} = \delta_{\alpha\alpha'}\,\delta_{\mu\mu'}\,\delta_{ll'} . \tag{3.7}
$$

Stumpf [10] postulates in his simplified crystal model that the local mode can be constructed by taking a linear combination of L.O.P. modes. We find it is more convenient to assume that the local mode is an additional independent mode of the crystal. As a direct consequence of Stumpf's hypothesis this mode will be orthogonal to all crystal modes except the L.O.P.'s where there will be a small overlap. However we are interested solely in the coupling of the local mode to the "continuous" spectrum of the acoustical phonons, where the orthogonality is still good. Hence our expansion of (3.5) into normal modes is:

$$
Y_\alpha\begin{pmatrix} l \\ \mu \end{pmatrix} = \xi_\alpha\begin{pmatrix} l & \\ \mu & 1 \end{pmatrix} Q_1 + \sum_{k\sigma} \xi\begin{pmatrix} l & k \\ \mu & \sigma \end{pmatrix} Q_{k\sigma} \tag{3.8}
$$

where the k summation is over the whole of the Brillouin zone and runs over all branches. We substitute (3.8) in (3.3) and select only those terms that contain the local mode and acoustic modes

$$
{}^{a}\mathscr{H} = \frac{1}{2} \sum_{\substack{ll'l'' \\ \mu\mu'\mu'' \\ \alpha\alpha'\alpha''}} \frac{G_{\alpha\alpha'\alpha''}\begin{pmatrix} l\ l'\ l'' \\ \mu\mu'\mu'' \end{pmatrix}}{(M_\mu M_{\mu'} M_{\mu''})^{1/2}} \left\{ \xi_\alpha\begin{pmatrix} l & \\ \mu & 1 \end{pmatrix} \sum_{\substack{kk' \\ \sigma\sigma'}} \xi_{\alpha'}\begin{pmatrix} l' & k \\ \mu' & \sigma \end{pmatrix} \xi_{\alpha''}\begin{pmatrix} l'' & k' \\ \mu'' & \sigma' \end{pmatrix} Q_1 Q_{k\sigma} Q_{k'\sigma} \right.
$$

$$
\left. + \xi_\alpha\begin{pmatrix} l & \\ \mu & 1 \end{pmatrix} \xi_{\alpha'}\begin{pmatrix} l' & \\ \mu' & 1 \end{pmatrix} \sum_{k\sigma} \xi_{\alpha''}\begin{pmatrix} l'' & k \\ \mu'' & \sigma \end{pmatrix} Q_1^2 Q_{k\sigma} \right\} . \tag{3.9}
$$

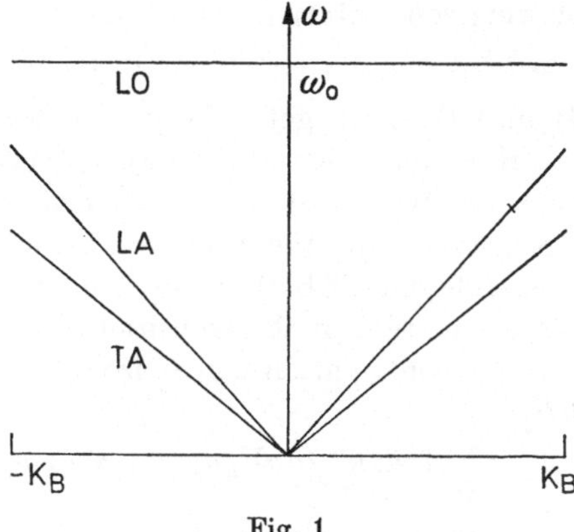

Fig. 1

Märtle [16] has solved the harmonic problem and gives for the acoustic eigenvectors:

$$\xi_\alpha\begin{pmatrix} l & k \\ \mu & \sigma \end{pmatrix} = N^{-1/2} \left(\frac{M_0}{M_{3-\mu}}\right)^{1/2} e_{\sigma\alpha}(k)\, e^{i\,k\cdot R_{l\mu}} \tag{3.10}$$

where

$$M_0 = M_1 M_2 (M_1 + M_2)^{-1}$$

and

$$e_4(k) = \begin{pmatrix} \sin\theta_k \cos\Phi_k \\ \sin\theta_k \sin\Phi_k \\ \cos\theta_k \end{pmatrix}; e_5(k) = \begin{pmatrix} -\sin\Phi_k \\ \cos\Phi_k \\ 0 \end{pmatrix}; e_6(k) = \begin{pmatrix} \cos\theta_k \cos\Phi_k \\ \cos\theta_k \sin\Phi_k \\ -\sin\theta_k \end{pmatrix} \tag{3.11}$$

are the polarization vectors for the longitudinal and two transverse modes respectively.

From (2.26) our local mode eigenvector is

$$\xi_\alpha\begin{pmatrix} 1 & k \\ & \varrho \end{pmatrix} = (-1)^{3-\varrho} F_{k\varrho}\, e_{4\alpha}(k\varrho) \tag{3.12}$$

and

$$F_{k\varrho} = \left| C_0^{-1} \nabla_{k\varrho} \int [\,|\Psi_F^1(r)|^2 - |\Psi_{CB}^1(r)|^2\,] C(r, R_{k\varrho})\, d\tau \right|. \tag{3.13}$$

From his calculations Märtle estimates an average L.O.P. frequency which is associated with our "local mode". We also assume that the dispersion relations for the acoustic branches can be represented by straight lines. The Brillouin zone is assumed iso-tropic (Fig. 1).

We now write the normal modes Q in terms of creation and annihilation operators

$$Q_1 = \left(\frac{\hbar}{2\,\omega_0}\right)^{1/2} (b + b^\dagger), \tag{3.14a}$$

$$Q_{\sigma k} = \left(\frac{\hbar}{2\,v_\sigma |k|}\right)^{1/2} (a_{k\sigma} + a_{\sigma,-k}^\dagger) \tag{3.14b}$$

where we have the following commutation relationships

$$[a_{\sigma k}, a_{\sigma' k'}] = \delta_{\sigma \sigma'} \, \delta_{k k'}; \quad [b, b^\dagger] = 1; \quad [b^\dagger, a_{\sigma k}^{(\dagger)}] = 0 . \tag{3.14c}$$

On substituting (3.14) into (3.9) we get a large number of possible operator combinations. However if we consider an energy conserving process, i.e. the calculation of the phonon line width, then only two combinations remain, being $b \, a_{\sigma k}^\dagger a_{\sigma' k'}^\dagger$; $b^\dagger a_{\sigma k} a_{\sigma' k'}$. Secondly with the proposed model of the Brillouin zone this energy balance is possible only if both acoustic modes come from the longitudinal branch. For the calculation of the transition rates all combinations are possible but since the major contribution comes from the above terms we take only these to have finally

$$^a\mathscr{H} = \sum_{k k'} A(k, k') \, \{b \, a_{-k}^\dagger a_{-k'}^\dagger + b^\dagger a_k a_{k'}\} \tag{3.15a}$$

and

$$A(k, k') = \left(\frac{\hbar^3}{8\,\omega_0 \, v^2 \, |k| \, |k'|}\right)^{1/2} \frac{M_0}{2 \, N \, M_1 \, M_2} \sum_{\substack{l l' l'' \\ \mu \mu' \mu'' \\ \alpha \alpha' \alpha''}} G_{\alpha \alpha' \alpha''} \begin{pmatrix} l & l' & l'' \\ \mu \mu' \mu'' \end{pmatrix}$$

$$\times (-1)^{3-\mu} \, e^{i k . R_{l' \mu'}} \, e^{i k . R_{l'' \mu''}} \, e_{4\alpha'}(k) \, e_{4\alpha''}(k') \, F_{l\mu} M_{l\mu}^{-1/2} e_{4\alpha}(l \mu) . \tag{3.15b}$$

The properties of the coupling constant G related to the crystal symmetry and to the potential (3.1) can be found in Leibfried and Ludwig [17] and also in [16].

Translational invariance of the lattice gives:

$$G_{\alpha \alpha' \alpha''} \begin{pmatrix} l & l' & l'' \\ \mu \mu' \mu'' \end{pmatrix} = G_{\alpha \alpha' \alpha''} \begin{pmatrix} 0 & l' - l & l'' - l \\ \mu & \mu' & \mu'' \end{pmatrix} \tag{3.16a}$$

and inversion symmetry

$$G_{\alpha \alpha' \alpha''} \begin{pmatrix} l & l' & l'' \\ \mu \mu' \mu'' \end{pmatrix} = - \, G_{\alpha \alpha' \alpha''} \begin{pmatrix} -l & -l' & -l'' \\ \mu & \mu' & \mu'' \end{pmatrix} . \tag{3.16b}$$

The potential requires that $R_{l\mu} = R_{l'\mu'}$ or $R_{l\mu} = R_{l''\mu''}$ and if we use a nearest neighbour interaction model then the G are defined by only two independent parameters

$$\Phi_1 = \frac{6 e^2}{d^4} - \eta(\eta + 1)(\eta + 2) \frac{b}{d^{\eta+3}} , \tag{3.17a}$$

$$\Phi_2 = - \frac{3 e^2}{d^4} + \eta(\eta + 2) \frac{b}{d^{\eta+3}} \tag{3.17b}$$

and we obtain the following table:

Table 1

$R_{l\mu} d^{-1}$	$G_{1\alpha'\alpha''} \begin{pmatrix} 00l \\ 112 \end{pmatrix}$			$G_{2\alpha'\alpha''} \begin{pmatrix} 00l \\ 112 \end{pmatrix}$			$G_{3\alpha'\alpha''} \begin{pmatrix} 0\,0\,l \\ 1\,1\,2 \end{pmatrix}$		
	Φ_1	0	0	0	Φ_2	0	0	0	Φ_2
[1, 0, 0]	0	Φ_2	0	Φ_2	0	0	0	0	0
	0	0	Φ_2	0	0	0	Φ_2	0	0
	0	Φ_2	0	Φ_2	0	0	0	0	0
[0, 1, 0]	Φ_2	0	0	0	Φ_1	0	0	0	Φ_2
	0	0	0	0	0	Φ_2	0	Φ_2	0
	0	0	Φ_2	0	0	0	Φ_2	0	0
[0, 0, 1]	0	0	0	0	0	Φ_2	0	Φ_2	0
	Φ_2	0	0	0	Φ_2	0	0	0	Φ_1

4. The Line Width

From (2.36) we have

$$\Gamma_{lm} = 2\pi \sum_{l'm'} |\langle lm | {}^{a}\mathcal{H} | l'm'\rangle|^2 \, \delta(E^n_{lm} - E^n_{l'm'}) \tag{4.1}$$

and (3.15) gives

$$^{a}\mathcal{H} = \sum_{kk'} A(k, k') \{ b\, a^{\dagger}_{-k} a^{\dagger}_{-k'} + b^{\dagger} a_k a_{k'} \}. \tag{4.2}$$

It is convenient to define a delta function of the form

$$\Delta = \prod_{t \neq k, k'} \delta_{m_t, m'_t} \tag{4.3}$$

where the index 't' runs over all acoustic modes with the exception of two. Then for a single matrix element of $^{a}\mathcal{H}$ we have

$$\langle lm | {}^{a}\mathcal{H} | l'm'\rangle = \sum_{kk'} A(k, k') \{ [l'(m'_k + 1)(m'_{k'} + 1)]^{1/2} \, \delta_{l', l+1} \, \delta_{m'_k, m_k-1}$$

$$\delta_{m'_{k'}, m_{k'}-1} + [(l'+1) m'_k m'_{k'}]^{1/2} \, \delta_{l', l-1} \, \delta_{m'_k, m_k+1} \, \delta_{m'_{k'}, m_{k'}+1} \} \, \Delta. \tag{4.4}$$

The summations over the complex conjugate k vectors must be done separately but this second sum is reduced to single terms for a fixed initial and final state so that (4.1) becomes

$$\Gamma_{lm} = 2\pi \sum_{l'm'} \sum_{kk'} |A(k, k')|^2 \{ l'(m'_k + 1)(m'_{k'} + 1) \, \delta_{l', l+1} \, \delta_{m'_k, m_k-1} \, \delta_{m'_{k'}, m_{k'}-1}$$

$$+ (l'+1) m'_k m'_{k'} \, \delta_{l', l-1} \, \delta_{m'_k, m_k+1} \, \delta_{m'_{k'}, m_{k'}+1} \} \, \Delta\, \delta(E^n_{lm} - E^n_{l'm'}). \tag{4.5}$$

We have for the energy of these phonon states

$$E^n_{lm} = (l + \tfrac{1}{2}) \hbar \omega_0 + \sum_k (m_k + \tfrac{1}{2}) \hbar v |k| \tag{4.6}$$

using this in (4.5) and performing the sum over the intermediate states immediately gives us

$$\Gamma_{lm} = 2\pi \sum_{kk'} |A(k, k')|^2 \{ (l+1) m_k m_{k'} + l(m_k + 1)(m_{k'} + 1) \}$$

$$\times \delta(\hbar \omega_0 - \hbar v(|k| + |k'|)). \tag{4.7}$$

We can proceed no further until we know the dependence of m_k on k. Hence we now consider a thermal average over all acoustic modes and define.

$$\Gamma_l = \mathcal{N} \sum_m \Gamma_{lm} \, e^{-\beta \sum_t m_t \hbar \omega_t} \tag{4.8a}$$

where

$$\mathcal{N} = \prod_{t \neq 1} (1 - e^{-\beta \hbar \omega_t}). \tag{4.8b}$$

These summations have the effect of replacing m_k in (4.7) by its thermal average, $\bar{m}_k = (e^{\beta \hbar v |k|} - 1)^{-1}$, which depends only on the modulus of k and (4.7) becomes

$$\Gamma_l = 2\pi \sum_{k,k'} |A(k, k')|^2 \{ (l+1) \bar{m}_k \bar{m}_{k'} + l(\bar{m}_k + 1)(\bar{m}_{k'} + 1) \}$$

$$\times \delta(\hbar \omega_0 - \hbar v(|k| + |k'|)). \tag{4.9}$$

The acoustic spectrum is a quasi continuum, so we replace the summations over k by integrations

$$\sum_k \rightarrow N (d/2\pi)^3 \int d^3k \tag{4.10}$$

82   S. Fraser

so that (4.9) becomes

$$\Gamma_l = 2\pi N^2 (d/2\pi)^6 \int d^3k \int d^3k' \, |A(\mathbf{k}, \mathbf{k}')|^2$$
$$\times \{(l+1)\bar{m}_{\mathbf{k}}\bar{m}_{\mathbf{k}'} + l(\bar{m}_{\mathbf{k}}+1)(\bar{m}_{\mathbf{k}'}+1)\}\,\delta(\hbar\omega_0 - \hbar v(|\mathbf{k}| + |\mathbf{k}'|)) \quad (4.11)$$

while (3.15 b) gives

$$|A(\mathbf{k}, \mathbf{k}')|^2 = \frac{\hbar^3}{32\,\omega_0 v^2 |\mathbf{k}||\mathbf{k}'|} \frac{M_0^2}{N^2 M_1^2 M_2^2} \sum_{\substack{ll'l''\\ \mu\mu'\mu''\\ \alpha\alpha'\alpha''}} \sum_{\substack{tt't''\\ \varrho\varrho'\varrho''\\ \beta\beta'\beta''}} \mathbf{G}_{\alpha\alpha'\alpha''}\begin{pmatrix} l\ l'\ l''\\ \mu\mu'\mu'' \end{pmatrix}$$

$$\times \mathbf{G}_{\beta\beta'\beta''}\begin{pmatrix} t\ t'\ t''\\ \varrho\varrho'\varrho'' \end{pmatrix} e^{i\,\mathbf{k}\,\cdot\,(\mathbf{R}_{l'\mu'} - \mathbf{R}_{t'\varrho'})}\,e^{i\,\mathbf{k}'\,\cdot\,(\mathbf{R}_{l''\mu''} - \mathbf{R}_{t''\varrho''})} \quad (4.12)$$

$$\times e_{4\alpha'}(\mathbf{k})\,e_{4\beta'}(\mathbf{k})\,e_{4\alpha''}(\mathbf{k}')\,e_{4\beta''}(\mathbf{k}')\,(-1)^{\mu+\varrho}\,e_{4\alpha}(l\,\mu)\,e_{4\beta}(t\,\varrho)$$

$$\times \frac{F_{l\mu}\,F_{t\varrho}}{M_\mu^{\frac{1}{2}}\,M_\varrho^{\frac{1}{2}}}\,.$$

We notice that all the angular $\mathbf{k}$ dependence occurs in (4.12). We choose our axes such that

$$\mathbf{k}\cdot(\mathbf{R}_{l\mu} - \mathbf{R}_{t\varrho}) = |\mathbf{k}|\,|\mathbf{R}_{l\mu} - \mathbf{R}_{t\varrho}|\,\cos\theta_k$$

and the φ_k dependence then lies solely in the polarization vectors. The θ_k integrations in (4.11) are now of the form

$$\int_0^\pi e^{i|\mathbf{k}||\mathbf{R}_{l\mu} - \mathbf{R}_{t\varrho}|\cos\theta_k} \left\{ \begin{array}{c} \sin^2\theta_k \\ \sin\theta_k \cos\theta_k \\ \cos^2\theta_k \end{array} \right\} \sin\theta_k\,d\theta_k$$

which throw up Bessel functions of first type with argument proportional to $|\mathbf{k}|\,|\mathbf{R}_{l\mu} - \mathbf{R}_{t\varrho}|$. These terms make the subsequent lattice summations and $|\mathbf{k}|$ integrations complicated as both have to be done by numerical means. Hence we take an approximation.

$$e^{i\,\mathbf{k}\,\cdot\,(\mathbf{R}_{l\mu} - \mathbf{R}_{t\varrho})} = \delta_{l\mu,\,t\varrho} \quad (4.13)$$

and all the angular $\mathbf{k}$ integrations become:

$$\int_0^\pi \sin\theta_k\,d\theta_k \int_0^{2\pi} d\,\varphi_k\,e_{4\alpha}(\mathbf{k})\,e_{4\beta}(\mathbf{k}) = \tfrac{4}{3}\pi\,\delta_{\alpha,\beta}\,. \quad (4.14)$$

Using this result in (4.11), (4.12) we obtain

$$\Gamma_l = \frac{\pi^3\hbar^3 M_0^2}{9\,\omega_0 v^2 M_1^2 M_2^2}\left(\frac{d}{2\pi}\right)^6 \int k'\,dk' \int k\,dk\,\{(l+1)\bar{m}_{\mathbf{k}}\bar{m}_{\mathbf{k}'} + l(\bar{m}_{\mathbf{k}}+1)(\bar{m}_{\mathbf{k}'}+1)\}$$

$$\times \delta(\hbar\omega_0 - \hbar v(k + k'))\sum_{\substack{ll'l''t\\ \mu\mu'\mu''\varrho\\ \alpha\alpha'\alpha''\beta}} \mathbf{G}_{\alpha\alpha'\alpha''}\begin{pmatrix} l\ l'\ l''\\ \mu\mu'\mu'' \end{pmatrix} \mathbf{G}_{\beta\alpha'\alpha''}\begin{pmatrix} t\ l'\ l''\\ \varrho\mu'\mu'' \end{pmatrix}(-1)^{\mu+\varrho} \quad (4.15)$$

$$\times e_{4\alpha}(l\,\mu)\,e_{4\beta}(t\,\varrho)\frac{F_{l\mu}\,F_{t\varrho}}{M_\mu^{1/2}\,M_\varrho^{1/2}}\,.$$

The remaining integrations are now separated from the lattice summations which is a vast simplification. We now use the properties of the coupling parameters to further simplify the summations.

Using the facts that we have a two point potential and inversion symmetry

$$\sum_{l'\mu'} G_{\alpha\alpha'\alpha''}\begin{pmatrix} l & l' & l'' \\ \mu & \mu' & \mu'' \end{pmatrix} G_{\beta\alpha'\alpha''}\begin{pmatrix} t & l' & l'' \\ \varrho & \mu' & \mu'' \end{pmatrix}$$

$$= G_{\alpha\alpha'\alpha''}\begin{pmatrix} l & l & l'' \\ \mu & \mu & \mu'' \end{pmatrix} G_{\beta\alpha'\alpha''}\begin{pmatrix} t & l & l'' \\ \varrho & \mu & \mu'' \end{pmatrix} + G_{\alpha\alpha'\alpha''}\begin{pmatrix} l & l'' & l'' \\ \mu & \mu'' & \mu'' \end{pmatrix} G_{\beta\alpha'\alpha''}\begin{pmatrix} t & l'' & l'' \\ \varrho & \mu'' & \mu'' \end{pmatrix}$$

$$= G_{\alpha\alpha'\alpha''}\begin{pmatrix} l & l & l'' \\ \mu & \mu & \mu'' \end{pmatrix} \left\{ G_{\beta\alpha'\alpha''}\begin{pmatrix} t & l & l'' \\ \varrho & \mu & \mu'' \end{pmatrix} - G_{\beta\alpha'\alpha''}\begin{pmatrix} t & l'' & l'' \\ \varrho & \mu'' & \mu'' \end{pmatrix} \right\}.$$

The first term is non-zero only if $t\varrho = l\mu$, $\alpha = \beta$ or $t\varrho = l''\mu''$; while the second must have $\varrho = \mu$

$$\sum_{l'\mu'} G_{\alpha\alpha'\alpha''}\begin{pmatrix} l & l' & l'' \\ \mu & \mu' & \mu'' \end{pmatrix} G_{\beta\alpha'\alpha''}\begin{pmatrix} t & l' & l'' \\ \varrho & \mu' & \mu'' \end{pmatrix} = G_{\alpha\alpha'\alpha''}^2 \begin{pmatrix} l & l & l'' \\ \mu & \mu & \mu'' \end{pmatrix} \delta_{\alpha\beta}\, \delta_{t\varrho,\, l\mu}$$

$$- G_{\alpha\alpha'\alpha''}\begin{pmatrix} l & l & l'' \\ \mu & \mu & \mu'' \end{pmatrix} \left\{ G_{\alpha''\beta\alpha'}\begin{pmatrix} l & l & l'' \\ \mu & \mu & \mu'' \end{pmatrix} \delta_{t\varrho,\, l''\mu''} - G_{\beta\alpha'\alpha''}\begin{pmatrix} t & t & l'' \\ \varrho & \varrho & \mu'' \end{pmatrix} \delta_{\mu\varrho} \right\}. \qquad (4.16)$$

Substitution of (4.16) into (4.15) gives:

$$\Gamma_l = \frac{\pi^3 \hbar^3 M_0^2}{9\,\omega_0\, v^2\, M_1^2\, M_2^2} \left(\frac{d}{2\pi}\right)^6 \int k'\, dk' \int k\, dk$$

$$\times \left\{ (l+1)\,\bar{m}_k\, \bar{m}_{k'} + l\,(\bar{m}_k + 1)\,(\bar{m}_{k'} + 1) \right\} \delta(\hbar\,\omega_0 - \hbar v (k + k'))$$

$$\times \left\{ \sum_{\substack{l'' \\ \mu\mu'' \\ \alpha\alpha'\alpha''}} G_{\alpha\alpha'\alpha''}^2 \begin{pmatrix} l & l & l'' \\ \mu & \mu & \mu'' \end{pmatrix} \frac{F_{l\mu}^2}{M_\mu}\, e_{4\alpha}^2(l\mu) + \sum_{\substack{l'' \\ \mu\mu'' \\ \alpha\beta\alpha'\alpha''}} G_{\alpha\alpha'\alpha''}\begin{pmatrix} l & l & l'' \\ \mu & \mu & \mu'' \end{pmatrix} \right. \qquad (4.17)$$

$$\times\, G_{\alpha''\beta\alpha'}\begin{pmatrix} l & l & l'' \\ \mu & \mu & \mu'' \end{pmatrix} e_{4\alpha}(l\mu)\, e_{4\beta}(l''\mu'') \frac{F_{l\mu} F_{l''\mu''}}{(M_1 M_2)^{1/2}} + \sum_{\substack{l''t \\ \mu\mu'' \\ \alpha\beta\alpha'\alpha''}} G_{\alpha\alpha'\alpha''}\begin{pmatrix} l & l & l'' \\ \mu & \mu & \mu'' \end{pmatrix}$$

$$\left. + G_{\beta\alpha'\alpha''}\begin{pmatrix} t & t & l'' \\ \mu & \mu & \mu'' \end{pmatrix} e_{4\alpha}(l\mu)\, e_{4\beta}(t\mu) \frac{F_{l\mu} F_{t\mu}}{M_\mu} \right\}.$$

From Table 1 we see that

$$\sum_{\substack{l''\mu'' \\ \alpha'\alpha''}} G_{\alpha\alpha'\alpha''}^2 \begin{pmatrix} l & l & l'' \\ \mu & \mu & \mu'' \end{pmatrix} = 2\,(\Phi_1^2 + 6\,\Phi_2^2)$$

and since $\displaystyle\sum_\alpha e_{4\alpha}^2(l\mu) = 1$,

(4.17) can be finally written as

$$\Gamma_l = (l+1)\,\gamma_1 + l\,\gamma_2 \qquad (4.18)$$

where

$$\gamma_1 = \text{LS.}\ I_\Gamma(m); \quad \gamma_2 = \text{LS.}\ I_\Gamma(m+1) \qquad (4.19)$$

$$\mathrm{LS} = \frac{\pi^3 \hbar^2 M_0^2 k_B^3}{9 \,\omega_0 v^3 M_1^2 M_2^2} \left(\frac{d}{2\pi}\right) \sum_{l\mu} \left\{ 2(\Phi_1^2 + 6\,\Phi_2^2) \frac{F_{l\mu}^2}{M_\mu}\right.$$

$$+ \sum_{\substack{l''\mu'' \\ \alpha\beta\alpha'\alpha''}} \left[\mathbf{G}_{\alpha\alpha'\alpha''} \begin{pmatrix} l & l & l'' \\ \mu & \mu & \mu'' \end{pmatrix} \mathbf{G}_{\alpha''\beta\alpha'} \begin{pmatrix} l & l & l'' \\ \mu & \mu & \mu'' \end{pmatrix} e_{4\alpha}(l\,\mu)\, e_{4\beta}(l''\,\mu'') \frac{F_{l\mu} F_{l''\mu''}}{(M_1 M_2)^{1/2}} \right.$$

$$\left. + \sum_t \mathbf{G}_{\alpha\alpha'\alpha''} \begin{pmatrix} l & l & l'' \\ \mu & \mu & \mu'' \end{pmatrix} \mathbf{G}_{\beta\alpha'\alpha''} \begin{pmatrix} t & t & l'' \\ \mu & \mu & \mu'' \end{pmatrix} e_{4\alpha}(l\,\mu)\, e_{4\beta}(t\,\mu) \frac{F_{l\mu} F_{t\mu}}{M_\mu} \right]\right\} \tag{4.20}$$

where $\mathbf{R}_{l''\mu''}$ is a nearest neighbour to both $\mathbf{R}_{l\mu}$ and $\mathbf{R}_{t\mu}$.

$$I_\Gamma(m) = k_B^{-3} \int\limits_0^{k_B} k'\,dk' \int\limits_0^{k_B} k\,dk\, \bar{m}_k \bar{m}_{k'}\, \delta\left(\frac{\omega_0}{v} - k - k'\right), \tag{4.21a}$$

$$I_\Gamma(m+1) = k_B^{-3} \int\limits_0^{k_B} k'\,dk' \int\limits_0^{k_B} k\,dk\,(\bar{m}_k + 1)\,(\bar{m}_{k'} + 1)\, \delta\left(\frac{\omega_0}{v} - k - k'\right). \tag{4.21b}$$

A full thermal average of (4.18) then becomes

$$\langle \Gamma_l \rangle = (1 - e^{-\beta\hbar\omega_0}) \sum_l \Gamma_l\, e^{-l\beta\hbar\omega_0} = (l+1)\,\gamma_1 + l\,\gamma_2, \tag{4.22a}$$

$$l = (e^{\beta\hbar\omega_0} - 1)^{-1}. \tag{4.22b}$$

Consequently

$$\langle \Gamma_{l+c} \rangle = (l + c + 1)\,\gamma_1 + (l + c)\,\gamma_2, \qquad c \geqq 0 \tag{4.23a}$$

$$\langle \Gamma_{l-1} \rangle = (e^{-\beta\hbar\omega_0}\, l + 1)\,\gamma_1 + e^{-\beta\hbar\omega_0}\, l\,\gamma_2. \tag{4.23b}$$

5. Non-Radiative Transition Matrix Elements

The form of the operator $^t\mathscr{K}$ is well known [18]. When written in normal coordinate representation we have for its non-diagonal matrix elements

$$\langle n'\,m' |\,^t\mathscr{K}\,| n\,m \rangle = -\frac{\hbar^2}{2} \sum_{t=1}^{3N+1} \left\{ \langle \Psi_{m'}^{n'} | \langle \Psi_{n'} | \frac{\partial^2}{\partial Q_t^2} | \Psi_n \rangle | \Psi_m^n \rangle \right.$$

$$\left. + 2 \langle \Psi_{m'}^{n'} | \langle \Psi_{n'} | \frac{\partial}{\partial Q_t} | \Psi_n \rangle \frac{\partial}{\partial Q_t} | \Psi_m^n \rangle \right\}. \tag{5.1}$$

But our electronic wave function (2.33) is a function of only the local mode and then only to first order. Hence (5.1) takes the much simpler form in our case

$$\langle n'\,m' |\,^t\mathscr{K}\,| n\,m \rangle = -\hbar^2 \langle \Psi_n | \frac{\partial}{\partial Q_1} | \Psi_{n'} \rangle \langle \Psi_m^n | \frac{\partial}{\partial Q_1} | \Psi_{m'}^{n'} \rangle \tag{5.2}$$

where we have used the hermitian properties of $^t\mathfrak{R}$. We now identify the states

$$| n, m \rangle = \Psi_{\mathrm{FC}}(r, Q_1)\, \Psi_m^{\mathrm{FC}}(Q_1), \tag{5.3a}$$

$$| n'\,m' \rangle = \Psi_{\mathrm{CB}}(r)\, \Psi_{m'}^{\mathrm{CB}}(Q_1)., \tag{5.3b}$$

We now transform to dimensionless normal coordinates

$$Q_t = (\hbar/\omega_0)^{1/2}\, q_t, \tag{5.4a}$$

$$A_t^{\mathrm{CB,FC}} = (2\,\hbar/\omega_0)^{1/2}\, \alpha\, \delta_{1t} \tag{5.4b}$$

and

$$\langle n'\,m'|\,{}^{t}\mathscr{K}\,|\,n\,m\rangle = -\,\hbar\,\omega_0\,\langle\varPsi_{\mathrm{CB}}|\frac{\partial}{\partial q_1}|\varPsi_{\mathrm{FC}}\rangle\,\langle\varPsi^{\mathrm{CB}}_{m}|\frac{\partial}{\partial q_1}|\varPsi^{\mathrm{FC}}_{m'}\rangle\,. \tag{5.5}$$

In particular we have

$$I_{11} = \langle\varPsi_{\mathrm{CB}}|\frac{\partial}{\partial q_1}|\varPsi_{\mathrm{FC}}\rangle = \left(\frac{\hbar}{\omega_0\,\Omega}\right)^{1/2}\int d\tau\,[C^{\mathrm{F}}_{21}\,\varPsi^{2}_{\mathrm{F}}(r) + C^{\mathrm{F}}_{31}\,\varPsi^{3}_{\mathrm{F}}(r)]\,. \tag{5.6}$$

The integration is over the microblock volume but this is sufficiently large
— $\sim 10^5\,d^3$ — that we replace it by an integration over all space. The wave
functions from (2.32) give:

$$\int d\tau\,\varPsi^{2}_{\mathrm{F}}(r) = -\,32\,\sqrt{\frac{2\,\pi}{\beta^3}}\,, \quad \int d\tau\,\varPsi^{3}_{\mathrm{F}}(r) = 72\,\sqrt{\frac{3\,\pi}{\beta^3}}\,. \tag{5.7}$$

Hence

$$I_{11} = \left(\frac{\hbar\,\pi}{\Omega\,\omega_0\,\beta^3}\right)^{1/2} 8\,(9\,\sqrt{3}\,C^{\mathrm{F}}_{31} - 4\,\sqrt{2}\,C^{\mathrm{F}}_{21})\,. \tag{5.8}$$

The coefficients C^{F}_{21}, C^{F}_{31} being defined in (2.30b) while the phonon-part of (5.5)
has the form

$$\langle\varPsi^{\mathrm{CB}}_{m}|\frac{\partial}{\partial q_1}|\varPsi^{\mathrm{FC}}_{m'}\rangle = \left\{\langle l\,m| + \sum_{j,k}\frac{\langle l\,m|\,{}^{a}\mathscr{K}\,|\,j\,k\rangle}{E_{lm} - E_{jk} - \dfrac{i}{2}\,\Gamma_{jk}}\,\langle j\,k|\right\}$$

$$\times\,\frac{\partial}{\partial q_1}\left\{|l'\,m'\rangle + \sum_{j'k'}\frac{\langle j'\,k'|\,{}^{a}\mathscr{K}\,|\,l'\,m'\rangle}{E_{l'm'} - E_{j'k'} + \dfrac{i}{2}\,\Gamma_{j'k'}}\,|j'\,k'\rangle\right\} \tag{5.9}$$

where we have used the dressed phonons of (2.35).

The partial transition rates (1.2) in this notation are

$${}^{t}W_{n'l'm',nlm} = \frac{2\,\pi}{\hbar}\,\hbar^2\,\omega_0^2\,I^2_{11}\,|\langle\varPsi^{\mathrm{CB}}_{m}|\frac{\partial}{\partial q_1}|\varPsi^{\mathrm{FC}}_{m'}\rangle|^2\,\delta(E^{\mathrm{CB}}_{lm} - E^{\mathrm{FC}}_{l'm'})\,. \tag{5.10}$$

For an exact energy balance in (5.10) we must use the continuous spectrum of
the acoustic phonons since in general the electronic energy change will not be
an integer multiple of the local mode energy. Thus in (5.9) we are forced to drop
the term diagonal in the harmonic phonons and take those linear in ${}^{a}\mathscr{K}$.

$$\langle\varPsi^{\mathrm{CB}}_{m}|\frac{\partial}{\partial q_1}|\varPsi^{\mathrm{FC}}_{m'}\rangle = \sum_{j,k}\frac{\langle l\,m|\,{}^{a}\mathscr{K}\,|j\,k\rangle}{E_{lm} - E_{jk} - \dfrac{i}{2}\,\Gamma_{jk}}\,\langle j\,k|\frac{\partial}{\partial q_1}|l'\,m'\rangle$$

$$+\,\sum_{j'k'}\frac{\langle j'\,k'|\,{}^{a}\mathscr{K}\,|l'\,m'\rangle\,\langle l\,m|\dfrac{\partial}{\partial q}|j'\,k'\rangle}{E_{l'm'} - E_{j'k'} + \dfrac{i}{2}\,\Gamma_{j'k'}}\,. \tag{5.11}$$

86 S. Fraser

We have immediately

$$\langle l\,m|\,\frac{\partial}{\partial q_1}\,|l'\,m'\rangle = \langle l|\,\frac{\partial}{\partial q}\,|l'\rangle\prod_t \delta_{m_{t'}m_{t'}} := D_l^{l'}\prod_t \delta_{m_{t'},\,m_t'} \qquad (5.12)$$

so that (5.11) now reads

$$\langle \Psi_m^{CB}|\,\frac{\partial}{\partial q_1}\,|\Psi_{m'}^{FC}\rangle = \sum_j \frac{\langle l\,m|\,{}^a\mathcal{H}\,|j\,m'\rangle}{E_{lm} - E_{jm'} - \dfrac{i}{2}\,\Gamma_{jm'}}\,D_j^{l'}$$

$$+ \sum_{j'} \frac{\langle j'\,m|\,{}^a\mathcal{H}\,|l'\,m'\rangle}{E_{l'm'} - E_{j'm} + \dfrac{i}{2}\,\Gamma_{j'm}}\,D_l^{j'} \qquad (5.13)$$

where the summation over the intermediate acoustic phonon states has been performed.

We now substitute our simplified form of the anharmonic operator from (3.15)

$$^a\mathcal{H} = \sum_{kk'} A\,(k,k')\,\{b\,a^\dagger_{-k}a^\dagger_{-k'} + b^\dagger a_k a_{k'}\} \qquad (5.14)$$

along with the definition (4.3), (5.13) becomes

$$\langle \Psi_m^{CB}|\,\frac{\partial}{\partial q_1}\,|\Psi_{m'}^{FC}\rangle = \sum_{kk'} A\,(k,k')\,\Bigg\{ \sum_j \frac{D_j^{l'}}{E_{lm} - E_{jm'} - \dfrac{i}{2}\,\Gamma_{jm'}}$$

$$\times\,[\{j\,(m_k' + 1)\,(m_{k'}' + 1)\}^{1/2}\,\delta_{j,\,l+1}\,\delta_{m_k' m_k - 1}\,\delta_{m_{k'}' m_{k'} - 1}$$

$$+ \{(j + 1)\,m_k'\,m_{k'}'\}^{1/2}\,\delta_{j,\,l-1}\,\delta_{m_k' m_k + 1}\,\delta_{m_{k'}' m_{k'} + 1}] \qquad (5.15)$$

$$+ \sum_{j'} \frac{D_{l'}^{j}}{E_{l'm'} - E_{j'm} + \dfrac{i}{2}\,\Gamma_{j'm}}$$

$$\times\,[\{l'\,(m_k' + 1)\,(m_{k'}' + 1)\}^{1/2}\,\delta_{l',\,j'+1}\,\delta_{m_k'\,,\,m_k - 1}\,\delta_{m_{k'}',\,m_{k'} - 1}$$

$$+ \{(l' + 1)\,m_k'\,m_{k'}'\}^{1/2}\,\delta_{l',\,j'-1}\,\delta_{m_k'\,m_k + 1}\,\delta_{m_{k'}',\,m_{k'} + 1}]\Bigg\}\,\Delta\,.$$

The summations over j, j' are easily performed. The line width in the denominator of the first term depends on the set $[m']$ which differs from the set $[m]$ in only two numbers of the set. We make the approximation, since the line widths consist of many terms most of which are the same, that

$$\Gamma_{lm'} = \Gamma_{lm}\,.$$

Then (5.15) becomes:

$$\langle \Psi_m^{CB}|\,\frac{\partial}{\partial q_1}\,|\Psi_{m'}^{FC}\rangle = \sum_{kk'} A\,(k,k')\,\Bigg\{ \frac{D_{l+1}^{l'}}{E_{lm} - E_{l+1,\,m'} - \dfrac{i}{2}\,\Gamma_{l+1,\,m}}$$

$$\times\,[(l + 1)\,m_k\,m_{k'}]^{1/2}\,\delta_{m_k',\,m_k - 1}\,\delta_{m_{k'}'\,m_{k'} - 1}$$

$$+ \frac{D_{l-1}^{l'}}{E_{lm} - E_{l-1,\,m} - \dfrac{i}{2}\,\Gamma_{l-1,\,m}}\,[l\,(m_k + 1)\,(m_{k'} + 1)]^{1/2}\,\delta_{m_k',\,m_k + 1}\,\delta_{m_{k'}'\,m_{k'} + 1}$$

$$+ \frac{D_l^{l'-1}}{E_{l'm'} - E_{l'-1,\,m} + \dfrac{i}{2}\, \Gamma_{l'-1,\,m}} \, [l'\, m_k\, m_{k'}]^{1/2}\, \delta_{m'_k,\,m_k-1}\, \delta_{m'_{k'},\,m_{k'}-1}$$

$$+ \frac{D_l^{l'+1}}{E_{l'm'} - E_{l'+1,\,m} + \dfrac{i}{2}\, \Gamma_{l'+1,\,m}}$$

$$\times\, [(l'+1)(m_k+1)(m_{k'}+1)]^{1/2}\, \delta_{m'_k,\,m_k+1}\, \delta_{m'_{k'},\,m_{k'}+1} \Bigg\}\, \Delta\,. \tag{5.16}$$

Thus (5.10) will have the form

$$^tW_{n'l'm',\,nlm} = 2\,\pi\,\hbar\,\omega_0^2\, I_{11}^2 \sum_{kk'} |\,A\,(k,\,k')\,|^2$$

$$\times \Bigg\{ \frac{|D_{l+1}^{l'}|^2}{(E_{lm} - E_{l+1,\,m'})^2 + \tfrac{1}{4}\, \Gamma_{l+1,m}^2}\, (l+1)\, m_k\, m_{k'}\, \delta_{m'_k\, m_k-1}\, \delta_{m'_{k'},\,m_{k'}-1}$$

$$+ \frac{|D_{l-1}^{l'}|^2}{(E_{lm} - E_{l-1,\,m'})^2 + \tfrac{1}{4}\, \Gamma_{l-1,m}^2}\, l\,(m_k+1)(m_{k'}+1)\, \delta_{m'_k,\,m_k+1}\, \delta_{m'_{k'},\,m_{k'}+1}$$

$$+ \frac{|D_l^{l'+1}|^2}{(E_{l'm'} - E_{l'+1,\,m})^2 + \tfrac{1}{4}\, \Gamma_{l'+1,m}^2}$$

$$\times\, (l'+1)(m_k+1)(m_{k'}+1)\, \delta_{m'_k,\,m_k+1}\, \delta_{m'_{k'},\,m_{k'}+1}$$

$$+ \frac{|D_{l'-1}^l|^2}{(E_{l'm'} - E_{l'-1,\,m})^2 + \tfrac{1}{4}\, \Gamma_{l'-1,m}^2}$$

$$\times\, l'\, m_k\, m_{k'}\, \delta_{m'_k,\,m_k-1}\, \delta_{m'_{k'},\,m_{k'}-1} \Bigg\}\, \Delta\,\delta\,(E_{lm}^{\mathrm{CB}} - E_{l'm'}^{\mathrm{FC}}) \tag{5.17}$$

because the independent k summations in the complex conjugate of the matrix element are reduced to single terms by the Kronecker delta functions that appear there.

To calculate the total non-radiative rates we have to perform a thermal summation over phonon states in (5.17). This is done in the next section.

6. The Thermal Weighting

(1.3) gives us

$$^tW_{n'n}(\beta) = \Re \sum_{\substack{lm\\l'm'}} {}^tW_{n'l'm',\,nlm}\, e^{-l\beta\hbar\omega_0 - \beta \sum_t m_t\hbar\omega_t} \tag{6.1}$$

and

$$\Re = (1 - e^{-\beta\hbar\omega_0}) \prod_{t \neq 1} (1 - e^{-\beta\hbar\omega_t})\,. \tag{6.2}$$

We first consider the summations over the acoustic phonon quantum numbers in (5.17)

$$\sum_{m'} {}^tW_{n'l'm',\,nlm} = 2\,\pi\,\hbar\,\omega_0^2\, I_{11}^2 \sum_{kk'} |\,A\,(k\,k')\,|^2$$

$$\times \Bigg\{ \frac{|D_{l+1}^{l'}|^2}{(\hbar\,\omega_0 - \hbar\,v(k+k'))^2 + \tfrac{1}{4}\, \Gamma_{l+1,m}^2} \cdot (l+1)\, m_k\, m_{k'}\, \delta\,(E_1)$$

$$+ \frac{|D^{l'}_{l-1}|^2}{(\hbar\,\omega_0 - \hbar\,v\,(k+k'))^2 + \frac{1}{4}\,\Gamma^2_{l-1,m}} \cdot l\,(m_k+1)\,(m_{k'}+1)\,\delta(E_2)$$

$$+ \frac{|D^{l}_{l'+1}|^2}{(\hbar\,\omega_0 - \hbar\,v\,(k+k'))^2 + \frac{1}{4}\,\Gamma^2_{l'+1,m}} \cdot (l'+1)\,(m_k+1)\,(m_{k'}+1)\,\delta(E_2)$$

$$+ \left. \frac{|D^{l'-1}_{l}|^2}{(\hbar\,\omega_0 - \hbar\,v\,(k+k'))^2 + \frac{1}{4}\,\Gamma^2_{l'-1,m}} \cdot l'\,m_k\,m_{k'}\,\delta(E_1)\right\} \tag{6.3}$$

where

$$\delta(E_1) = \delta(E^{CB}_{lm} - E^{FC}_{l',m_1\cdots m_k-1\cdots m_{3N}})\,, \tag{6.4a}$$

$$\delta(E_2) = \delta(E^{CB}_{lm} - E^{FC}_{l',m_1\cdots m_k+1\cdots m_{3N}})\,. \tag{6.4b}$$

We now write the electronic energy change as:

$$E^{CB}_{lm} - E^{FC}_{lm} = \hbar\,\omega_0\,(r_0 + \delta) \tag{6.5}$$

with r_0 integer and $0 \leq \delta < 1$.

The case $\delta = 0$ implies we have exact resonance conditions when an energy fit could be obtained using harmonic phonons only. Substituting in (6.4) we have:

$$\delta(E_1) = \delta(\hbar\,\omega_0\,(l + r_0 + \delta - l') + \hbar\,v\,(k+k'))\,, \tag{6.6a}$$

$$\delta(E_2) = \delta(\hbar\,\omega_0\,(l + r_0 + \delta - l') - \hbar\,v\,(k+k'))\,. \tag{6.6b}$$

However there is a maximum energy that two acoustic modes involved can be excited by $\varepsilon_{max} = 2\,\hbar\,v\,k_B < 2\,\hbar\,\omega_0$, so for the delta functions in (6.6) to be satisfied we must have $0 \leq |l + r_0 + \delta - l'| < 2$. Let $l' = l + r_0 + s$ where $s = 0, \pm 1, \pm 2\ldots$ then we must have for (6.6)

$$\delta(E_1) = \sum_{s=0}^{1}\delta_{l',l+r_0+s+1}\,\delta(\hbar\,\omega_0\,(\delta - s - 1) + \hbar\,v\,(k+k'))\,, \tag{6.7a}$$

$$\delta(E_2) = \sum_{s=0}^{1}\delta_{l',l+r_0+s-1}\,\delta(\hbar\,\omega_0\,(\delta - s + 1) - \hbar\,v\,(k+k'))\,. \tag{6.7b}$$

Thus the final local mode state is closely defined to one of two levels. We substitute (6.7) into (6.3) and use the Dirac delta functions to remove the k, k' dependence from the denominators.

$$\sum_{m'}{}^t W_{n'l'm',\,nlm} = 2\,\pi\,\hbar\,\omega_0^2\,I_{11}^2 \sum_{s=0}^{1}\sum_{kk'}|A\,(k,k')|^2$$

$$\times \left\{ \frac{(l+1)\,|D^{l'}_{l+1}|^2}{\hbar^2\,\omega_0^2\,(s-\delta)^2 + \frac{1}{4}\,\Gamma^2_{l+1,m}}\,\delta_{l',l+r_0+s+1}\,m_k\,m_{k'}\,\delta(\hbar\,\omega_0\,(s+1-\delta) - \hbar\,v\,(k+k'))\right.$$

$$+ \frac{l\,|D^{l'}_{l-1}|^2}{\hbar^2\,\omega_0^2\,(s-\delta)^2 + \frac{1}{4}\,\Gamma^2_{l-1,m}}$$

$$\times \delta_{l',l+r_0+s-1}(m_k+1)\,(m_{k'}+1)\,\delta(\hbar\,\omega_0\,(s-1-\delta) + \hbar\,v\,(k+k'))$$

$$+ \frac{(l'+1)\,|D^{l'+1}_{l}|^2}{\hbar^2\,\omega_0^2\,(s-\delta)^2 + \frac{1}{4}\,\Gamma^2_{l'+1,m}}$$

$$\times \delta_{l',l+r_0+s-1}(m_k+1)\,(m_{k'}+1)\,\delta(\hbar\,\omega_0\,(s-1-\delta) + \hbar\,v\,(k+k'))$$

$$+ \frac{l' \, | D_l^{l'-1} |^2}{\hbar^2 \, \omega_0^2 (s-\delta)^2 + \tfrac{1}{4} \, \Gamma_{l'-1,m}^2}$$

$$\times \, \delta_{l',\,l+r_0+s+1} \, m_k \, m_{k'} \, \delta \left(\hbar \, \omega_0 (s+1-\delta) - \hbar \, v(k+k') \right) \Bigg\} . \tag{6.8}$$

The summation in (6.1) over the initial acoustic modes is not possible because of their occurrence in the line width. The same applies to the local mode summation. Hence we approximate (6.8) by replacing Γ_{l_m} by $\langle \Gamma_l \rangle$. The summations over m then reduce to replacing m_k by its thermal average $\bar{m}_k = (e^{\beta \hbar vk} - 1)^{-1}$. (6.1) then becomes:

$$^t W_{n',\,n}(\beta) = 2 \, \pi \, \hbar \, \omega_0^2 \, I_{11}^2 (1 - e^{-\beta \hbar \omega_0}) \sum_{s=0}^{1} \sum_{ll'} \sum_{kk'} | A(k\,k') |^2 \, e^{-\beta l \hbar \omega_0}$$

$$\times \left\{ \frac{(l+1) \, | D_{l+1}^{l'} |^2}{\hbar^2 \, \omega_0^2 (s-\delta)^2 + \tfrac{1}{4} \langle \Gamma_{l+1} \rangle^2} \right.$$

$$\times \, \delta_{l',\,l+r_0+s+1} \, \bar{m}_k \, \bar{m}_{k'} \, \delta \left(\hbar \, \omega_0 (s+1-\delta) - \hbar \, v(k+k') \right)$$

$$+ \frac{l \, | D_{l-1}^{l'} |^2}{\hbar^2 \, \omega_0^2 (s-\delta)^2 + \tfrac{1}{4} \langle \Gamma_{l-1} \rangle^2}$$

$$\times \, \delta_{l',\,l+_0 r+s-1} \, (\bar{m}_k + 1) \, (\bar{m}_{k'} + 1) \, \delta \left(\hbar \, \omega_0 (s-1-\delta) + \hbar \, v(k+k') \right)$$

$$+ \frac{(l'+1) \, | D_l^{l'+1} |^2}{\hbar^2 \, \omega_0^2 (s-\delta)^2 + \tfrac{1}{4} \langle \Gamma_{l'+1} \rangle^2}$$

$$\times \, \delta_{l',\,l+r_0+s-1} \, (\bar{m}_k + 1) \, (\bar{m}_{k'} + 1) \, \delta \left(\hbar \, \omega_0 (s-1-\delta) + \hbar \, v(k+k') \right)$$

$$+ \frac{l' \, | D_l^{l'-1} |^2}{\hbar^2 \, \omega_0^2 (s-\delta)^2 + \tfrac{1}{4} \langle \Gamma_{l'-1} \rangle^2}$$

$$\left. \times \, \delta_{l',\,l+r_0+s+1} \, \bar{m}_k \, \bar{m}_{k'} \, \delta \left(\hbar \, \omega_0 (s+1-\delta) - \hbar \, v(k+k') \right) \right\} . \tag{6.9}$$

The k summations here are very similar to those that appeared in the line width and we treat them in a likewise manner.

Analogous to (4.2) we define the integrals:

$$I_w^s(m) = k_B^{-3} \int_0^{k_B} k' \, dk' \int_0^{k_B} k \, dk \, \bar{m}_k \, \bar{m}_{k'} \, \delta \left(\frac{\omega_0}{v} (1-\delta+s) - k - k' \right), \tag{6.10a}$$

$$I_w^s(m+1) = k_B^{-3} \int_0^{k_B} k' \, dk' \int_0^{k_B} k \, dk \, (\bar{m}_k + 1)(\bar{m}_{k'} + 1) \, \delta \left(\frac{\omega_0}{v} (1+\delta-s) - k - k \right) \tag{6.10b}$$

and using the lattice constant of (4.20) we obtain for (6.9)

$$^t W_{n'n}(\beta) = \omega_0 \, I_{11}^2 \, \hbar \, \omega_0 \, LS \sum_{s=0}^{1} (1 - e^{-\beta \hbar \omega_0}) \sum_{l,l'} e^{-\beta l \hbar \omega_0}$$

$$\times \left\{ \frac{l \, | D_{l-1}^{l'} |^2 \, \delta_{l',\,l+r_0+s-1}}{\hbar^2 \, \omega_0^2 (s-\delta)^2 + \tfrac{1}{4} \langle \Gamma_{l-1} \rangle^2} \, I_w^s(m+1) + \frac{(l+1) \, | D_{l+1}^{l'} |^2 \, \delta_{l',\,l+r_0+s+1}}{\hbar^2 \, \omega_0^2 (s-\delta)^2 + \tfrac{1}{4} \langle \Gamma_{l+1} \rangle^2} \, I_w^s(m) \right.$$

$$+ \frac{(l'+1) \, | D_l^{l'+1} |^2 \, \delta_{l',\,l+r_0+s-1}}{\hbar^2 \, \omega_0^2 (s-\delta)^2 + \tfrac{1}{4} \langle \Gamma_{l+r_0+s} \rangle^2} \, I_w^s(m+1)$$

$$\left. + \frac{l' \, | D_l^{l'-1} |^2 \, \delta_{l',\,l+r_0+s+1}}{\hbar^2 \, \omega_0^2 (s-\delta)^2 + \tfrac{1}{4} \langle \Gamma_{l+r_0+s} \rangle^2} \, I_w^s(m) \right\} . \tag{6.11}$$

7 *

Only the local mode summations remain. These are of the form:

$$FC_1(s) = (1 - e^{-\beta\hbar\omega_0}) \sum_{ll'} e^{-\beta l\hbar\omega_0}\, l \left| D_{l-1}^{l'} \right|^2 \delta_{l', l+r_0+s-1}, \tag{6.12a}$$

$$FC_2(s) = (1 - e^{-\beta\hbar\omega_0}) \sum_{ll'} e^{-\beta l\hbar\omega_0}\, (l+1) \left| D_{l+1}^{l'} \right|^2 \delta_{l', l+r_0+s+1}, \tag{6.12b}$$

$$FC_3(s) = (1 - e^{-\beta\hbar\omega_0}) \sum_{ll'} e^{-\beta l\hbar\omega_0}\, (l'+1) \left| D_{l}^{l'+1} \right|^2 \delta_{l', l+r_0+s-1}, \tag{6.12c}$$

$$FC_4(s) = (1 - e^{-\beta\hbar\omega_0}) \sum_{ll'} e^{-\beta l\hbar\omega_0}\, l' \left| D_{l}^{l'-1} \right|^2 \delta_{l', l+r_0+s+1}. \tag{6.12d}$$

They can be evaluated exactly using the techniques of Heinzel [14]. The method is an extension of the theory of Koide [19]. With the following notation

$$B_\nu = e^{-\alpha^2(2\bar{l}+1)}\, e^{\nu\beta\hbar\omega_0}\, I_\nu\left(\frac{2\alpha^2\, e^{\beta(\hbar/2)\omega_0}}{e^{\beta\hbar\omega_0} - 1} \right), \tag{6.13a}$$

$$\bar{l} = (e^{\beta\hbar\omega_0} - 1)^{-1} \tag{6.13b}$$

and with α defined in (5.4b) a long but simple calculation gives us:

$$\begin{aligned}
FC_1(s) = \tfrac{1}{2}\, \bar{l}\{&\alpha^4 \bar{l}(\bar{l}+1)^2\, B_{r_0+s-3} + \alpha^2(\bar{l}+1)\,[4\bar{l}+1 - 2\alpha^2(3\bar{l}+2)]\, B_{r_0+s-2} \\
&+ [2\bar{l}+1 - 2\alpha^2(7\bar{l}^2 + 7\bar{l} + 1) + \alpha^4\bar{l}(15\bar{l}^2 + 20\bar{l} + 6)]\, B_{r_0+s-1} + \alpha^2[20\bar{l}^2 + 13\bar{l} \\
&+ 1 - 4\alpha^2\bar{l}(5\bar{l}^2 + 5\bar{l} + 1)]\, B_{r_0+s} + \bar{l}[2 - 2\alpha^2(7\bar{l}+2) + \alpha^4(15\bar{l}^2 + 10\bar{l} + 1)] \\
&\times B_{r_0+s+1} + 2\alpha^2\bar{l}^2[2 - \alpha^2(3\bar{l}+1)]\, B_{r_0+s+2} + \alpha^4\bar{l}^3\, B_{r_0+s+3}\},
\end{aligned} \tag{6.14a}$$

$$\begin{aligned}
FC_2(s) = \tfrac{1}{2}(\bar{l}+1)\,\{&\alpha^4(\bar{l}+1)^3\, B_{r_0+s-3} + 2\alpha^2(\bar{l}+1)^2\,[2 - \alpha^2(3\bar{l}+2)]\, B_{r_0+s-2}, \\
&+ (\bar{l}+1)[2 - 2\alpha^2(7\bar{l}+5) + \alpha^4(15\bar{l}^2 + 20\bar{l} + 6)]\, B_{r_0+s-1} \\
&+ \alpha^2[20\bar{l}^2 + 27\bar{l} + 8 - 4\alpha^2(\bar{l}+1)(5\bar{l}^2 + 5\bar{l} + 1)]\, B_{r_0+s} \\
&+ [2\bar{l}+1 - 2\alpha^2(7\bar{l}^2 + 7\bar{l} + 1) + \alpha^4(\bar{l}+1)(15\bar{l}^2 + 10\bar{l} + 1)]\, B_{r_0+s+1} \\
&+ \alpha^2[4\bar{l}+3 - 2\alpha^2(\bar{l}+1)(3\bar{l}+1)]\, B_{r_0+s+2} + \alpha^4\bar{l}^2(\bar{l}+1)\, B_{r_0+s+3}\},
\end{aligned} \tag{6.14b}$$

$$\begin{aligned}
FC_3(s) = \tfrac{1}{2}\{&\alpha^4(\bar{l}+1)^4\, B_{r_0+s-3} + \alpha^2(\bar{l}+1)^2\,[4\bar{l}+3 - 2\alpha^2(\bar{l}+1)(3\bar{l}+1)]\, B_{r_0+s-2} \\
&+ [(\bar{l}+1)(2\bar{l}+1) - 2\alpha^2(\bar{l}+1)(7\bar{l}^2 + 7\bar{l} + 1) + \alpha^4(\bar{l}+1)^2(15\bar{l}^2 + 10\bar{l} + 1)] \\
&\times B_{r_0+s-1} + \alpha^2\bar{l}[20\bar{l}^2 + 27\bar{l} + 8 - 4\alpha^2(\bar{l}+1)(5\bar{l}^2 + 5\bar{l} + 1)]\, B_{r_0+s} \\
&+ \bar{l}^2[2 - 2\alpha^2(7\bar{l}+5) + \alpha^4(15\bar{l}^2 + 20\bar{l} + 6)]\, B_{r_0+s+1} \\
&+ 2\alpha^2\bar{l}^3[2 - \alpha^2(3\bar{l}+2)]\, B_{r_0+s+2} + \alpha^4\bar{l}^4\, B_{r_0+s+3}\},
\end{aligned} \tag{6.14c}$$

$$\begin{aligned}
FC_4(s) = \tfrac{1}{2}\{&\alpha^4(\bar{l}+1)^4\, B_{r_0+s-3} + 2\alpha^2(\bar{l}+1)^3\,[2 - \alpha^2(3\bar{l}+1)]\, B_{r_0+s-2} \\
&+ (\bar{l}+1)^2[2 - 2\alpha^2(7\bar{l}+2) + \alpha^4(15\bar{l}^2 + 10\bar{l} + 1)]\, B_{r_0+s-1} \\
&+ \alpha^2(\bar{l}+1)[20\bar{l}^2 + 13\bar{l} + 1 - 4\alpha^2\bar{l}(5\bar{l}^2 + 5\bar{l} + 1)]\, B_{r_0+s} \\
&+ [\bar{l}(2\bar{l}+1) - 2\alpha^2\bar{l}(7\bar{l}^2 + 7\bar{l} + 1) + \alpha^4\bar{l}^2(15\bar{l}^2 + 20\bar{l} + 6)]\, B_{r_0+s+1} \\
&+ \alpha^2\bar{l}^2[4\bar{l}+1 - 2\alpha^2\bar{l}(3\bar{l}+2)]\, B_{r_0+s+2} + \alpha^4\bar{l}^4\, B_{r_0+s+3}\}.
\end{aligned} \tag{6.14d}$$

Substituting (6.14) in (6.11) we obtain the final result

$$\begin{aligned}
{}^tW_{\mathrm{FC,CB}}(\beta) = \omega_0\, I_{11}^2\, \hbar\,\omega_0\, \mathrm{LS} \sum_{s=0}^{1} \Bigg\{ & \frac{FC_1(s)\, I_w^s(m+1)}{\hbar^2\,\omega_0^2(s-\delta)^2 + \tfrac{1}{4}\langle\Gamma_{l-1}\rangle^2} \\
& + \frac{FC_2(s)\, I_w^s(m)}{\hbar^2\,\omega_0^2(s-\delta)^2 + \tfrac{1}{4}\langle\Gamma_{l+1}\rangle^2} + \frac{FC_3(s)\, I_w^s(m+1)}{\hbar^2\,\omega_0^2(s-\delta)^2 + \tfrac{1}{4}\langle\Gamma_{l+r_0+s}\rangle^2}
\end{aligned}$$

$$+ \frac{FC_4\, I_w^s(m)}{\hbar^2\, \omega_0^2(s-\delta)^2 + \frac{1}{4}\langle \Gamma_{l+r_0+s}\rangle^2}\bigg\} \, . \tag{6.15}$$

7. Numerical Calculations

We take the lattice parameters from Renn [12] while the electronic variational parameter in (2.32) is from Schmid [11]. There were, however, no accurate calculations of the energy difference between the F-centre ground state and the bottom of the conduction band available. Consequently we obtain an estimate of this quantity from experimental data.

If E_a; and E_e; are the energies corresponding to the maxima of the absorption and emission spectra for radiative transitions between the $1s$ and $2p$ states of the F-centre we have:

$$E_{lm}^{2p} - E_{lm}^{1s} = \tfrac{1}{2}(E_a + E_e) \tag{7.1}$$

and if ε_t is the thermal ionisation energy of the excited state then we have our required energy difference:

$$E_{lm}^{CB} - E_{lm}^{FC} = \tfrac{1}{2}(E_a + E_e) + \varepsilon_t \, . \tag{7.2}$$

The spectral data comes from Gebhardt and Kühnert [20] while the ionisation energies are from Swank and Brown [21]. We can also estimate the zero-point shift of the coupled oscillator coordinate between the $1s$- and $2p$-electronic states. This is

$$\Delta q^{1s-2p} = \left(\frac{E_a - E_e}{\hbar\,\omega_0}\right)^{1/2} \, . \tag{7.3}$$

From (2.23) we see that if the excited state is very diffuse compared with the ground state then the zero point shift will have only a small dependence on the exact form of the excited state wave function. Consequently one expects the zero-point shifts to be similar for all the excited states of the F-centre and this seems to be so [22, 23].

The lattice summations were performed using the procedure of Renn [12] which gives rapid convergence of charge dependant summations. The phonon

Table 2

	NaCl	NaBr	KCl	KBr	KI
β [Å^{-1}]	0.40	0.40	0.40	0.40	0.35
d [Å]	2.820	2.989	3.147	3.298	3.533
α^e	0.902	1.044	0.820	0.922	1.087
α^g	1.886	1.901	1.563	1.571	1.582
$\omega_0 \times 10^{-13}$ [sec^{-1}]	4.0	3.4	3.2	2.6	2.1
$v \times 10^{-5}$ [cm · sec^{-1}]	5.5	4.3	5.1	4.1	3.1
$E_{lm}^{CB} - E_{lm}^{1s}$ [eV]	1.95	1.8†	1.91	1.63	1.46
$r_0 + \delta \cdot$	73.93	80.43	90.87	94.96	105.70

† Estimated from Löffler [24].

β	electronic variational parameter,	d	lattice constant,
α^e, α^g	polarisation parameters,	ω_0	local mode frequency,
v	velocity of sound		

integrations were evaluated using a standard Gauss integration method. In Table 3 we compare the zero point shifts calculated from the wave functions (2.32a) and (2.31) with those estimated from (7.3).

Table 3

Δq	NaCl	NaBr	KCl	KBr	KI
(2:32a)	4.32	4.75	4.86	5.42	5.19
(2:31)	5.48	5.91	5.89	6.46	6.42
(7:3)	8.02	7.2	7.22	8.19	8.71

We immediately see the strong dependence of the zero point shift on the actual form of the ground state wave function. However the large discrepancy between these values and those from (7.3) cannot be explained. While a more exact derivation of equation (2.14) might bring some improvement the discrepancy is more likely to stem from our assumption that the local mode frequency is independent of the electronic state. From the spectral data, effective frequencies of the local mode associated with emission and absorption are calculated in [20]. These frequencies are different from each other and also from our effective frequency. Because of the extreme sensitivity of the rates to the absolute value of the zero-point shift we use the values given by (7.3) in our calculations.

In Tables 4 and 5 we give the linewidth parameters γ_1, and γ_2 from (4.18) as a function of temperature. For our temperature range we have $\Gamma_l \ll \hbar\omega_0$ for all l in the range $0-1000$ so that we can see from the size of δ in Table 2 that we are far from a resonance transition.

Table 4

T (°K)	NaCl	NaBr	KCl	KBr	KI
0	0.0	0.0	0.0	0.0	0.0
50	$1.50 \cdot 10^{-22}$	$1.88 \cdot 10^{-22}$	$4.61 \cdot 10^{-22}$	$7.00 \cdot 10^{-22}$	$1.49 \cdot 10^{-21}$
100	$4.88 \cdot 10^{-21}$	$4.14 \cdot 10^{-21}$	$9.17 \cdot 10^{-21}$	$9.74 \cdot 10^{-21}$	$1.56 \cdot 10^{-20}$
150	$2.05 \cdot 10^{-20}$	$1.55 \cdot 10^{-20}$	$3.34 \cdot 10^{-20}$	$3.20 \cdot 10^{-20}$	$4.74 \cdot 10^{-20}$
200	$4.89 \cdot 10^{-20}$	$3.53 \cdot 10^{-20}$	$7.49 \cdot 10^{-20}$	$6.83 \cdot 10^{-20}$	$9.73 \cdot 10^{-20}$
250	$9.07 \cdot 10^{-20}$	$6.36 \cdot 10^{-20}$	$1.34 \cdot 10^{-19}$	$1.19 \cdot 10^{-19}$	$1.66 \cdot 10^{-19}$
300	$1.46 \cdot 10^{-19}$	$1.01 \cdot 10^{-19}$	$2.10 \cdot 10^{-19}$	$1.84 \cdot 10^{-19}$	$2.52 \cdot 10^{-19}$

$\gamma_1(T)$ (erg).

Table 5

T (°K)	NaCl	NaBr	KCl	KBr	KI
0	$5.84 \cdot 10^{-20}$	$2.85 \cdot 10^{-20}$	$4.71 \cdot 10^{-20}$	$2.59 \cdot 10^{-20}$	$2.29 \cdot 10^{-20}$
50	$6.78 \cdot 10^{-20}$	$3.40 \cdot 10^{-20}$	$6.12 \cdot 10^{-20}$	$3.72 \cdot 10^{-20}$	$3.69 \cdot 10^{-20}$
100	$1.04 \cdot 10^{-19}$	$5.56 \cdot 10^{-20}$	$1.06 \cdot 10^{-19}$	$7.09 \cdot 10^{-20}$	$7.77 \cdot 10^{-20}$
150	$1.57 \cdot 10^{-19}$	$8.78 \cdot 10^{-20}$	$1.71 \cdot 10^{-19}$	$1.20 \cdot 10^{-19}$	$1.38 \cdot 10^{-19}$
200	$2.25 \cdot 10^{-19}$	$1.29 \cdot 10^{-19}$	$2.54 \cdot 10^{-19}$	$1.84 \cdot 10^{-19}$	$2.17 \cdot 10^{-19}$
250	$3.08 \cdot 10^{-19}$	$1.80 \cdot 10^{-19}$	$3.56 \cdot 10^{-19}$	$2.63 \cdot 10^{-19}$	$3.14 \cdot 10^{-19}$
300	$4.05 \cdot 10^{-19}$	$2.39 \cdot 10^{-19}$	$4.75 \cdot 10^{-19}$	$3.56 \cdot 10^{-19}$	$4.30 \cdot 10^{-19}$

$\gamma_2(T)$ (erg).

This will invariably be the case for transitions between discrete electronic levels. Consequently our approximation in (6.8) where the linewidth is replaced by its thermal average is good. Numerical values for the rates from the band bottom to the F-centre ground state, calculated for a microblock volume $\Omega = 10^5 d^3$, are given in Table 6, columns (a). The relatively small values obtained can largely be attributed to the nonresonance character of the transitions. We performed a similar calculation where we increased fractionally the energy difference and thereby moved towards the resonance condition. The resulting rates are given in Table 6, columns (b). These are much greater than the non-resonance rates. However they must be regarded as estimates only as the approximation of (6.8) is no longer good.

Table 6

T	NaCl		NaBr	
(°K)	(a)	(b)	(a)	(b)
50	.25	$.11 \cdot 10^8$	$.88 \cdot 10^{-10}$	2.1
100	1.1	$.97 \cdot 10^9$	$.56 \cdot 10^{-8}$	$.57 \cdot 10^3$
150	11	$.62 \cdot 10^{10}$	$.59 \cdot 10^{-6}$	$.26 \cdot 10^5$
200	110	$.18 \cdot 10^{11}$	$.36 \cdot 10^{-4}$	$.41 \cdot 10^6$
250	520	$.41 \cdot 10^{11}$	$.11 \cdot 10^{-2}$	$.38 \cdot 10^7$
300	3700	$.77 \cdot 10^{11}$	.017	$.22 \cdot 10^8$

KCl		KBr		KI	
(a)	(b)	(a)	(b)	(a)	(b)
$.23 \cdot 10^{-13}$	$.20 \cdot 10^{-4}$	$.64 \cdot 10^{-8}$	1.5	$.24 \cdot 10^{-10}$	4.3
$.19 \cdot 10^{-11}$	.012	$.14 \cdot 10^{-5}$	$.92 \cdot 10^3$	$.62 \cdot 10^{-7}$	$.58 \cdot 10^3$
$.62 \cdot 10^{-9}$	1.1	$.41 \cdot 10^{-3}$	$.51 \cdot 10^5$	$.50 \cdot 10^{-4}$	$.66 \cdot 10^5$
$.12 \cdot 10^{-6}$	54	.037	$.10 \cdot 10^7$	$.61 \cdot 10^{-2}$	$.17 \cdot 10^7$
$.10 \cdot 10^{-4}$	$.13 \cdot 10^4$	1.2	$.88 \cdot 10^7$	.20	$.18 \cdot 10^8$
$.37 \cdot 10^{-3}$	$.17 \cdot 10^5$	18	$.48 \cdot 10^8$	3.0	$.15 \cdot 10^9$

$^t W_{\text{FC·CB}}(\beta) \cdot (\text{sec}^{-1})$.

For a near resonance transition the electron must have an energy of the order $\frac{1}{2}\hbar\omega_0 \sim 10$ meV above the band bottoms. This is the same as $\varkappa T$ at 125°K. If we consider the electronic excited state as a localized polaron, which is more realistic, then we have a discrete set of states instead of a band. There will be a thermal distribution of the electronic energy over these levels and thus near resonance transitions become possible at finite temperatures. By summing over all polaron states with a suitable thermal weighting factor one can obtain a total non-radiative capture rate for the vacancy. In practice the summation can be restricted to energy levels close to resonance as the rates fall drastically when one moves away from this range.

We can now consider a simple model for the quenching of F-luminescence and compare it with experiment. The F-centre excited state lies very close to the bottom of the band and if this state is a mixture of s- and p-type wave functions as is postulated in [23] then the electronic overlap integrals for the two transitions

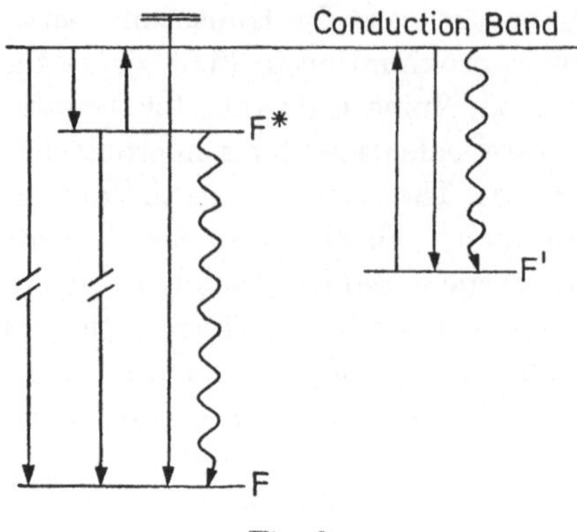

Fig. 2

will be similar. Hence we can say that non-radiative transitions from the excited state to the ground state are of the same order as those from the band minima. These rates are very small compared with the radiative transition rate which is approximately 10^6 sec^{-1}. Thus to quench the F-luminescence we must excite our electron into the band and then have a fast resonance transition or trap the electron in a meta-stable level — the F'-centre. Fig. 2 is suggested.

In [23] it is found that the quantum yield of F-luminescence is controlled by the thermal excitation from the excited state into the band. Capture of the freed electron by the F'-centre is important as is shown by Markham [25] in his analysis of Pick's data [26, 27]. However the F'-centre becomes thermally unstable as room temperature is approached and in this region, $T > 170°K$, the "resonance transition" becomes increasingly important.

In conclusion we feel that a tractable method of treating non-radiative multiphonon transitions has been found and that it is particularily relevant to transitions between discrete electronic levels. However the sensitivity of the resulting rates on r_0 and Δq requires accurate knowledge of these quantities. In this respect transitions to shallow defect levels should be less susceptible to inaccuracies in Δq and so capture of a polaron to form an F'-centre or an excited F-centre would be worth treating. Complete reaction kinetics for the problem could then be considered.

Acknowledgement. I am grateful to Prof. H. Stumpf and Prof. H. Motz who made my visit to the University of Tübingen possible. I would like to thank Prof. Stumpf for suggesting this topic and for the encouragement he has given me throughout my stay. I also thank W. Heinzel for many enlightening discussions.

I am indebted to the Science Research Council for financial support.

Appendix A

(1) *Calculation of the Matrix* **A** (2.4)

$$\mathbf{A}_{ij} = \left(\frac{\partial}{\partial \alpha_i} \frac{\partial}{\partial \alpha_j} U_n \right)_s . \qquad (A.1)$$

The total electronic energy is given by Stumpf [10]

$$U_n = \frac{1}{2\,m}\,P_n^2 + \sum_{k\varrho}{}' z_{k\varrho}\,P_{k\varrho}\int \xi_n^s(r)\,C(r,R_{k\varrho})\,d\tau + H_{\mathrm{F}}^{\mathrm{ex}} + \sum_{k\varrho} H_{k\varrho}^0$$

$$- \frac{1-\alpha^e}{1+\alpha^e}\int \xi_n(r_1)\,\xi_n(r_2)\,F(r_1,r_2)\,d\tau_1\,d\tau_2 - \sum_{k\varrho}{}' P_{k\varrho}\,b\,C(00,k\varrho)^\eta$$

$$+ \tfrac{1}{2}\sum_{\substack{l\mu \\ \neq k\varrho}} G_{l\mu,\,k\varrho}[z_{k\varrho}\,z_{l\mu}\,C(k\varrho,l\mu) + b\,C(k\varrho,l\mu)^\eta] \qquad (A.2)$$

where

$$\xi_n^s(r) = (1+\alpha^e)^{-1/2}\,\xi_n(r) = (1+\alpha^e)^{-1/2}\,[e\,|\Psi_n(r)|^2 - a_{00}\,\delta(r-R_{00})]\,,$$
$$z_{k\varrho} = (1+\alpha^e)^{-1/2}\,a_{k\varrho}\,,$$
$$P_{k\varrho} = 1 + \mathfrak{M}_{k\varrho}\nabla_{k\varrho}\,,$$
$$F(r_1,r_2) = \frac{d^3}{2\pi}\sum_{k\varrho}\nabla_{k\varrho}P_{k\varrho}C(r_1,R_{k\varrho})\cdot\nabla_{k\varrho}P_{k\varrho}C(r_2,R_{k\varrho})\,.$$

The notation has been taken en-bloc from [10]. The terms appearing are the kinetic energy of the defect electron, electron interaction with the lattice ions, exchange interaction with all other crystal electrons, self energy and the interaction energy of the remaining closed ion shells. The only terms involving the electronic wave function are the 1st, 2nd and 5th. Since the ionic displacements $\mathfrak{M}_{k\varrho}$ fall of very rapidly [11] we approximate by setting $P_{k\varrho} = 1$.

Only the first two terms are considered as the fifth contains a lattice summation of the form $\sum_{k\varrho} R_{k\varrho}^{-4}$ which falls off much more rapidly than the summation in the second term.

$$U_n = \frac{1}{2\,m}\,P_n^2 + \sum_{k\varrho}{}'(1-\alpha^e)^{-1}\,e\,a_{k\varrho}\int |\Psi_n(r)|^2\,C(r,R_{k\varrho})\,d\tau + U_n^\Phi \qquad (A.3)$$

where U_n^Φ is all the terms independent of $\Psi_n(r)$.

With the expansion (2.1) we have

$$\Psi_n(r) = \left(\sum_{k=1}^{n}\alpha_k^2\right)^{-1/2}\sum_{k=1}^{n}\alpha_k\,\Psi_n^k(r) \qquad (A.4)$$

we immediately get:

$$\left(\frac{\partial}{\partial\alpha_i}\,\Psi_n(r)\right)_{\!s} = \Psi_n^i(r) - \delta_{1i}\,\Psi_n^1(r)\,, \qquad (A.5\,a)$$

$$\left(\frac{\partial^2}{\partial\alpha_i\,\partial\alpha_j}\,\Psi_n(r)\right)_{\!s} = (3\,\delta_{1i}\,\delta_{1j} - \delta_{ij})\,\Psi_n^1(r) - \delta_{1i}\,\Psi_n^j(r) - \delta_{1j}\,\Psi_n^i(r)\,, \qquad (A.5\,b)$$

$$\left(\frac{\partial^2}{\partial\alpha_i\,\partial\alpha_j}\,|\Psi_n(r)|^2\right)_{\!s} = 2\,\Psi_n^1(r)\left(\frac{\partial^2}{\partial\alpha_i\,\partial\alpha_j}\,\Psi_n(r)\right)_{\!s} + \left(\frac{\partial}{\partial\alpha_i}\,\Psi_n(r)\right)_{\!s}\!\left(\frac{\partial}{\partial\alpha_j}\,\Psi_n(r)\right)_{\!s}.$$
$$(A.5\,c)$$

It can be readily seen that all terms in (A.2) contribute nothing to the element A_{11}. This is why we are forced to take a minimum of three wave functions in the expansion (A.4) to avoid a non-trivial result in (2.30 b).

If we define the integrals

$$F_{ij} = \int \Psi_n^i(\mathbf{r})\, \nabla^2 \Psi_n^j(\mathbf{r})\, d\tau \,, \tag{A.6a}$$

$$f_{ij} = \int \Psi_n^i(\mathbf{r})\, C(\mathbf{r}, \mathbf{R}_{k\varrho})\, \Psi_n^j(\mathbf{r})\, d\tau \,. \tag{A.6b}$$

Then the matrix $\mathbf{A}$ has elements

$$\mathbf{A}_{11} = 0 \,,$$

$$\mathbf{A}_{12} = \frac{\hbar^2}{2\,m}\, [F_{12} + F_{21}] - 2(1 + \alpha^e)^{-1}\, e^2 \sum_{k\varrho}{}' \frac{a_{k\varrho}}{e}\, f_{12} \,,$$

$$\mathbf{A}_{13} = \frac{\hbar^2}{2\,m}\, [F_{13} + F_{31}] - 2(1 + \alpha^e)^{-1}\, e^2 \sum_{k\varrho}{}' \frac{a_{k\varrho}}{e}\, f_{13} \,,$$

$$\mathbf{A}_{22} = \frac{\hbar^2}{m}\, [F_{11} - F_{22}] + 2(1 + \alpha^e)^{-1}\, e^2 \sum_{k\varrho}{}' \frac{a_{k\varrho}}{e}\, (f_{22} - f_{11}) \,,$$

$$\mathbf{A}_{23} = - \frac{\hbar^2}{2\,m}\, [F_{23} + F_{32}] + 2(1 + \alpha^e)^{-1}\, e^2 \sum_{k\varrho}{}' \frac{a_{k\varrho}}{e}\, f_{23} \,,$$

$$\mathbf{A}_{33} = \frac{\hbar^2}{m}\, [F_{11} - F_{33}] + 2(1 + \alpha^e)^{-1}\, e^2 \sum_{k\varrho}{}' \frac{a_{k\varrho}}{e}\, (f_{33} - f_{11}) \,. \tag{A.7}$$

The matrix is symmetric.

The integrals (A.6) are

$$F_{11} = -\beta^2 \,, \quad F_{12} = F_{21} = -8\sqrt{2}/27\,\beta^2 \,, \quad F_{13} = F_{31} = -\sqrt{3/8}\,\beta^2 \,,$$

$$F_{22} = -\tfrac{1}{4}\beta^2 \,, \quad F_{23} = F_{32} = -\sqrt{\tfrac{2}{3}}\,\tfrac{552}{3125}\,\beta^2 \,, \quad F_{33} = -\tfrac{1}{9}\beta^2 \,, \tag{A.8}$$

$$f_{11} = \frac{1}{R_{k\varrho}}\, [1 - e^{-2\beta R_{k\varrho}}(1 + \beta R_{k\varrho})] \,, \tag{A.9}$$

$$f_{12} = \frac{2\sqrt{2}}{9}\, \beta\, e^{-\frac{3}{2}\beta R_{k\varrho}}[\beta R_{k\varrho} + \tfrac{2}{3}] \,,$$

$$f_{13} = \frac{1}{3\sqrt{3}}\, \beta\, e^{-4/3\,\beta R_{k\varrho}}\, [-\tfrac{1}{6}(\beta R_{k\varrho})^2 + \tfrac{3}{4}(\beta R_{k\varrho}) + \tfrac{9}{16}] \,,$$

$$f_{22} = \frac{1}{R_{k\varrho}}\, [1 - e^{-\beta R_{k\varrho}}[1 + \tfrac{3}{4}\beta R_{k\varrho} + \tfrac{1}{4}(\beta R_{k\varrho})^2 + \tfrac{1}{8}(\beta R_{k\varrho})^3]] \,,$$

$$f_{23} = \frac{1}{3\sqrt{3}}\, \beta\, e^{-5/6\,\beta R_{k\varrho}}\, [\tfrac{4}{75}(\beta R_{k\varrho})^3 - \tfrac{28}{375}(\beta R_{k\varrho})^2 + \tfrac{138}{625}\beta R_{k\varrho} + \tfrac{828}{3125}] \,,$$

$$f_{33} = \frac{1}{R_{k\varrho}}\, [1 - \tfrac{4}{27}e^{-4/3\,\beta R_{k\varrho}}(\tfrac{27}{4} + \tfrac{15}{4}\beta R_{k\varrho} + (\beta R_{k\varrho})^2 + \tfrac{1}{3}(\beta R_{k\varrho})^3$$
$$- \tfrac{1}{27}(\beta R_{k\varrho})^4 + \tfrac{1}{81}(\beta R_{k\varrho})^5)] \,.$$

$$(2)\ \textit{Vector Coefficients in (2.30)}$$

(2.26) gives

$$C_0\, \xi(1|k\varrho) = \frac{a_{k\varrho}}{e}\, \nabla_{k\varrho} \int [|\Psi_{\mathrm{F}}^1(\mathbf{r})|^2 - |\Psi_{\mathrm{CB}}^1(\mathbf{r})|^2]\, C(\mathbf{r}, \mathbf{R}_{k\varrho})\, d\tau \,.$$

The conduction band wave function is very diffuse and we set it equal to zero in this equation.

Then

$$C_0\,\xi(1|k\varrho) = \frac{a_{k\varrho}}{e}\,\nabla_{k\varrho}\int |\Psi_{\mathrm{F}}^1(r)|^2\,C(r,R_{k\varrho})\,d\tau = \frac{a_{k\varrho}}{e}\,\nabla_{k\varrho}f_{11}\,. \qquad (A.10)$$

Normalization implies

$$C_0^2 = \sum_{k\varrho}|\nabla_{k\varrho}f_{11}|^2\,. \qquad (A.11)$$

(2.25) gives directly:

$$\frac{a_{k\varrho}}{e}\,\nabla_{k\varrho}f_{1r} = \sum_{m=1}^{h}{}^{n}C_m^r\,\xi(m|k\varrho) \qquad (A.12)$$

so the overlap of this vector with (A.10) is just

$${}^{n}C_1^r = C_1^{-1}\sum_{k\varrho}\nabla_{k\varrho}f_{11}\,\nabla_{k\varrho}f_{1r}\,. \qquad (A.13)$$

References

1. Markham, J. J.: Solid State Phys., Suppl. 8, Academic Press 1966
2. Davydov, A. S.: Phys. Status Solidi **30**, 357 (1968)
3. Evans, B. D., Kemp, J. C.: Phys. Rev. B **2**, 4179 (1970)
4. Tewordt, L.: Z. Physik **137**, 604 (1954)
5. Kubo, R., Toyozawa, Y.: Prog. Theor. Phys. **13**, 160 (1955)
6. Stumpf, H.: Z. Naturforsch. **12a**, 465 (1957)
7. Stumpf, H.: Quantentheorie der Ionenrealkristalle. Berlin-Göttingen-Heidelberg: Springer 1961.
8. Stumpf, H.: Z. Physik **229**, 488 (1969)
9. Stumpf, H.: Phys. kondens. Materie **13**, 9 (1971)
10. Stumpf, H.: Phys. kondens. Materie **13**, 101 (1971)
11. Schmid, J.: Phys. kondens. Materie **15**, 119 (1972)
12. Renn W.: Dissertation, Univ. Tübingen 1973
13. Schwarz, G.: Dissertation, Univ. Tübingen 1973
14. Heinzel, W.: Dissertation, Univ. Tübingen, in preparation
15. Rampacher, H.: Z. Naturforsch. **23a**, 401 (1968)
16. Märtl, H.: Diplomarbeit, Univ. München 1967
17. Leibfried, G., Ludwig, W.: Solid State Phys. **12**, 275, Academic Press 1961
18. Ziman, J.: Principles of the Theory of Solids. Cambridge Univ. Press 1969
19. Koide, S.: Z. Naturforsch. **15a**, 123 (1960).
20. Gebhardt, W., Kühnert, H.: Phys. Letters **11**, 15 (1964)
21. Swank, R. K., Brown, F. C.: Phys. Rev. **130**, 34 (1963)
22. Wood, R. F., Öpik, U.: Phys. Rev. **179**, 783 (1969)
23. Honda, S., Tomura, M.: J. Phys. Soc. Japan **33**, 1003 (1972)
24. Löffler, A.: Z. Naturforsch. **24a**, 516 (1969)
25. Markham, J. J.: Phys. Rev. **88**, 500 (1952)
26. Pick, H.: Ann. Physik **31**, 365 (1938)
27. Pick, H.: Ann. Physik **37**, 421 (1940)

S. Fraser
Department of Engineering Science
Parks Road
Oxford, England

Phys. cond. Matter 17, 99—111 (1974)
© by Springer-Verlag 1974

An Algebraic Treatment of Franck-Condon-Transitions

W. Heinzel

Institut für Theoretische Physik, Universität Tübingen

Received July 11, 1973

A purely algebraic evaluation of Franck-Condon-Integrals, of transition rates for energy conserving Franck-Condon-transitions and of the related thermal averaged quantities is performed.

1. Introduction

In a recent paper W. Witschel [1] used a purely algebraic method to calculate Franck-Condon-matrixelements. The algebraic method has the advantage that it is independent of the related representations of the algebra in contrast to the methods used by Wagner [2], Ansbacher [3] and others. Already Koide [4] has made use of algebraic methods but only partially however and it was a theoretical progress to treat Franck-Condon transition probabilities with the algebra of oscillator-operators alone [1]. A short historical review on the different treatments of two-centre harmonic oscillator integrals (Franck-Condon-integrals) is given by Witschel [1].

In the present paper we also want to use algebraic methods for Franck-Condon transitions. But our method will be different from Witschel's one. Moreover our methods will enable us to perform thermal averaging of the respective transition probabilities. Especially we will get an evaluation of the thermal averaged probabilities for energy-conserving Franck-Condon-processes. We will discuss all that for one-dimensional harmonic oscillators. It can easily be extended to multidimensional harmonic oscillators.

2. Fundamentals

One has two harmonic oscillators with different frequencies ω and ω' and with oscillator coordinates q and q' being related by a shift:

$$q' = q + \Delta q. \tag{2.1}$$

The canonical conjugates are

$$p = -i\frac{\partial}{\partial q} = -i\frac{\partial}{\partial q'} = p' \qquad (\hbar = 1). \tag{2.2}$$

The hamiltonians are:

$$H = \tfrac{1}{2}(p^2 + \omega^2 q^2), \tag{2.3a}$$

$$H' = \tfrac{1}{2}(p^2 + \omega'^2 q'^2). \tag{2.3b}$$

The eigenstates of H are denoted by

$$|0, \omega; m\rangle = |m\rangle \tag{2.4}$$

100 W. Heinzel

and the eigenstates of H' by

$$|\Delta q, \omega'; m'\rangle .\tag{2.5}$$

Defining the oscillator creation and annihilation operators a^+ and a by

$$p = i \sqrt{\omega/2}\,(a^+ - a) ,\tag{2.6a}$$

$$q = \sqrt{\frac{1}{2\,\omega}}\,(a^+ + a)\tag{2.6b}$$

then Koide shows that the states (2.5) are obtained from the states (2.4) by the following transformation:

$$|\Delta q, \omega'; m'\rangle = C_{m'}^{-1}\, U V\,|m'\rangle ,\tag{2.7a}$$

$$\langle \Delta q, \omega'; m'| = C_{m'}\,\langle m'|\, V^{-1}\, U^{-1}\tag{2.7b}$$

with

$$U = e^{-(\alpha/2)^2}\, e^{-\alpha a^+}\, e^{\alpha a} = e^{ip\Delta q} ,\tag{2.8}$$

$$V = e^{-Aa^2}\, e^{B(a^+)^2} ,\tag{2.9a}$$

$$V^{-1} = e^{-B(a^+)^2}\, e^{Aa^2} ,\tag{2.9b}$$

$$\alpha = \sqrt{\frac{\omega}{2}}\, \Delta q ,\tag{2.10}$$

$$A = \frac{\omega - \omega'}{2\,(\omega + \omega')}\tag{2.11a}$$

$$B = \frac{\omega^2 - \omega'^2}{8\,\omega\,\omega'} .\tag{2.11b}$$

U is a unitary transformation, whereas V is not unitary; therfore:

$$C_{m'}^2 = \langle m'|\, V^+\, U^+\, U V\,|m'\rangle = \langle m'|\, V^+ V\,|m'\rangle \neq 1 .\tag{2.12}$$

The problem is proposed to evaluate the overlaps (Franck-Condon-integrals).

$$\langle 0, \omega; m\,|\,\Delta q, \omega', m'\rangle = C_{m'}^{-1}\langle m|\, U V\,|m'\rangle ,\tag{2.13a}$$

$$\langle \Delta q, \omega', m'\,|\,0, \omega; m\rangle = C_{m'}\langle m'|\, V^{-1}\, U^{-1}\,|m\rangle\tag{2.13b}$$

respectively their squares:

$$|\langle \Delta q, \omega', m'\,|\,0, \omega, m\rangle|^2 = \langle m|\, U V\,|m'\rangle\,\langle m'|\, V^{-1}\, U^{-1}\,|m\rangle .\tag{2.14}$$

The occupation number operator

$$\hat{m} = a^+ a\tag{2.15}$$

has the eigenstates (2.4). We are interested in the evaluation of (2.13), (2.14) as well as of the following quantities:

$$\langle 0, \omega; m\,|\, q^\nu\, p^\mu\,|\,\Delta q, \omega'; m'\rangle\tag{2.16}$$

or

$${}^1 I_{mm'} = C_{m'}\langle 0, \omega; m\,|\, e^{q\eta}\, e^{ip\delta}\,|\,\Delta q, \omega'; m'\rangle ,\tag{2.17a}$$

$${}^2 I_{mm'} = C_{m'}^{-1}\langle m', \omega', \Delta q\,|\, e^{-ip\delta}\, e^{q\eta}\,|\,0, \omega, m\rangle .\tag{2.17b}$$

Additionally we are interested in

$$\gamma_m(\Omega) = 2\pi \sum_{m'} |\langle 0, \omega, m | e^{q\eta} e^{ip\delta} | \Delta q, \omega', m'\rangle|^2 \, \delta(\omega[m+\tfrac{1}{2}] - \omega'[m'+\tfrac{1}{2}] + \Omega)$$

(2.18)

which represents a general transition probability for energy balanced Franck-Condon-transitions from the state $|m\rangle$, and we want to calculate the thermal averaged quantity:

$$\gamma(\beta, \Omega) = \mathfrak{n} \sum_m e^{-\beta\omega m} \gamma_m(\Omega)$$

(2.19)

where:

$$\mathfrak{n}^{-1} = \sum_m e^{-\beta\omega m}, \qquad \beta = 1/kT.$$

(2.20)

All interesting quantities are known if we have evaluated $^1I_{mm'}$, $^2I_{mm'}$, $\gamma_m(\Omega)$, $\gamma(\beta, \Omega)$ which will all be done by algebraic methods, introduced in the next section[1].

3. Algebra and "Hyper-Algebra"

To evaluate (2.17), we use (2.7):

$$^1I_{mm'} = \langle m | e^{q\eta} e^{ip\delta} U V | m'\rangle,$$

(3.1 a)

$$^2I_{mm'} = \langle m' | V^{-1} U^{-1} e^{-ip\delta} e^{q\eta} | m\rangle.$$

(3.1 b)

We have the oscillator algebra

$$[a, a^+] = 1$$

(3.2)

and

$$a|0\rangle = \langle 0| a^+ = 0,$$

(3.3 a)

$$\hat{m}|m\rangle = m|m\rangle.$$

(3.3 b)

In (3.1) we need a reordering of the operators in order to make use of (3.3). Especially we need a normal-ordering of e.g. $V = e^{-Aa^2} e^{B(a^+)^2}$. That normal ordering is generally not possible in a closed form but only within states. To avoid rather complicated operator relations one can use the following representation:

$$V = e^{-A(\partial/\partial\alpha_1)^2} e^{\alpha_1 a} e^{B(\partial/\partial\alpha_2)^2} e^{\alpha_2 a^+}\big|_{\alpha_1 = \alpha_2 = 0}$$

(3.4)

with the condition that α_1 and α_2 have to be equalized to zero after having applied the differentiation operators $\partial/\partial\alpha_1$ and $\partial/\partial\alpha_2$. The advantage of (3.4) is, that we have no longer any square of the operators a^+ or a in the exponents and get thus a normal-ordering for the operators of the oscillator-algebra simply by the Hausdorff-formula (see Sect. 4). The parameters α_1, α_2 and the differentiation operators $\partial/\partial\alpha_1$, $\partial/\partial\alpha_2$ build also algebras, which we will call "hyper-algebras":

$$[\partial/\partial\alpha_i, \alpha_i] = 1.$$

(3.5)

---

[1] In this paper instead of $\gamma(\beta, \Omega)$ we treat its Fouriertransformation:

$$\tilde{\gamma}(\beta, \tau) = \frac{1}{2\pi} \int d\Omega \, e^{-i\tau\Omega} \gamma(\beta, \Omega) \quad \text{(see Sect. 9).}$$

The problem of algebra-elements appearing squared as an exponent is shifted
from the oscillator-algebra to the hyper-algebras. But there, it will turn out to
be easier handled. The reason for that is the splitting of one algebra $(a,\, a^+)$ into
two (hyper-)algebras $(\partial/\partial\alpha_1,\, \alpha_1)$ and $(\partial/\partial\alpha_2,\, \alpha_2)$ or even more than two (see Sect. 7).

4. Hausdorff-Formula and Related Formulas

From (2.15) and (3.2) one gets

$$\hat{m}\,a = a\,(\hat{m} - 1)\,, \tag{4.1a}$$

$$\hat{m}\,a^+ = a^+\,(\hat{m} + 1)\,. \tag{4.1b}$$

From that we get

$$e^{c\hat{m}}\,a = a\,e^{c(\hat{m}-1)} = a\,e^{-c}\,e^{c\hat{m}}\,, \tag{4.2a}$$

$$e^{c\hat{m}}\,a^+ = a^+\,e^{c(\hat{m}+1)} = a^+\,e^{c}\,e^{c\hat{m}} \tag{4.2b}$$

and from that we can conclude

$$e^{c_1\hat{m}}\,e^{c_2 a^+} = e^{c_2 a^+\,\exp c_1}\,e^{c_1\hat{m}}\,, \tag{4.3a}$$

$$e^{c_1\hat{m}}\,e^{c_2 a} = e^{c_2 a\,\exp(-c_1)}\,e^{c_1\hat{m}}\,. \tag{4.3b}$$

The Hausdorff-formula reads:

$$e^{c_1 a}\,e^{c_2 a^+} = e^{c_2 a^+}\,e^{c_1 a}\,e^{c_1 c_2} \tag{4.4a}$$

or

$$e^{c_1 a + c_2 a^+} = e^{c_2 a^+}\,e^{c_1 a}\,e^{\frac{1}{2}c_1 c_2}\,. \tag{4.4b}$$

Also the following formulae will be needed:

$$e^{ca}\,(a^+)^m = (c + a^+)^m\,e^{ca}\,, \tag{4.5a}$$

$$e^{ca^2}\,(a^+)^m = (a^+ + 2\,c\,a)^m\,e^{ca^2}\,, \tag{4.5b}$$

$$a^m\,e^{ca^+} = e^{ca^+}\,(c + a)^m\,, \tag{4.5c}$$

$$a^m\,e^{c(a^+)^2} = e^{c(a^+)^2}\,(a + 2\,c\,a^+)^m\,. \tag{4.5d}$$

(4.5a) and (4.5c) follow directly from the formula (4.4a) and (4.5b, d) follow from

$$e^{c_1 a}\,e^{c_2(a^+)^2} = e^{c_2(a^+ + c_1)^2}\,e^{c_1 a}\,, \tag{4.6a}$$

$$e^{c_1 a^2}\,e^{c_2 a^+} = e^{c_2 a^+}\,e^{c_1(a + c_2)^2}\,. \tag{4.6b}$$

We can prove (4.6) with hyper-algebra-technique:

$$
\begin{aligned}
e^{c_1 a}\,e^{c_2(a^+)^2} &= e^{c_2(\partial/\partial\alpha_1)^2}\,e^{c_1 a}\,e^{\alpha_1 a^+}\Big|_{\alpha_1=0} \\
&= e^{c_2(\partial/\partial\alpha_1)^2}\,e^{c_1\alpha_1}\,e^{\alpha_1 a^+}\,e^{c_1 a}\Big|_{\alpha_1=0} \\
&= e^{c_2(\partial/\partial\alpha_1)^2}\,e^{\alpha_1(c_1 + a^+)}\,e^{c_1 a}\Big|_{\alpha_1=0} \\
&= e^{c_2(c_1 + a^+)^2}\,e^{c_1 a}\,.
\end{aligned}
\qquad \text{q.e.d.}
$$

In the next sections we will not always note explicitly that the parameters are
to be set equal zero after differentiation.

5. Evaluation of $^2I_{mm'}$

From (2.17b) and (2.7b):

$$^2I_{mm'} = \langle m'|\, V^{-1}\, U^{-1} e^{-ip\delta}\, e^{q\eta}\, |m\rangle\,. \tag{5.1}$$

Using the abbreviations:

$$\Theta = \sqrt{\frac{1}{2\omega}}\,\eta\,, \tag{5.2a}$$

$$\Delta = \sqrt{\frac{\omega}{2}}\,\delta \tag{5.2b}$$

we obtain by use of (2.6), (2.8) and (2.9b) from (5.1)

$$^2I_{mm'} = \langle m'|\, e^{-B(a^+)^2}\, e^{A\,a^2}\, e^{(a^+-a)\,(\alpha+\Delta)}\, e^{(a^++a)\,\Theta}\, |m\rangle\,. \tag{5.3}$$

With (4.4b) we get:

$$^2I_{mm'} = \langle m'|\, e^{-B(a^+)^2}\, e^{A\,a^2}\, e^{a(\alpha+\Delta)}\, e^{-a(\alpha+\Delta)}\, e^{a^+\Theta}\, e^{a\Theta}\, |m\rangle \cdot e^{-\frac{1}{2}(\alpha+\Delta)^2}\, e^{\frac{1}{2}\Theta^2}\,. \tag{5.4}$$

A normal ordering of the operator within the brackets is achieved by use of (4.6b) and (4.4a)

$$^2I_{mm'} = \langle m'|\, e^{-B(a^+)^2}\, e^{c_1 a^+}\, e^{A\,a^2}\, e^{c_2 a}\, |m\rangle\, e^{c_3} \tag{5.5}$$

with the abbreviations:

$$c_1 = \alpha + \Delta + \Theta\,, \tag{5.6a}$$

$$c_2 = 2A\,c_1 - (\alpha + \Delta) + \Theta\,, \tag{5.6b}$$

$$c_3 = -\tfrac{1}{2}(\alpha + \Delta)^2 + \tfrac{1}{2}\Theta^2 - (\alpha + \Delta)\,\Theta + A\,c_1^2\,. \tag{5.6c}$$

Using $|m\rangle = (a^+)^m/\sqrt{m!}\,|0\rangle$ and (4.5) we arrive at:

$$^2I_{mm'} = \langle 0|\, (a - 2\,B\,a^+ + c_1)^{m'}\, (a^+ + 2\,A\,a + c_2)^m\, |0\rangle\, \frac{e^{c_3}}{\sqrt{m!}\,\sqrt{m'!}}\,. \tag{5.7}$$

With hyperalgebras we may write

$$^2I_{mm'} = \frac{e^{c_3}}{\sqrt{m!}\,\sqrt{m'!}} \left(\frac{\partial}{\partial\alpha_1}\right)^{m'} \left(\frac{\partial}{\partial\alpha_2}\right)^m \langle 0|\, e^{\alpha_1(a-2Ba^++c_1)}\, e^{\alpha_2(a^++2Aa+c_2)}\, |0\rangle\,. \tag{5.8}$$

By use of (4.4) and (3.3) one gets

$$^2I_{mm'} = \frac{e^{c_3}}{\sqrt{m!}\,\sqrt{m'!}} \left(\frac{\partial}{\partial\alpha_1}\right)^{m'} \left(\frac{\partial}{\partial\alpha_2}\right)^m e^{\alpha_1(c_1-B)}\, e^{\alpha_2(c_2+A)}\, e^{\alpha_1\alpha_2}\,. \tag{5.9}$$

Now one uses the algebraic relations (4.5c, d) for the hyperalgebras (replace a by $\partial/\partial\alpha_i$ and a^+ by α_i). Thus we get:

$$\begin{aligned}
^2I_{mm'} &= \frac{e^{c_3}}{\sqrt{m!}\,\sqrt{m'!}} \left(\frac{\partial}{\partial\alpha_1} + c_1 - B\right)^{m'} \left(\frac{\partial}{\partial\alpha_2} + \alpha_1 + c_2 + A\right)^m \\
&= \frac{e^{c_3}}{\sqrt{m!}\,\sqrt{m'!}} \sum_{\nu=0}^{\min(m,m')} \binom{m'}{\nu}\binom{m}{\nu} (c_1 - B)^{m'-\nu}(c_2 + A)^{m-\nu}\,\nu!
\end{aligned} \tag{5.10}$$

104 W. Heinzel

That is Witchel's result for the quantity $^2I_{mm'}$. It's identity with Koide's result can be shown rather simply.

6. Evaluation of $^1I_{mm'}$

From (2.17a) and (2.7a):

$$^1I_{mm'} = \langle m |\, e^{qn} e^{ip\delta}\, U V\, | m' \rangle. \tag{6.1}$$

Analogous to (5.3) we have

$$^1I_{mm'} = \langle m |\, e^{(a^+ + a)\,\Theta}\, e^{(a^+ - a)\,(-\alpha - \varDelta)}\, e^{-A a^2}\, e^{B(a^+)^2}\, | m' \rangle. \tag{6.2}$$

The complication in contrast to $^2I_{mm'}$ is, that $e^{-A a^2}$ and $e^{B(a^+)^2}$ must still be normal ordered, whereas in (5.3) those operators appear in a normal ordered form from the beginning. For the normal ordering of the operator in (6.2) the hyper-algebra-technique will show its power. First we write for (6.2):

$$^1I_{mm'} = \langle m |\, e^{d_1 a^+}\, e^{d_2 a}\, e^{-A a^2}\, e^{B(a^+)^2}\, | m' \rangle\, e^{\frac{1}{2} d_3} \tag{6.3}$$

with

$$d_1 = -\alpha - \varDelta + \Theta, \tag{6.4a}$$

$$d_2 = \alpha + \varDelta + \Theta, \tag{6.4b}$$

$$\tfrac{1}{2} d_3 = \tfrac{1}{2}\Theta^2 - \tfrac{1}{2}(\varDelta + \alpha)^2 + \Theta(-\varDelta - \alpha). \tag{6.4c}$$

Now use the hyper-algebra-representation (3.4)

$$^1I_{mm'} = e^{-A(\partial/\partial\alpha_1)^2}\, e^{B(\partial/\partial\alpha_2)^2}\, \langle m |\, e^{d_1 a^+} e^{d_2 a}\, e^{\alpha_1 a}\, e^{\alpha_2 a^+}\, | m' \rangle\, e^{\frac{1}{2} d_3}. \tag{6.5}$$

Normal ordering of the operator in the brackets is easily done with the aid of the Hausdorff-formula and we get by use of (4.5a, c) and (3.3a):

$$^1I_{mm'} = \frac{1}{\sqrt{m!}\,\sqrt{m'!}}\, e^{-A(\partial/\partial\alpha_1)^2}\, e^{B(\partial/\partial\alpha_2)^2}$$
$$\cdot \langle 0 |\, (d_1 + \alpha_2 + a)^m\, (d_2 + \alpha_1 + a^+)^{m'}\, | 0 \rangle\, e^{\alpha_2(\alpha_1 + d_2)}\, e^{\frac{1}{2} d_3}. \tag{6.6}$$

Using again a hyper-algebra, (6.6) may be written:

$$^1I_{mm'} = \frac{1}{\sqrt{m!}\,\sqrt{m'!}}\, \left(\frac{\partial}{\partial j_1}\right)^m \left(\frac{\partial}{\partial j_2}\right)^{m'}\, e^{-A(\partial/\partial\alpha_1)^2}\, e^{B(\partial/\partial\alpha_2)^2}$$
$$\cdot \langle 0 |\, e^{j_1(d_1 + \alpha_2 + a)}\, e^{j_2(d_2 + \alpha_1 + a^+)}\, | 0 \rangle\, e^{\alpha_2(\alpha_1 + d_2)}\, e^{\frac{1}{2} d_3}. \tag{6.7}$$

With (4.4a) we get by regarding (3.3a):

$$^1I_{mm'} = \frac{1}{\sqrt{m!}\,\sqrt{m'!}}\, \left(\frac{\partial}{\partial j_1}\right)^m \left(\frac{\partial}{\partial j_2}\right)^{m'}\, e^{j_1 j_2}\, e^{j_1 d_1}\, e^{j_2 d_2}$$
$$\cdot e^{-A(\partial/\partial\alpha_1)^2}\, e^{B(\partial/\partial\alpha_2)^2}\, e^{\alpha_1 j_2}\, e^{\alpha_2(j_1 + d_2)}\, e^{\alpha_1 \alpha_2}\, e^{\frac{1}{2} d_3} \Big|_{j_1 = j_2 = \alpha_1 = \alpha_2 = 0}. \tag{6.8}$$

Now we treat the algebra $(\alpha_1, \partial/\partial\alpha_1)$, i.e., we intend a normal ordering, which means all operators containing α_1 being to the left of those operators containing $\partial/\partial\alpha_1$. Having obtained that normal ordering with respect to α_1 and $\partial/\partial\alpha_1$, both are set zero (that is the analogue to (3.3a) for the oscillator-algebra). Now we use

(4.6b) for our hyper-algebra i.e. replace in (4.6b) a by $\partial/\partial\alpha_1$ and a^+ by α_1, then we get from (6.8):

$$^1I_{mm'} = \frac{1}{\sqrt{m!}\,\sqrt{m!'}}\left(\frac{\partial}{\partial j_1}\right)^m\left(\frac{\partial}{\partial j_2}\right)^{m'} e^{j_1 j_2}\,e^{j_1 d_1}\,e^{j_2 d_2}\,e^{-A j_2{}^2}\,e^{B(\partial/\partial\alpha_2)^2}\,e^{-A\alpha_2{}^2}$$
$$\cdot\, e^{\alpha_2(j_1 - 2A j_2 + d_2)}\,e^{\frac{1}{2}d_3}\,. \tag{6.9}$$

Now we treat the hyper-algebra $(\alpha_2, \partial/\partial\alpha_2)$.

By use of the formula (see (A.11) in Appendix A)

$$e^{B(\partial/\partial\alpha_2)^2}\,e^{-A\alpha_2{}^2}\,e^{\alpha_2(j_1 - 2A j_2 + d_2)} = e^{B(j_1 - 2A j_2 + d_2)^2(1 + 4AB)^{-1}}\,(1 + 4AB)^{-\frac{1}{2}}. \tag{6.10}$$

we get from (6.9)

$$^1I_{mm'} = \frac{1}{\sqrt{m!}\,\sqrt{m!}}\left(\frac{\partial}{\partial j_1}\right)^m\left(\frac{\partial}{\partial j_2}\right)^{m'} e^{j_1 j_2 D_0}\,e^{j_1 D_1}\,e^{j_2 D_2}\,e^{j_1{}^2 F_1}e^{j_2{}^2 F_2}\,e^{F_3}\cdot Z_{|j_1 = j_2 = 0} \tag{6.11}$$

with the constants:

$$D_0 = \frac{1}{1 + 4AB}, \tag{6.12a}$$

$$D_1 = d_1 + \frac{2B d_2}{1 + 4AB}, \tag{6.12b}$$

$$D_2 = \frac{d_2}{1 + 4AB}. \tag{6.12c}$$

$$F_1 = \frac{B}{1 + 4AB}, \tag{6.12d}$$

$$F_2 = -\frac{A}{1 + 4AB}, \tag{6.12e}$$

$$F_3 = \frac{B d_2^2}{1 + 4AB} + \frac{1}{2}d_3, \tag{6.12f}$$

$$Z = \frac{1}{\sqrt{1 + 4AB}}. \tag{6.12g}$$

Using (4.5c) for the hyper-algebra $(j_i, \partial/\partial j_i)$, we can put $e^{j_1 j_2 D_0}$ to the left of the operators $\partial/\partial j_i$ and get:

$$^1I_{mm'} = \frac{1}{\sqrt{m!}\,\sqrt{m'!}}\left(\frac{\partial}{\partial j_1}\right)^m\left(\frac{\partial}{\partial j_2} + j_1 D_0\right)^{m'} e^{j_1 D_1}\,e^{j_2 D_2}\,e^{j_1{}^2 F_1}\,e^{j_2{}^2 F_2}\,e^{F_3}\cdot Z_{|j_i = 0}\,. \tag{6.13}$$

Now bring $(\partial/\partial j_2 + j_1 D_0)^{m'}$ to the left of $(\partial/\partial j_1)^m$ by use of

$$\left(\frac{\partial}{\partial j_1}\right)^m j_1^{m'} = \sum_{\mu=0}^{\min(m,m')}\binom{m}{\mu}\binom{m'}{\mu}\mu!\,j_1^{m'-\mu}\left(\frac{\partial}{\partial j_1}\right)^{m-\mu} \tag{6.14}$$

then we get from (6.13):

$$^1I_{mm'} = \frac{1}{\sqrt{m!}\,\sqrt{m'!}}\sum_{\mu=0}^{\min(m,m')} D_0^\mu \binom{m}{\mu}\binom{m'}{\mu}\mu!\left(\frac{\partial}{\partial j_1}\right)^{m-\mu}\left(\frac{\partial}{\partial j_2}\right)^{m'-\mu}$$
$$\cdot\, e^{j_1 D_1}\,e^{j_1{}^2 F_1}\,e^{j_2 D_2}\,e^{j_2{}^2 F_2}\,e^{F_3}\cdot Z\,. \tag{6.15}$$

If we now make use of (4.5c, d) for the hyper-algebras $(j_i,\ \partial/\partial j_i)$ we get directly Witchel's [1] result for the quantity $^1I_{mm'}$. If we make use of the following formula for Hermite-polynomials

$$H_n(x) = e^{x^2}\left[\frac{\partial^n}{\partial j^n}\ e^{-(x-j)^2}\right]_{j=0} \tag{6.16}$$

we get directly Koide's result [4].

Using (4.5c, d) we get from (6.15) (if we replace j_i, $\partial/\partial j_i$ by a^+, a)

$$\begin{aligned}
^1I_{mm'} = \frac{1}{\sqrt{m!}\,\sqrt{m'!}} \sum_{\mu=0}^{\min(m,\,m')} D_0^\mu \binom{m}{\mu}\binom{m'}{\mu}\mu! \\
\cdot \langle 0|\,(a+D_1)^{m-\mu}\,(a+2F_1 a^+)^{m-\mu}\,|0\rangle \\
\cdot \langle 0|\,(a+D_2)^{m'-\mu}\,(a+2F_2 a^+)^{m'-\mu}\,|0\rangle\, e^{F_3}\cdot Z\,.
\end{aligned} \tag{6.17}$$

That result is rather simple to deal with. Using binomial expansion and using

$$\langle 0|\,a^k\,(a^+)^{k'}\,|0\rangle = k!\,\delta_{k,k'} \tag{6.18}$$

one can perform normal-ordering in the matrix-elements of (6.17), which brings a lengthy but finite multinomial series of the c-numbers D_0, D_1, D_2, F_1, F_2 factored by $1/\sqrt{m!}\,\sqrt{m'!}\,e^{F_3}\cdot Z\,.$

7. Evaluation of the Transition-Probability γ_m (see (2.18))

From (2.18) and (2.17) we get

$$\gamma_m = 2\pi \sum_{m'} {}^1I_{mm'}\,{}^2I_{mm'}\,\delta(\omega[m+\tfrac{1}{2}] - \omega'(m'+\tfrac{1}{2}) + \Omega)\,. \tag{7.1}$$

Because we want quite a general treatment we won't make any specification of the energy Ω. E.g. it may be the energy of any other oscillator. For a general treatment we use the Fourier-representation of the δ-function

$$\delta(x) = \frac{1}{2\pi}\int_{-\infty}^{\infty} d\tau\,e^{i\tau x} \tag{7.2}$$

and instead of (7.1) we will treat the Fourier-transform:

$$\tilde{\gamma}_m(\tau) = \frac{1}{2\pi}\int_{-\infty}^{\infty} d\Omega\,e^{-i\tau\Omega}\,\gamma_m(\Omega) = \sum_{m'} {}^1I_{mm'}\,{}^2I_{mm'}\,e^{i\tau(\omega[m+\frac{1}{2}] - \omega'[m'+\frac{1}{2}])}\,. \tag{7.3}$$

The final results for $^1I_{mm'}$, $^2I_{mm'}$, obtained in Sects. 5 and 6 (see (5.10)), are too long-winded for the evaluation of (7.3). Therefore we will use the original forms (6.3) and (5.4) for $^1I_{mm'}$ and $^2I_{mm'}$, respectively:

$$\begin{aligned}
\tilde{\gamma}_m(\tau) = \sum \langle m|\,e^{d_1 a^+}\,e^{d_2 a}\,e^{-A a^2}\,e^{B(a^+)^2}\,|m'\rangle \\
\cdot \langle m'|\,e^{-B(a^+)^2}\,e^{A a^2}\,e^{d_2 a^+}\,e^{d_1 a}\,|m\rangle\,e^{i\tau(\omega[m+\frac{1}{2}] - \omega'[m'+\frac{1}{2}])}\,e^{d_3}\,.
\end{aligned} \tag{7.4}$$

Using $\hat{m}\,|m'\rangle = m'\,|m'\rangle$ and regarding that the oscillator states form a complete orthonormal set, we may write

$$\begin{aligned}
\tilde{\gamma}_m(\tau) = \langle m|\,e^{d_1 a^+}\,e^{d_2 a}\,e^{-A a^2}\,e^{B(a^+)^2}\,e^{-i\tau\omega'\hat{m}}\,e^{-B(a^+)^2}\,e^{A a^2}\,e^{d_2 a^+}\,e^{d_1 a}\,|m\rangle \\
\cdot e^{i\tau\omega m}\,e^{i\tau(\omega-\omega')/2}\,e^{d_3}\,.
\end{aligned} \tag{7.5}$$

Now we want to commute the operator $e^{-i\tau\omega'\hat{m}}$ to the states. That can be achieved by use of (4.3):

$$\tilde{\gamma}_m(\tau) = \langle m \,|\, e^{a^+ d_1 \exp i\tau(\omega'/2)} \, e^{a d_2 \exp - i\tau(\omega'/2)} \, e^{-A a^2 \exp - i\tau\omega'} \, e^{B(a^+)^2 \exp i\tau\omega'}$$

$$\cdot \, e^{-B(a^+)^2 \exp - i\tau\omega'} \, e^{A a^2 \exp i\tau\omega'} \, e^{a^+ d_2 \exp - i\tau(\omega'/2)} \, e^{a d_1 \exp i\tau(\omega'/2)} \,|\, m \rangle$$

$$\cdot \, e^{i\tau(\omega-\omega')m} \, e^{i\tau(\omega-\omega')/2} \, e^{d_3} \, . \tag{7.6}$$

The next step is a complete normal-ordering of the operator between the states, for which we use the hyper-algebra technique, by substituting:

$$e^{-A a^2 \exp - i\tau\omega'} = e^{-A(\partial/\partial\alpha_1)^2} \, e^{\alpha_1 a \exp - i\tau(\omega'/2)} \, , \tag{7.7a}$$

$$e^{B(a^+)^2 (\exp i\tau\omega' - \exp - i\tau\omega')} = e^{\bar{B}(\partial/\partial\alpha_2)^2} \, e^{\alpha_2 a^+} \, , \tag{7.7b}$$

$$e^{A a^2 \exp i\tau\omega'} = e^{A(\partial/\partial\alpha_3)^2} \, e^{\alpha_3 a \exp i\tau(\omega'/2)} \tag{7.7c}$$

with

$$\bar{B} = B(\exp i\tau\omega' - \exp - i\tau\omega') \tag{7.8}$$

we get from (7.6) after normal-ordering

$$\tilde{\gamma}_m(\tau) = e^{-A(\partial/\partial\alpha_1)^2} \, e^{\bar{B}(\partial/\partial\alpha_2)^2} \, e^{A(\partial/\partial\alpha_3)^2} \, \langle m \,|\, e^{a^+ r_1} \, e^{a r_2} \,|\, m \rangle \, e^{r_3} \, e^{i\tau m(\omega-\omega')} \, e^{i\tau(\omega-\omega')/2} \tag{7.9}$$

with the abbreviations:

$$r_1 = d_1 \exp i\tau(\omega'/2) + d_2 \exp - i\tau(\omega'/2) + \alpha_2 \, , \tag{7.10a}$$

$$r_2 = (d_2 + \alpha_1) \exp - i\tau(\omega'/2) + (d_1 + \alpha_3) \exp i\tau(\omega'/2) \, , \tag{7.10b}$$

$$r_3 = (d_2\alpha_2 + \alpha_1\alpha_2) \exp - i\tau(\omega'/2) + (d_2^2 + \alpha_1 d_2) \exp - i\tau\omega' + \alpha_3 d_2 + d_3 \, . \tag{7.10c}$$

(7.9) may be written

$$\tilde{\gamma}_m(\tau) = e^{-A(\partial/\partial\alpha_1)^2} \, e^{\bar{B}(\partial/\partial\alpha_2)^2} \, e^{A(\partial/\partial\alpha_3)^2} \, \frac{1}{m!} \, \langle 0 \,|\, (r_1 + a)^m \, (r_2 + a^+)^m \,|\, 0 \rangle$$

$$\cdot \, e^{r_3} \, e^{i\tau m(\omega-\omega')} \, e^{i\tau(\omega-\omega')/2} \, . \tag{7.11}$$

The evaluation of (7.11) is straight forward. First the hyper-algebra $(\alpha_3, \partial/\partial\alpha_3)$ is treated and then one gets a form for $\hat{\gamma}_m(\tau)$ which is quite similar to $^1 I_{mm'}$, as given in (6.6). Therefore we do not discuss it in detail. Instead of that we give a detailed evaluation of the related thermal averaged quantity

$$\tilde{\gamma}(\tau) = \mathfrak{n} \sum_{m=0} e^{-\beta\omega m} \, \tilde{\gamma}_m(\tau) \, .$$

8. Auxiliary Algebraic Formulae

For the thermal averaging we are interested in the following quantity:

$$M(\varkappa) = \sum_{m=0} \varkappa^m \, \frac{1}{m!} \, \langle 0 \,|\, (r_1 + a)^m \, (r_2 + a^+)^m \,|\, 0 \rangle \tag{8.1}$$

with $|\varkappa| < 1$. (8.1) may be written

$$M(\varkappa) = \sum_{m=0} \frac{\varkappa^m}{m!} \left(\frac{\partial}{\partial j_1}\right)^m \left(\frac{\partial}{\partial j_2}\right)^m \langle 0 \,|\, e^{j_1(r_1+a)} \, e^{j_2(r_2+a^+)} \,|\, 0 \rangle_{|j_1=j_2=0} \, . \tag{8.2}$$

After normal-ordering in the matrix-element one gets by use of (3.3a)

$$M(\varkappa) = \sum_{m=0} \frac{\varkappa^m}{m!} \left(\frac{\partial}{\partial j_1}\right)^m \left(\frac{\partial}{\partial j_2}\right)^m e^{j_1 r_1} \, e^{j_2 r_2} \, e^{j_1 j_2}{}_{|j_1=j_2=0} \tag{8.3}$$

108   W. Heinzel

which is identical with

$$M(\varkappa) = e^{\varkappa(\partial/\partial j_1)(\partial/\partial j_2)}\, e^{j_1 r_1}\, e^{j_2 r_2}\, e^{j_1 j_2}\big|_{j_1=j_2=0}\,. \tag{8.4}$$

Using now the Hausdorff-formula (4.4a) for the hyper-algebras we get the following equalities:

$$\begin{aligned}
M(\varkappa) &= e^{j_1 r_1}\, e^{\varkappa(\partial/\partial j_1)(\partial/\partial j_2)}\, e^{\varkappa r_1(\partial/\partial j_2)}\, e^{j_2 r_2}\, e^{j_1 j_2}\\
&= e^{\varkappa(\partial/\partial j_1)(\partial/\partial j_2)}\, e^{j_2 r_2}\, e^{\varkappa r_1(\partial/\partial j_2)}\, e^{j_1 j_2}\, e^{\varkappa r_1 r_2}\\
&= e^{\varkappa(\partial/\partial j_1)(\partial/\partial j_2)}\, e^{j_2 r_2}\, e^{j_1 j_2}\, e^{\varkappa r_1(\partial/\partial j_2)}\, e^{\varkappa r_1 j_1}\, e^{\varkappa r_1 r_2}\\
&= e^{\varkappa(\partial/\partial j_1)(\partial/\partial j_2)}\, e^{j_1 \varkappa r_1}\, e^{j_2 r_2}\, e^{j_1 j_2}\, e^{\varkappa r_1 r_2}\\
&= e^{\varkappa(\partial/\partial j_1)(\partial/\partial j_2)}\, e^{j_1 \varkappa r_1}\, e^{j_2 \varkappa r_2}\, e^{j_1 j_2}\, e^{\varkappa r_1 r_2}\, e^{\varkappa^2 r_1 r_2}
\end{aligned} \tag{8.5}$$

That result can be interpreted as follows:

In (8.4) $e^{j_1 r_1}$ may be replaced by $e^{j_1 \varkappa r_1}\, e^{\varkappa r_1 r_2}$, then $e^{j_1 \varkappa r_1}$ may be replaced by $e^{j_1 \varkappa^2 r_1}\, e^{\varkappa^2 r_1 r_2}$ and so on. One gets for instance

$$M(\varkappa) = e^{\varkappa(\partial/\partial j_1)(\partial/\partial j_2)}\, e^{j_1 r_1 \varkappa^n}\, e^{j_2 r_2 \varkappa^m}\, e^{j_1 j_2}\, e^{\sum\limits_{\nu=1}^{n+m} \varkappa^\nu r_1 r_2} \tag{8.6}$$

for arbitrary n and m. Provided $|\varkappa| < 1$ we thus get:

$$M(\varkappa) = e^{\varkappa(\partial/\partial j_1)(\partial/\partial j_2)}\, e^{j_1 j_2}\, e^{r_1 r_2 \varkappa/(1-\varkappa)}\,. \tag{8.7}$$

Additionally we get:

$$e^{\varkappa(\partial/\partial j_1)(\partial/\partial j_2)}\, e^{j_1 j_2}\big|_{j=j_2=0} = \sum_{\nu_1=0} \frac{\varkappa^{\nu_1}}{\nu_1!}\left(\frac{\partial}{\partial j_1}\frac{\partial}{\partial j_2}\right)^{\nu_1}\sum_{\nu_2=0}\frac{1}{\nu_2!}(j_1 j_2)^{\nu_2} = \sum_{\nu=0}\varkappa^\nu = \frac{1}{1-\varkappa}\,. \tag{8.8}$$

Using (8.8) we obtain from (8.7):

$$M(\varkappa) = \frac{1}{1-\varkappa}\, e^{r_1 r_2 \varkappa/(1-\varkappa)}\,. \tag{8.9}$$

9. Thermal Averaged Transition Probabilities

We treat the quantity

$$\tilde{\gamma}(\beta,\tau) = \mathfrak{n}\sum_{m=0} e^{-\beta\omega m}\,\tilde{\gamma}_m(\tau) \quad (\mathfrak{n}^{-1} = \sum_{m=0} e^{-\beta\omega m}) \tag{9.1}$$

where $\gamma_m(\tau)$ may be given in the form (7.11). To make use of Sect. 8 we must set:

$$\varkappa = e^{-\beta\omega}\, e^{i\tau(\omega-\omega')} \qquad (|\varkappa| < 1)\,. \tag{9.2}$$

Using (8.9) and regarding (7.11), (8.1) und (9.2), we get

$$\tilde{\gamma}(\beta,\tau) = \mathfrak{n}\,\frac{1}{1-\varkappa}\, e^{i\tau(\omega-\omega')/2}\, e^{-A(\partial/\partial\alpha_1)^2}\, e^{\bar{B}(\partial/\partial\alpha_2)^2}\, e^{A(\partial/\partial\alpha_3)^2}\, e^{(r_1 r_2 \varkappa/(1-\varkappa)+r_3)}\,. \tag{9.3}$$

With the abbreviations (7.10) we may write

$$r_1 r_2 \frac{\varkappa}{1-\varkappa} + r_3 = L_1\alpha_1\alpha_2 + L_3\alpha_2\alpha_3 + M_1\alpha_1 + M_2\alpha_2 + M_3\alpha_3 + N \tag{9.4}$$

with

$$L_1 = \frac{1}{1-\varkappa}\, e^{-i\tau(\omega'/2)}, \tag{9.5a}$$

$$L_3 = \frac{\varkappa}{1-\varkappa}\, e^{i\tau(\omega'/2)}, \tag{9.5b}$$

$$M_1 = \frac{\varkappa}{1-\varkappa}\, d_1 + \frac{1}{1-\varkappa}\, d_2\, e^{-i\tau\omega'}, \tag{9.5c}$$

$$M_2 = \frac{\varkappa}{1-\varkappa}\, d_1\, e^{i\tau(\omega'/2)} + \frac{1}{1-\varkappa}\, d_2\, e^{-i\tau(\omega'/2)}, \tag{9.5d}$$

$$M_3 = \frac{\varkappa}{1-\varkappa}\, d_1\, e^{i\tau\omega'} + \frac{1}{1-\varkappa}\, d_2, \tag{9.5e}$$

$$N = \frac{\varkappa}{1-\varkappa}\, 2\, d_1\, d_2 + \frac{\varkappa}{1-\varkappa}\, d_1^2\, e^{i\tau\omega'} + \frac{1}{1-\varkappa}\, d_2^2\, e^{-i\tau\omega'} + d_3. \tag{9.5f}$$

Using formula (A.1) after insertion of (9.4) into (9.3) we get

$$\tilde{\gamma}(\beta,\tau) = \mathfrak{n}\, \frac{1}{1-\varkappa}\, e^{i\tau(\omega-\omega')/2}\, e^N\, e^{-AM_1^2}\, e^{AM_3^2}\, e^{\bar{B}(\partial/\partial\alpha_2)^2}$$
$$\cdot\, e^{(-AL_1^2 + AL_3^2)\alpha_2^2}\, e^{(-2AM_1L_1 + 2AM_3L_3 + M_2)\alpha_2}\big|_{\alpha_2=0}. \tag{9.6}$$

The treatment of the algebra $(\alpha_2, \partial/\partial\alpha_2)$ is done by use of formula (A.9)

$$\tilde{\gamma}(\beta,\tau) = \mathfrak{n}\, \frac{1}{1-\varkappa}\, e^{i\tau(\omega-\omega')/2}\, e^N\, e^{-AM_1^2}\, e^{AM_3^2}\, e^{\bar{B}D^2 \sum_{\nu=0}(4\bar{B}C)^\nu}\, \sum_{n=0}^{\infty} \frac{(2n)!}{n!\,n!}\, (\bar{B}C)^n \tag{9.7}$$

where $C\alpha_2^2$ and $D\alpha_2^2$ are just the exponents of the last two exponentials in (9.6):

$$C = -A\,L_1^2 + A\,L_3^2, \tag{9.8a}$$

$$D = -2\,A\,M_1\,L_1 + 2\,A\,M_3\,L_3 + M_2. \tag{9.8b}$$

In (9.7) all algebraic operations have come to an end. One may replace

$$\sum_{\nu=0} (4\,\bar{B}C)^\nu \quad \text{by} \quad \frac{1}{1-4\,\bar{B}C} \quad \text{and} \quad \sum_{n=0} \frac{(2n)!}{n!\,n!}\, (\bar{B}C)^n \quad \text{by} \quad \frac{1}{\sqrt{1-4\,\bar{B}C}}$$

for $|4\,\bar{B}C| < 1$ but that is not very handsome, because the quantity

$$\bar{B}C = \frac{A\,B}{(1-\varkappa)^2}\, (\varkappa^2\, e^{2i\tau\omega'} + e^{-2i\tau\omega'}) - A\,B\, \frac{1+\varkappa^2}{(1-\varkappa)^2} \tag{9.9}$$

is intended to appear not in any denominator for a Fourier-transformation of $\tilde{\gamma}(\beta,\tau)$ to $\gamma(\beta,\Omega) = \int d\tau\, e^{i\tau\Omega}\, \tilde{\gamma}(\beta,\tau)$ which will be treated in a next paper, where the result (9.7) is applied to electronic Franck-Condon-transitions of Colour-centre-electrons in ionic crystals. In general the Fourier transformation of $\tilde{\gamma}(\beta,\tau)$ to $\gamma(\beta,\Omega)$ is not possible exactly except for the special case of no frequency-shift ($\omega = \omega'$, $A = 0$, $B = 0$). We will give some approximate solutions for the general case in the next paper.

Finally we write down an explicit form of $-A\,M_1^2 + A\,M_3^2$, $\mathring{B}C$, D, N:

$$-A\,M_1^2 + A\,M_3^2 = \frac{A}{(1-\varkappa)^2}\,(e^{i\tau\omega'} - e^{-i\tau\omega'})\,(d_1\varkappa\,e^{i\tau(\omega'/2)} + d_2\,e^{-i\tau(\omega'/2)})^2,$$

$$\tag{9.10a}$$

$$\mathring{B}C = \frac{A\,B}{(1-\varkappa)^2}\,(e^{i\tau\omega'} - e^{-i\tau\omega'})\,(\varkappa^2 e^{i\tau\omega'} - e^{-i\tau\omega'})\,,\tag{9.10b}$$

$$D = \frac{2\,A}{(1-\varkappa)^2}\,(\varkappa\,e^{i\tau\omega'} - e^{-i\tau\omega'} + \frac{1-\varkappa}{2\,A}\,(d_1\varkappa\,e^{i\tau(\omega'/2)} + d_2\,e^{-i\tau(\omega'/2)})\,,\tag{9.10c}$$

$$N = \frac{\varkappa}{1-\varkappa}\,2\,d_1 d_2 + \frac{\varkappa}{1-\varkappa}\,d_1^2\,e^{i\tau\omega'} + \frac{1}{1-\varkappa}\,d_2^2\,e^{-i\tau\omega'} + d_3\,.\tag{9.10d}$$

Regard: $\qquad\qquad\qquad\qquad\qquad \varkappa = e^{-\beta\omega}\,e^{i\tau(\omega-\omega')}\,.$

The approximations for the Fourier transform of (9.7) will essentially be based on the smallness of A, B (i.e. smallness of ω-ω' compared to ω), for a termination of the respective series'.

Appendix A

We prove the following identity

$$e^{K_1(\partial/\partial\alpha_1)^2}\,e^{K_2(\partial/\partial\alpha_2)^2}\,e^{K_3(\partial/\partial\alpha_3)^2}\,e^{L_1\alpha_1\alpha_2}\,e^{L_3\alpha_2\alpha_3}\,e^{M_1\alpha_1}\,e^{M_2\alpha_2}\,e^{M_3\alpha_3}{}_{|\alpha_i=0}$$

$$= e^{K_1 M_1^2}\,e^{K_3 M_3^2}\,e^{K_2(\partial/\partial\alpha_2)^2}\,e^{C\alpha_2^2}\,e^{D\alpha_2}\tag{A.1}$$

with

$$C = K_1 L_1^2 + K_3 L_3^2\,,\tag{A.2a}$$

$$D = 2\,M_1 K_1 L_1 + 2\,M_3 K_3 L_3 + M_2\,.\tag{A.2b}$$

The proof makes use of the commutation (4.6b) for the hyper-algebras $(\alpha_i,\ \partial/\partial\alpha_i)$. First we treat the algebra $(\alpha_1,\ \partial/\partial\alpha_1)$:

$$e^{K_1(\partial/\partial\alpha_1)^2}\,e^{L_1\alpha_1\alpha_2}\,e^{M_1\alpha_1}{}_{|\alpha_1=0} = e^{K_1(\partial/\partial\alpha_1)^2}\,e^{(L_1\alpha_2+M_1)\alpha_1}{}_{|\alpha_1=0} = e^{K_1 L_1^2\alpha_2^2}\,e^{2\,K_1 L_1 M_1\alpha_2}\,e^{K_1 M_1^2}.$$

$$\tag{A.3}$$

Analogous we get:

$$e^{K_3(\partial/\partial\alpha_3)^2}\,e^{L_3\alpha_3\alpha_2}\,e^{M_3\alpha_3}{}_{|\alpha_3=0} = e^{K_3 L_3^2\alpha_2^2}\,e^{2\,K_3 L_3 M_3\alpha_2}\,e^{K_3 M_3^2}\,.\tag{A.4}$$

Insert (A.3) and (A.4) in the left-hand side of (A.1) and one gets the right-hand side.
Now we prove

$$e^{B(\partial/\partial\alpha_2)^2}\,e^{C\alpha_2^2}\,e^{D\alpha_2} = e^{BD^2 1/(1-4BC)}\sum_{n=0}^{\infty}\frac{(2\,n)!}{n!\,n!}\,(B\,C)^n\,.\tag{A.5}$$

Making repeatedly use of the commutations (4.4a) and (4.6) for the hyper-algebra $(\alpha_2,\ \partial/\partial\alpha_2)$, we get:

$$e^{B(\partial/\partial\alpha_2)^2}\,e^{C\alpha_2^2}\,e^{D\alpha_2} = e^{BD^2}\,e^{B(\partial/\partial\alpha_2)^2}\,e^{2\,BD(\partial/\partial\alpha_2)}\,e^{C\alpha_2^2} = e^{BD^2}\,e^{B(\partial/\partial\alpha_2)^2}\,e^{C(\alpha_2 + 2\,BD)^2}$$

$$= e^{BD^2}\,e^{BD^2 4\,BC}\,e^{B(\partial/\partial\alpha_2)^2}\,e^{C\alpha_2^2}\,e^{D4\,BC\alpha_2}\,.$$

Repeating that procedure n-times, we get:

$$e^{B(\partial/\partial\alpha_2)^2}e^{C\alpha_2{}^2}e^{D\alpha_2} = e^{BD^2\sum\limits_{\nu=0}^{n}(4BC)^{\nu}}\,e^{B(\partial/\partial\alpha_2)^2}\,e^{C\alpha_2{}^2}\,e^{D(4BC)^n\alpha_2}\,.$$

If $|4BC| < 1$ we get in the limit $n \to \infty$:

$$e^{B(\partial/\partial\alpha_2)^2}\,e^{C\alpha_2{}^2}\,e^{D\alpha_2} = e^{BD^2\sum\limits_{\nu=0}^{\infty}(4BC)^{\nu}}\,e^{B(\partial/\partial\alpha_2)^2}\,e^{C\alpha_2{}^2}\,. \tag{A.6}$$

For $|4BC| < 1$ we may write

$$\sum_{\nu=0}^{\infty}(4BC)^{\nu} = \frac{1}{1-4BC}\,. \tag{A.7}$$

Additionally we get

$$e^{B(\partial/\partial\alpha_2)^2}\,e^{C\alpha_2{}^2} = \sum_{n_1=0}\frac{B^{n_1}}{n_1!}\left(\frac{\partial}{\partial\alpha_2}\right)^{2n_1}\sum_{n_2=0}\frac{C^{n_2}}{n_2!}\,\alpha_2^{2n_2}|_{\alpha_2=0} = \sum_{n=0}\frac{(2n)!}{n!\,n!}(BC)^n \tag{A.8}$$

with (A.8) we get from (A.6):

$$e^{B(\partial/\partial\alpha_2)^2}\,e^{C\alpha_2{}^2}\,e^{D\alpha_2} = e^{BD^2\sum\limits_{\nu=0}^{\infty}(4BC)^{\nu}}\sum_{n=0}\frac{(2n)!}{n!\,n!}(BC)^n\,. \tag{A.9}$$

One can show that:

$$\sum_{n=0}\frac{(2n)!}{n!\,n!}(BC)^n = \frac{1}{\sqrt{1-4BC}} \tag{A.10}$$

provided $|4BC| < 1$.

Using (A.7) and (A.10) one gets from (A.9):

$$e^{B(\partial/\partial\alpha_2)^2}\,e^{C\alpha_2{}^2}\,e^{D\alpha_2} = e^{BD^2 1/(1-4BC)}\frac{1}{\sqrt{1-4BC}}\,. \tag{A.11}$$

Acknowledgement. I am very much obliged to Prof. Dr. H. Stumpf for helpful advices. I am also grateful to Mr. Fraser for reading the manuscript.

References

1. Witschel, W.: J. Phys. B: Atom. Molec. Phys. **6**, 527 (1973)
2. Wagner, M.: Z. Naturforsch. **14a**, 81 (1959)
3. Ansbacher, F.: Z. Naturforsch. **14a**, 889 (1959)
4. Koide, S.: Z. Naturforsch. **15a**, 123 (1960)

Wolfram Heinzel
Institut für Theoretische Physik
der Universität Tübingen
D-74 Tübingen
Auf der Morgenstelle 14
Federal Republic of Germany

Phys. cond. Matter 17, 113—124 (1974)
© by Springer-Verlag 1974

Spektroskopische Untersuchungen an DyAsO$_4$

Dieter Wappler

Physikalisches Institut der Universität Karlsruhe

Eingegangen am 25. Juli 1973

Spectroscopic Investigations on DyAsO$_4$

DyAsO$_4$ undergoes a crystallographic phase transition at $T_D = 11.2$ K which is induced by a cooperative Jahn-Teller-effect. As deduced from the optical absorption spectra the distance between the two lowest lying Kramers doublets of the Dy^{3+} ion is increased from (6.1 $\pm$ 0.5) cm^{-1} above T_D to (25.0 $\pm$ 0.5) cm^{-1} at 4.2 K. Below T_D the splitting factor of the lowest doublet becomes nearly uniaxial with a maximum value of $g_b = 17.5 \pm 1.0$ along the crystallographic b-axis. At $T_N = 2.44$ K the crystals order antiferromagnetically. The absorption lines of Er^{3+} ions in DyAsO$_4$ show already a splitting immediately below T_D which is explained by magnetic short range ordering of the Dy^{3+} ions in the temperature range $T_N < T < T_D$.

DyAsO$_4$ erfährt bei $T_D = 11,2$ K einen durch einen kooperativen Jahn-Teller-Effekt induzierten kristallographischen Phasenübergang. Dabei vergrößert sich, wie aus den optischen Absorptionsspektren folgt, der Abstand der beiden tiefsten Kramers-Dubletts des Dy^{3+} von (6,1 $\pm$ 0,5) cm^{-1} oberhalb T_D mit sinkender Temperatur auf (25,0 $\pm$ 0,5) cm^{-1} bei 4,2 K. Unterhalb T_D ist der g-Faktor des tiefsten Dubletts stark anisotrop und besitzt seinen maximalen Wert von 17,5 $\pm$ 1,0 in Richtung der b-Achse des Kristalls. Bei $T_N = 2,44$ K ordnet die Substanz antiferromagnetisch, die Dubletts des Dy^{3+} spalten im inneren Magnetfeld auf. In mit Er^{3+} dotierten DyAsO$_4$-Kristallen zeigen die Er-Linien bereits unterhalb T_D eine Aufspaltung, die auf eine Nahordnung der Dy-Momente im Temperaturbereich $T_N < T < T_D$ zurückzuführen ist.

1. Einleitung

Die Verbindungen der Seltenen Erden mit Zirkon-Struktur sind wegen ihrer besonderen magnetischen und kristallographischen Eigenschaften bei tiefen Temperaturen Gegenstand zahlreicher Untersuchungen geworden.

Ein besonders ausgeprägtes magnetisches Verhalten zeigen das Phosphat, Vanadat und Arsenat des Dyprosiums, die im Temperaturbereich zwischen 2 K und 4 K antiferromagnetisch ordnen [1—9]. Während DyPO$_4$ in der tetragonalen Symmetrie der Zirkon-Struktur in einer einfachen antiferromagnetischen Zwei-Untergitter-Struktur mit den magnetischen Momenten parallel zur tetragonalen c-Achse ordnet, erfahren DyVO$_4$ und DyAsO$_4$ zunächst einen kristallographischen Phasenübergang von tetragonaler zu orthorhombischer Symmetrie mit Verzerrungen in Richtung der a- und b-Achsen und ordnen dann in einer antiferromagnetischen Vier-Untergitter-Struktur mit den Momenten senkrecht zur c-Achse. Aus Neutronenbeugungsuntersuchungen folgt, daß im DyVO$_4$ die Dy^{3+}-Momente parallel zur b-Achse des Kristalls liegen, während sie im DyAsO$_4$ einen Winkel von 22° zur b-Achse bilden [8, 9]. Der kristallographische Phasenübergang kann als ein kooperativer Jahn-Teller-Effekt interpretiert werden. Bei DyAsO$_4$ beträgt die Temperatur des kristallographischen Phasenübergangs $T_D = (11,17$

$\pm$ 0,05) K und die antiferromagnetische Ordnungstemperatur $T_N = (2{,}437 \pm 0{,}005)$ K [7].

In den bereits vorliegenden Untersuchungen an $DyAsO_4$ [7—15], die sich vor allem mit den kristallographischen und magnetischen Eigenschaften dieser Verbindung beschäftigen, konnten mit Ausnahme der Neutronenbeugungsmessungen weder Hinweise noch eine Erklärung für die spezielle Ordnungsrichtung der Dy^{3+}-Momente gefunden werden.

In der vorliegenden Arbeit wird über weitere optisch-spektroskopische Messungen an $DyAsO_4$ berichtet, die der Aufklärung der magnetischen Eigenschaften der Substanz dienen sollen.

Dabei wurden insbesondere untersucht: Die Lage der Kristallfeldkomponenten des Grundterms $^6H_{15/2}$ des Dy^{3+} für Temperaturen $T > T_D$ (Abschnitt 4.1), der Temperaturverlauf der energetisch tiefsten Kristallfeldkomponenten des Dy^{3+} unterhalb der Temperatur T_D (Abschnitt 4.2) und der Zeeman-Effekt dieser Kristallfeldkomponenten (Abschnitt 4.3), ferner das Spektrum von Er-Ionen, die in kleiner Konzentration in $DyAsO_4$ eingebaut waren (Abschnitt 5). Aus diesem Spektrum kann auf eine ausgeprägte antiferromagnetische Nahordnung der Dy-Momente im Temperaturbereich zwischen T_D und T_N geschlossen werden.

Ein Einfluß des magnetischen Ordnungswinkels von 22° auf die optischen Spektren konnte nicht festgestellt werden.

2. Experimentelles

Die für die Untersuchungen verwendeten Einkristalle von $DyAsO_4$ und $(Dy, Er)AsO_4$ (100:1) wurden nach einem Flußmittelverfahren [16] gezogen. Für die Messungen wurden Kristalle mit den typischen Abmessungen von $2 \times 2 \times 8\,mm^3$ verwendet.

Die optischen Absorptionsspektren wurden mit der in [17] beschriebenen Apparatur aufgenommen. Die Aufnahmen im fernen Infrarot wurden mit einem Fourier-Spektrometer am Physikalischen Institut der Universität Freiburg von Herrn Prettl durchgeführt.

3. Theoretische Vorbemerkungen

Die Kristalle des $DyAsO_4$ und des $(Dy, Er)AsO_4$ kristallisieren in der tetragonalen Zirkon-Struktur mit der Raumsymmetrie $I4_1/amd$. Die Punktsymmetrie am Ort eines Seltenen-Erd-Ions ist oberhalb der Temperatur T_D der kristallographischen Phasenumwandlung $\bar{4}m2$, unterhalb T_D höchstens $mm2$ mit der zweizähligen Achse parallel c [11, 18].

Die Zustände eines freien Seltenen-Erd-Ions werden durch die Quantenzahlen J und M des Gesamtdrehimpulses und seiner z-Komponente gekennzeichnet. In der Näherung des statischen Einzelionenmodells spalten die Zustände sowohl des Dy^{3+} als auch des Er^{3+} unter der Wirkung eines elektrischen Kristallfeldes der Symmetrie $\bar{4}m2$ oder $mm2$ in $(2J+1)/2$ Kramers-Dubletts auf [19]. Den Dubletts werden bei der Punktsymmetrie $\bar{4}m2$ die Kristallquantenzahlen $\bar{\mu} = \pm \frac{3}{2}$ oder $\bar{\mu} = \pm \frac{1}{2}$ (Quantelungsachse parallel c) zugeordnet, bei der Punktsymmetrie $mm2$ die Kristallquantenzahlen $\bar{\mu} = \pm \frac{1}{2}$ (Quantelungsachse parallel b). Die Auswahlregeln für elektrische Dipolstrahlung bei der Punktsymmetrie $\bar{4}m2$ sind in [20], bei $mm2$ in [19] angegeben. Wird bei der Punktsymmetrie $mm2$ ein äußeres Magnetfeld parallel zur a- oder b-Achse angelegt, so reduziert sich die Punktsymmetrie auf m, und die Übergänge in elektrischer Dipolstrahlung sind dann entweder parallel oder senkrecht zur Magnetfeldrichtung polarisiert.

4. Spektren des Dy^{3+} in DyAsO$_4$

Das optische Absorptionsspektrum von DyAsO$_4$ wurde im Temperaturbereich von 1,8 K bis 15,0 K in Magnetfeldern bis zu 20 kOe untersucht. Im einzelnen wurden die Übergänge vom Grundterm $^6H_{15/2}$ des Dy^{3+} zu den angeregten Termen $^6F_{3/2}$, $^4F_{9/2}$, $^4I_{15/2}$ und $^4G_{11/2}$ vermessen. In Emission wurden Übergänge vom Term $^4F_{9/2}$ zum Grundterm $^6H_{15/2}$ beobachtet.

4.1 Spektren in der tetragonalen Phase

Für Temperaturen oberhalb der kristallographischen Phasenumwandlung $(T > T_D)$ folgt aus den Spektren der erwähnten Übergänge, daß im Grundterm $^6H_{15/2}$ zwei Kramers-Dubletts als tiefste Kristallfeldkomponenten relativ dicht beieinander liegen. Das unterste Dublett bei 0,0 cm^{-1} hat die Kristallquantenzahlen $\bar{\mu} = \pm \frac{3}{2}$, das erste angeregte Dublett bei $\Delta (T > T_D) = (6,1 \pm 0,5)$ cm^{-1} $\bar{\mu} = \pm \frac{1}{2}$. Das nächsthöhere Dublett liegt bereits bei (97 ± 5) cm^{-1}. Die Energien der beobachteten Grundtermkomponenten, ihre Kristallquantenzahlen, sowie berechnete Energien sind in Tabelle 1 angegeben.

Tabelle 1. Beobachtete und berechnete Energien der Grundtermkomponenten des Dy^{3+} in DyAsO$_4$

Energie experimentell (cm^{-1})	Energie berechnet (cm^{-1})	Kristall-quantenzahl 2μ
0,0	0,0	± 3
$6,1 \pm 0,5$	6,0	± 1
$97,0 \pm 5,0$	113,2	± 3
$140,0 \pm 10,0$	131,6	± 3
	181,3	± 1
$243,0 \pm 10,0$	250,4	± 1
	308,0	± 3
	311,0	± 1

Für die Rechnungen wurde der Kristallfeldparameter B_2^0 direkt aus der Aufspaltung des Terms $^6F_{3/2}$ des Dy^{3+} im DyAsO$_4$ experimentell ermittelt. Die übrigen Parameter, die für die Rechnungen verwandt wurden, waren die des Er^{3+} in YAsO$_4$ [21], die mit Hilfe der Mittelwerte $\langle r^k \rangle$ $(k = 2, 4, 6)$ von Freeman und Watson [22] für das Dy^{3+} korrigiert wurden. Die α-, β- und γ-Werte wurden der Arbeit von Washimiya [23] entnommen. Da die Korrekturen große Unsicherheiten enthalten, können die berechneten Energien nur als relativ grobe Näherung gelten. Trotzdem ist die Übereinstimmung zwischen berechneten und experimentellen Energien erfreulich gut.

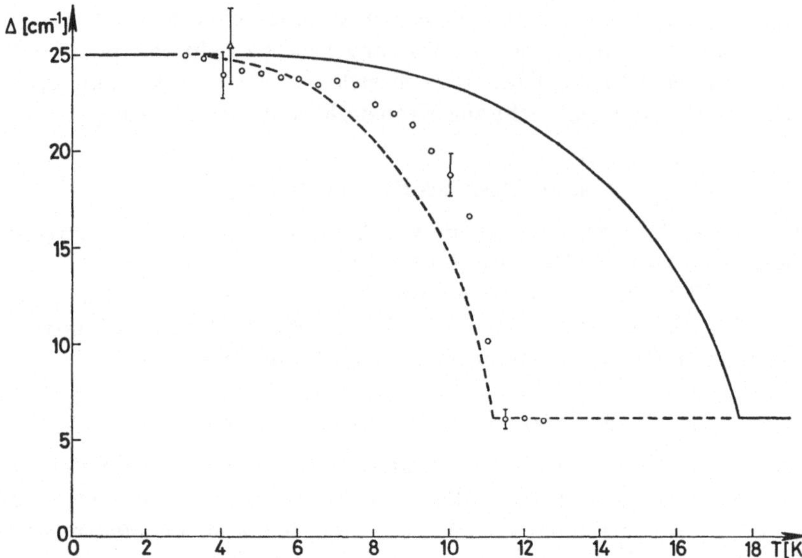

Fig. 1. Temperaturabhängigkeit des Abstandes Δ der beiden tiefsten Kramers-Dubletts des Dy^{3+} in $DyAsO_4$. Die Kreise stellen die spektroskopisch gemessenen Werte dar, das Dreieck bezeichnet die Messung mit Hilfe der Fourier-Spektroskopie. Die ausgezogene Kurve ist nach der Molekularfeldtheorie von Elliott *et al.* [15] berechnet; die gestrichelte Kurve ist gegenüber der ausgezogenen um den Faktor 0,63 in Abszissenrichtung gestaucht (siehe Text)

Aus diesen Rechnungen folgen als Eigenfunktionen für das tiefste Dublett (I) bei 0,0 cm^{-1}:

$$|\bar{\mu} = \pm \tfrac{3}{2}\rangle = 0{,}406\,|\mp \tfrac{9}{2}\rangle + 0{,}595\,|\mp \tfrac{1}{2}\rangle + 0{,}687\,|\pm \tfrac{7}{2}\rangle + 0{,}09\,|\pm \tfrac{15}{2}\rangle$$

und für das erste angeregte Dublett (II) bei $(6{,}1 \pm 0{,}5)$ cm^{-1}:

$$|\bar{\bar{\mu}} = \pm \tfrac{1}{2}\rangle = 0{,}219\,|\mp \tfrac{11}{2}\rangle + 0{,}652\,|\mp \tfrac{3}{2}\rangle + 0{,}717\,|\pm \tfrac{5}{2}\rangle + 0{,}113\,|\pm \tfrac{13}{2}\rangle .$$

4.2 Kristallographischer Phasenübergang

Unterhalb T_D nimmt der Abstand Δ der beiden tiefsten Dubletts mit sinkender Temperatur zu, er beträgt $\Delta(4{,}2\ \mathrm{K}) = (25{,}0 \pm 0{,}5)$ cm^{-1} bei $T = 4{,}2$ K. Die aus den Spektren ermittelte Temperaturabhängigkeit des Abstands Δ ist in Fig. 1 wiedergegeben. Bei 4,2 K konnte der Abstand Δ zusätzlich mit Hilfe der Fourier-Spektroskopie direkt zu $(25{,}5 \pm 2{,}0)$ cm^{-1} bestimmt werden. Dieser Meßpunkt ist in Fig. 1 durch ein Dreieck gekennzeichnet. Die Übereinstimmung der Ergebnisse dieser beiden Meßmethoden und der Raman-Streuung [15] ist gut. In allen Fällen wurden keine Korrekturen der Meßwerte wegen möglicher Einflüsse einer magnetischen Nahordnung vorgenommen.

Die Temperaturabhängigkeit des Abstands Δ hat im $DyAsO_4$ ziemlich genau denselben Verlauf wie im $DyVO_4$. Auch die kristallographische Verzerrung unterhalb T_D ist in beiden Fällen von demselben Symmetrietyp Γ_4^+. Daher sollte die von Elliott u. Mitarb. [15] für das $DyVO_4$ entwickelte Molekularfeldtheorie des kooperativen Jahn-Teller-Effekts das Verhalten des $DyAsO_4$ wenigstens so gut wie das des DyVO$_4$ beschreiben. Unter Berücksichtigung der Werte $\Delta(T \to 0) = 25\ \mathrm{cm}^{-1}$ und $\Delta(T > T_D) = 6{,}1\ \mathrm{cm}^{-1}$ (dieser Wert war Elliott u. Mitarb. [15] nicht bekannt) ergibt sich nach dieser Theorie der in Fig. 1 mit der ausgezogenen Kurve eingezeichnete Verlauf des Abstands Δ der Dubletts. Die berechnete Temperatur des Phasenübergangs beträgt $T_{D\,\mathrm{ber}} = 17{,}64\ \mathrm{K}$ im Gegensatz zum gemessenen Wert $T_{D\,\mathrm{gem}} = 11{,}17\ \mathrm{K}$. Berücksichtigt man bei der Molekularfeld-theorie lediglich die Kopplung mit den vier nächsten Nachbarn (nn), so ergibt sich für das Diamantgitter nach einer Reihenentwicklung ein T_D, das um den Faktor 0,68 kleiner ist, vgl. z.B. Domb [24], in unserem Fall also $T_{D\,\mathrm{ber\,(nn)}} = 12{,}00\ \mathrm{K}$. Verkleinert man den Korrekturfaktor auf 0,63, so erhält man zwar eine gute Übereinstimmung der theoretischen und der gemessenen Übergangstempera-tur T_D, der Temperaturverlauf des Abstands Δ der beiden Dubletts, der in Fig. 1 gestrichelt eingezeichnet ist, liegt aber deutlich unterhalb der Meßpunkte. Auch die modifizierte Molekularfeldtheorie kann also den experimentell bestimmten Verlauf von Δ nicht vollständig wiedergeben. Offenbar wird die Verzerrung des Gitters unterhalb T_D mit steigender Temperatur schneller abgebaut, als dies von einer Molekularfeldtheorie beschrieben wird.

4.3 Messung des Zeeman-Effekts

Die Untersuchung des Zeeman-Effekts galt vor allem der Bestimmung der Anisotropie des g-Faktors des tiefsten Grundterm-Dubletts ($I\ ^6H_{15/2}$) in der (a, b)-Ebene unterhalb T_D. Der g-Faktor, der mit dem magnetischen Moment μ einer Termkomponente durch die Beziehung $\mu = \pm\frac{1}{2}g\mu_B$ verknüpft ist ($\mu_B = $ Bohrsches Magneton), wurde aus der Steigung der tiefsten Zeeman-Komponente des Dubletts (I) für Magnetfelder über 10 kOe ermittelt. Die bei einer Temperatur von 1,8 K gemessenen Werte sind in Fig. 2 in Abhängigkeit vom Azimutwinkel φ zwischen b-Achse und Magnetfeld angegeben. Bei dieser Temperatur ist zwar die antiferromagnetische Ordnungstemperatur T_N bereits unterschritten, bei einem inneren Magnetfeld von etwa 2 kOe wird die Ordnung aber aufgebrochen, und bei äußeren Feldern über 10 kOe ist für Kristalle der benutzten Form sicher die magnetische Sättigung erreicht, so daß die Verschiebung der Zeeman-Komponen-ten linear mit der Stärke des äußeren Feldes erfolgt.

Der Verlauf des g-Faktors mit dem Winkel φ zeigt eine $\pi/2$-Periode. Der Grund hierfür ist, wie beim DyVO$_4$ [4], der, daß ein genügend starkes äußeres Magnetfeld die kristallographische Verzerrung des Kristalls beeinflußt und die a- und b-Achsen vertauscht (magnetic switching) [7]. Im Bereich $-\pi/4 \leqq \varphi \leqq \pi/4$ folgen die Meßpunkte gut einer Kosinus-Funktion, die in Fig. 2 als ausgezogene Kurve eingezeichnet ist. Aus diesem Kurvenverlauf folgt $g_{(I)}(\varphi = \pi/2) = g_{(I)a} = 0$; bei Berücksichtigung der Fehlergrenzen der Meßpunkte ergibt sich $g_{(I)a} = 0 + 3$. Für $g_{(I)b}$ erhält man als Mittelwert vieler (z.T. bei unterschiedlichen Tempera-turen durchgeführter) Messungen: $g_{(I)b} = 17{,}5 \pm 1{,}0$. Messungen bei $\varphi = 22°$ ergaben in jedem Fall einen Wert für $g_{(I)}(\varphi = 22°)$, der kleiner war als $g_{(I)b}$.

Bildet man aus den Eigenfunktionen $|\bar{\mu} = \pm\frac{3}{2}\rangle$ und $|\bar{\mu} = \pm\frac{1}{2}\rangle$ der beiden tiefsten Dubletts (vgl. Abschnitt 4.1) unter Berücksichtigung des Abstands $\Delta(T > T_D)$ Linearkombinationen, die der orthorhombischen Symmetrie unter-halb T_D mit $\Delta(T \to 0)$ angepaßt sind, so ergeben sich folgende theoretische g-Faktoren:

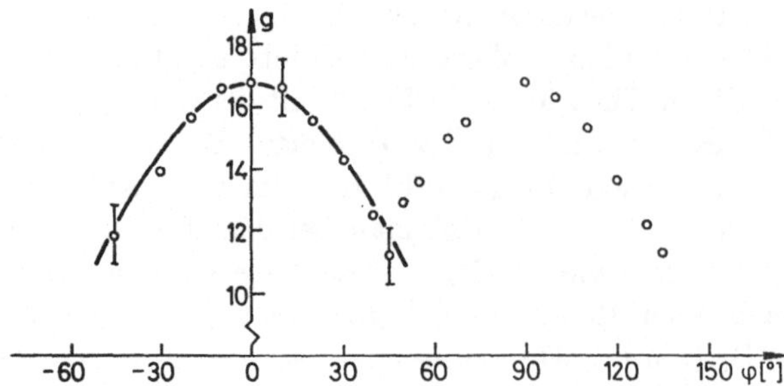

Fig. 2. g-Faktor des tiefsten Dubletts des Dy^{3+} in $DyAsO_4$ als Funktion des Azimutwinkels φ in der (a, b)-Ebene. Die Kreise stellen die gemessenen Werte dar, die ausgezogene Kurve entspricht einer Kosinusfunktion

für das tiefste Dublett $\qquad g_{(I)b} = 18{,}8,\qquad g_{(I)a} = 0{,}1$

für das erste angeregte Dublett $g_{(II)b} = 0{,}1,\qquad g_{(II)a} = 19{,}1.$

Die Abweichungen vom experimentellen Wert $g_{(I)b} = 17{,}5 \pm 1{,}0$ sind recht groß, was angesichts der verwendeten Näherungen aber verständlich ist.

4.4 Beobachtungen im magnetisch geordneten Zustand von $DyAsO_4$

Beim Einsetzen der antiferromagnetischen Ordnung bei T_N spalten die Absorptionslinien des Dy^{3+} aufgrund von Austausch- und Dipolwechselwirkung im allgemeinen in mehrere Linienkomponenten auf, wobei die Linienkomponenten an der Seite niederer Energie mit sinkender Temperatur schnell an Intensität verlieren. Außerdem treten in der direkten Umgebung der intensiven Linien mehrere Satelliten-Linien auf.

Legt man unterhalb T_N an einen Kristall das erstemal ein äußeres Magnetfeld parallel zur a- oder b-Achse an, so sind die Spektren bis etwa 2 kOe äußerst linienreich und kompliziert. Der Grund hierfür ist der, daß der Kristall zunächst aus verschiedenen kristallographischen Domänen besteht, deren b-Achsen um $90°$ gegeneinander gedreht sind. Für die beiden verschiedenen Domänensorten liegt das äußere Magnetfeld entweder parallel zur a- oder parallel zur b-Achse der einzelnen Domäne. Die beobachteten Spektren sind eine Überlagerung der im Zeeman-Effekt verschiedenen Spektren der beiden Domänensorten. In Feldern zwischen 2 und 3 kOe wandelt sich der Kristall in einen Eindomänenkristall mit der b-Achse parallel zum Magnetfeld um. Ein Kristall, der parallel b vormagnetisiert war, bleibt auch nach Abschalten des Magnetfeldes bei 1,8 K weitgehend ein Eindomänenkristall. Die Spektren im Zeeman-Effekt eines solchen Kristalls sind bereits bei kleinen Magnetfeldern einfach und übersichtlich.

Die Existenz der kristallographischen Domänen bei $DyAsO_4$ ist inzwischen durch kristalloptische Beobachtungen von Kasten und Becker [25] gesichert.

Der Zeeman-Effekt einer typischen Linie bei 20945 cm^{-1}, wie er bei $T = 1{,}8$ K gemessen wurde, ist in Fig. 3 wiedergegeben. Die Linie entspricht dem Übergang vom tiefsten Dublett (I) des Grundterms $^6H_{15/2}$ zum tiefsten Dublett (a) des Terms $^4F_{9/2}$. Die unterschiedliche Polarisation der Linienkomponenten parallel bzw. senkrecht zur Magnetfeldrichtung ist durch offene bzw. geschlossene Kreise angegeben. Die Meßpunkte für die intensiven (Haupt-)Linien sind mit ausgezogenen Kurven verbunden.

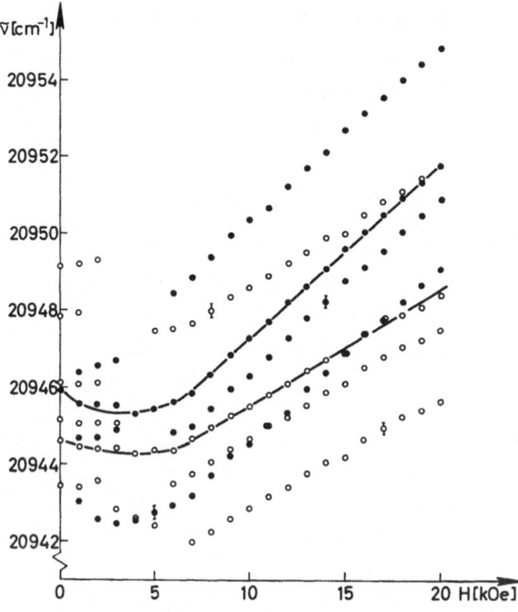

Fig. 3. Zeeman-Effekt der Linie $I\,^6H_{15/2} \to a\,^4F_{9/2}$ des Dy^{3+} mit dem Magnetfeld parallel zur b-Achse, $T = 1,8$ K. Die offenen Kreise entsprechen dem parallel zur b-Achse, die geschlossenen dem senkrecht zur b-Achse polarisierten Spektrum. Die ausgezogenen Kurven bezeichnen die Hauptlinien des Spektrums

Der Abstand der Hauptlinien bei $H = 0$ entspricht der Aufspaltung des angeregten Dubletts im effektiven Magnetfeld, das gleich der Summe von Austausch- und Dipolfeld ist. Aus dem Zeeman-Effekt dieses Übergangs ergibt sich der g-Faktor des Dubletts $(a\,^4F_{9/2})$ zu $g_{(a)a} = g_{(a)b} = 3,58 \pm 0,10$, er ist also isotrop in der (a, b)-Ebene. Mit diesem Wert erhält man als effektives Feld für dieses Dublett $H_{\text{eff}}(a\,^4F_{9/2}) = (7,6 \pm 0,3)$ kOe.

Bei $T = 2,1$ K konnte auch die Aufspaltung des tiefsten Grundtermdubletts beobachtet werden. Daraus wurde die Komponente des effektiven Feldes in Richtung der b-Achse für dieses Dublett zu $H_{\text{eff}\,b}(I\,^6H_{15/2}) = (7,7 \pm 0,4)$ kOe bestimmt.

Aus dem Zeeman-Effekt der Fig. 3 folgt, daß der Übergang von der antiferromagnetischen in die quasiferromagnetische Struktur im äußeren Magnetfeld, der zwischen etwa 1 kOe und 7 kOe abläuft, so erfolgt, daß das auf das einzelne Dy^{3+}-Ion wirkende gesamte Magnetfeld annähernd konstant bleibt. Eine detaillierte Untersuchung der Linienlagen im Bereich zwischen 0 und 7 kOe, wie sie Wright und Moos durchgeführt haben [12], war wegen der großen Linienbreiten und der Satellitenlinien nicht möglich.

Die Satellitenlinien verschieben sich für Magnetfelder über 7 kOe parallel zu den Hauptlinien mit gleicher Polarisation. Möglichkeiten zur Deutung der Satellitenlinien werden im Zusammenhang mit den Spektren des Er^{3+} in DyAsO$_4$ in Abschnitt 5 diskutiert.

5. Spektren des Er^{3+} in $DyAsO_4$

Zu weiteren Untersuchungen der Eigenschaften von $DyAsO_4$ wurden Kristalle der Zusammensetzung $(Dy, Er)AsO_4$ (100:1) hergestellt. Es wurden Übergänge vom Grundterm $^4I_{15/2}$ des Er^{3+} zu den angeregten Termen $^4S_{3/2}$, $^2H_{11/2}$ und $^4F_{5/2}$ untersucht.

Die Absorptionslinien des Er^{3+} spalten mit sinkender Temperatur bereits dicht unterhalb T_D auf. Die Aufspaltung bleibt mit sinkender Temperatur konstant und ändert sich auch bei T_N nicht wesentlich.

Als Beispiel ist in Fig. 4 das Verhalten des Übergangs vom tiefsten Grundtermdublett ($I\,^4I_{15/2}$) zum Dublett (b) des Terms $^4F_{5/2}$ angegeben. Die Linienaufspaltung muß als Zeeman-Aufspaltung im effektiven Magnetfeld des $DyAsO_4$ gedeutet werden. Offenbar bildet sich bereits direkt unterhalb der kristallographischen Phasenumwandlung eine antiferromagnetische Nahordnung der Dy-Momente. Dabei kann man nicht entscheiden, ob die Er^{3+}-Ionen die Nahordnung begünstigen oder ob sie sie aufgrund unterschiedlicher Zeitkonstanten eher zu messen vermögen als die Dy^{3+}-Ionen selbst.

Die Unsymmetrie der Aufspaltung in Fig. 4 konnte noch nicht eindeutig erklärt werden.

Aus dem Zeeman-Effekt des Übergangs $I\,^4I_{15/2} \to b\,^4S_{3/2}$ im starken äußeren Magnetfeld parallel b wurden die g-Faktoren des tiefsten Grundtermdubletts (I) zu $g_{(I)b} = 5,8 \pm 0,2$ und des Dubletts (b) des Terms $^4S_{3/2}$ zu $g_{(b)b} = 0 + 0,2$ bestimmt. Mit diesen Werten ergibt sich aus der Nullfeldaufspaltung dieses Übergangs die Komponente des effektiven Feldes in Richtung der b-Achse für das Grundtermdublett zu $H_{eff\,b}(I\,^4I_{15/2}) = (7,5 \pm 0,5)$ kOe.

Im Gegensatz zu den meisten Dy^{3+}-Linien zeigen die Er^{3+}-Linien im Magnetfeld parallel zur c-Achse einen großen Zeeman-Effekt, der bei kleinen Feldern nichtlinear ist und bei großen Feldern asymptotisch in einen linearen Effekt übergeht. In Fig. 5 ist dieses für eine Linie des Übergangs $^4I_{15/2} \to {}^2H_{11/2}$ bei $T = 1,8$ K dargestellt. Der nichtlineare Verlauf rührt daher, daß sich das auf das Er^{3+} wirkende gesamte Magnetfeld aus dem äußeren Feld parallel zur c-Achse und dem effektiven Feld, das senkrecht zur c-Achse liegt, zusammensetzt.

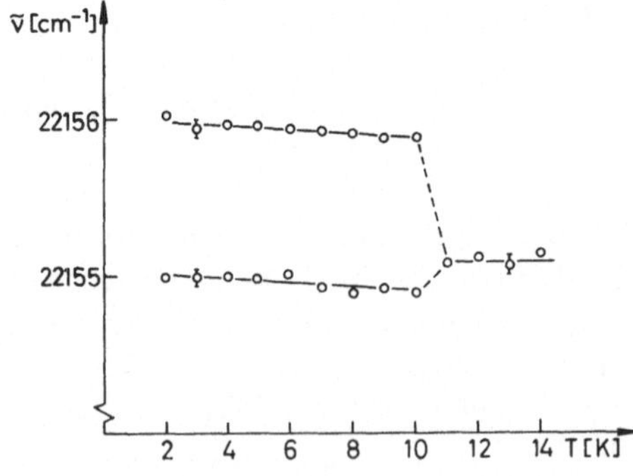

Fig. 4. Linienlage des Übergangs $I\,^4I_{15/2} \to b\,^4F_{5/2}$ des Er^{3+} in $DyAsO_4$ in Abhängigkeit von der Temperatur

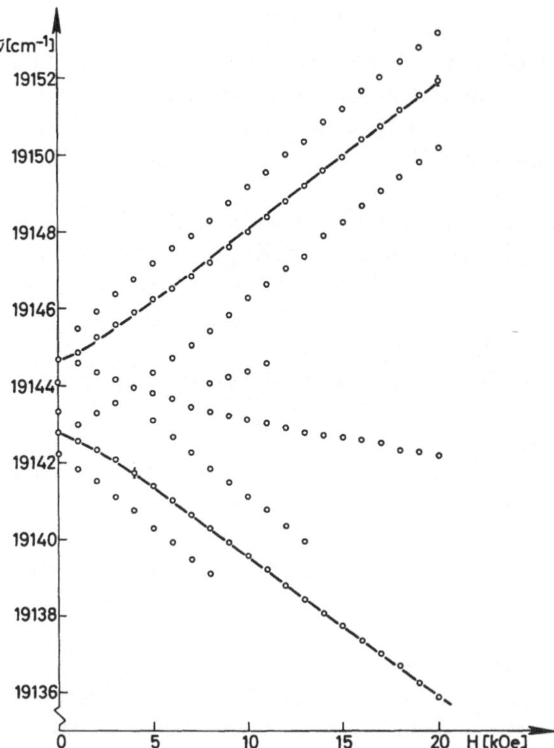

Fig. 5. Zeeman-Effekt des Er^{3+} in DyAsO$_4$ mit dem Magnetfeld parallel zur c-Achse, $T = 1,8$ K, Übergang $^4I_{15/2} \rightarrow\ ^2H_{11/2}$. Die Hauptlinien sind durch die durchgezogenen Kurven gekennzeichnet

Auch im Er^{3+}-Spektrum treten zahlreiche Satellitenlinien auf, die sich im Zeeman-Effekt $H \parallel c$ für Feldstärken über 6 kOe parallel zu den Hauptlinien verschieben, vgl. Fig. 5. Aus dieser Tatsache folgt, daß die Satelliten nicht durch magnetisch nichtäquivalente Plätze der Er-Ionen im DyAsO$_4$ (mit verschieden großen effektiven Feldern H_{eff} in der Ebene senkrecht zu c, d. h. unterschiedlicher Nahordnung) zustande kommen. Denn sonst müßten die Satellitenlinien im großen Feld $H \parallel c$ (mit $H \gg H_{eff}$) asymptotisch in die Hauptlinien einmünden, statt parallel zu ihnen zu verlaufen.

Eine mögliche Erklärung für die Satelliten, sowohl im Er^{3+}- als auch im Dy^{3+}-Spektrum, ist die folgende: Es sind im DyAsO$_4$ Fremdatome oder Fehlstellen in kleiner Konzentration auf definierten Plätzen eingebaut, die zur Folge haben, daß das auf die benachbarten Gitterplätze der Seltenen Erden wirkende Kristallfeld um einen bestimmten Betrag abweicht von dem Kristallfeld an einem Gitterplatz mit regulärer Umgebung. Das veränderte Kristallfeld führt, wie Hellwege gezeigt hat [26], zu definiert geänderten Termaufspaltungen und Absorptionsspektren, aber wenig oder gar nicht veränderten magnetischen Momenten. Göbel hat beobachtet, daß sich bei DyAsO$_4$ das stöchiometrische Verhältnis von Dy zu As durch Tempern verschieben kann [27]. Daher sind Störungen der AsO$_4^{3-}$-Komplexe denkbar, die zu den beobachteten Satelliten führen könnten.

6. Diskussion

$DyAsO_4$ zeigt bei tiefen Temperaturen Eigenschaften, die denen des $DyVO_4$ sehr ähnlich sind. Beide Substanzen erfahren einen kristallographischen Phasenübergang mit Verzerrung in Richtung der a- und der b-Achse, der durch einen kooperativen Jahn-Teller-Effekt induziert ist. Die von Elliott u. Mitarb. [15] entwickelte Molekularfeldtheorie dieses Effektes gibt, im Gegensatz zum $TbVO_4$ und $TbAsO_4$, keine sehr gute Übereinstimmung mit den experimentellen Ergebnissen. Wie weit dieses mit einer Kopplung an optische bzw. akustische Eigenschwingungen zu tun hat, soll hier nicht diskutiert werden, siehe hierzu z. B. [15].

Die Spektren des in $DyAsO_4$ eingebauten Er^{3+} zeigen bereits dicht unterhalb T_D die magnetische Nahordnung der Dy-Momente. Daraus folgt, daß der kristallographische Phasenübergang entscheidende Voraussetzung für die magnetische Ordnung des $DyAsO_4$ ist.

Die antiferromagnetische Struktur des $DyAsO_4$ ist nach den Neutronenbeugungsuntersuchungen von Will [9] verschieden von der des $DyVO_4$. Aus den vorliegenden optischen Untersuchungen konnte allerdings kein Hinweis auf den Ordnungswinkel der magnetischen Momente von 22° gegen die b-Achse gefunden werden. Wenn tatsächlich $g_{(I)a} = 0$ ist, müssen bei den normalen Austausch- und Dipol-Wechselwirkungen die Dy-Momente parallel zur b-Achse liegen. Bei $g_{(I)a} \neq 0$ wäre eine Drehung der Momente möglich.

Es ist allerdings bisher nicht ausgeschlossen, daß der Ordnungswinkel, der sich bei den Neutronenbeugungsmessungen ergeben hat, auf eine schlechte Qualität der untersuchten $DyAsO_4$-Proben zurückzuführen ist. An den von Will benutzten Kristallpulvern konnte der kristallographische Phasenübergang durch Röntgenbeugung teilweise nicht nachgewiesen werden [27]. Bei den Kristallen derselben Zucht, wie die von Will benutzten Proben, konnte auch im Raman-Effekt kein kristallographischer Phasenübergang beobachtet werden [14].

Mein besonderer Dank gilt Herrn Prof. Dr. H. G. Kahle für die Anregung zu dieser Arbeit, seine stete Förderung und für wertvolle Diskussionen. Herrn Dr. G. Müller-Vogt und Herrn Dr. L. Klein danke ich für die Züchtung der Einkristalle, Herrn Dipl.-Phys. R. Herb für die Ausführung der numerischen Rechnungen und Herrn Dr. W. Prettl, jetzt beim Max-Planck-Institut für Festkörperforschung, Stuttgart, für die Messungen am Fourier-Spektrometer. Die Arbeit wurde teilweise von der Deutschen Forschungsgemeinschaft unterstützt.

Literatur

1. Wright, J. C., Moos, H. W., Colwell, J. H., Mangum, B. W., Thornton, D. D.: Phys. Rev. B **3**, 843 (1971)
2. Koonce, C. S., Mangum, B. W., Thornton, D. D.: Phys. Rev. B **4**, 4054 (1971)
3. Scharenberg, W., Will, G.: Intern. J. Magnetism 1, 277 (1971)
4. Cooke, A. H., Ellis, C. J., Gehring, K. A., Leask, M. J. M., Martin, D. M., Wanklyn, B. M., Wells, M. R., White, R. L.: Solid State Commun. 8, 689 (1970)
5. Cooke, A. H., Martin, D. M., Wells, M. R.: Solid State Commun. 9, 519 (1971)
6. Becker, P. J., Dummer, G., Kahle, H. G., Klein, L., Müller-Vogt, G., Schopper, H. C.: Phys. Letters **31** A, 499 (1970)
7. Kahle, H. G., Klein, L., Müller-Vogt, G., Schopper, H. C.: Phys. status solidi (b) **44**, 619 (1971)
8. Will, G., Schäfer, W., Scharenberg, W., Göbel, H.: Z. angew. Phys. **32**, 122 (1971)

9. Schäfer, W., Will, G.: J. Phys. C, Solid State Phys. **4**, 3224 (1971)
10. Göbel, H., Will, G.: Phys. status solidi (b) **50**, 147 (1972)
11. Göbel, H., Will, G.: Phys. Letters **39** A, 79 (1972)
12. Wright, J. C., Moos, H. W.: Phys. Rev. B **4**, 163 (1971)
13. Hudson, R. P., Mangum, B. W.: Phys. Letters **36** A, 157 (1971)
14. Brüesch, P., Kalbfleisch, H.: Phys. status solidi (b) **45**, K 103 (1971)
15. Elliott, R. J., Harley, R. T., Hayes, W., Smith, S. R. P.: Proc. Roy. Soc. A **328**, 217 (1972)
16. Hintzmann, W., Müller-Vogt, G.: J. Crystal Growth 5, 274 (1969)
17. Wüchner, W., Böhm, W., Kahle, H. G., Kasten, A., Laugsch, J.: Phys. status solidi (b) **54**, 273 (1972)
18. Laugsch, J.: Private Mitteilung
19. Hellwege, K. H.: Ann. Phys. (6) **4**, 95, 127, 136, 143, 150, 357 (1949)
20. Kuse, D.: Z. Physik **203**, 49 (1967)
21. Kahle, H. G., Klein, L.: Phys. status solidi **42**, 479 (1970)
22. Freeman, A. J., Watson, R. E.: Phys. Rev. **127**, 2058 (1962)
23. Washimiya, S.: J. Phys. Soc. Japan **27**, 56 (1969)
24. Domb, C.: Magnetism, Vol. II A (ed. Rado, G. T., Suhl, H.): New York, Academic Press 1965
25. Kasten, A., Becker, P. J.: Intern. J. Magnetism, im Druck
26. Hellwege, K. H.: Phys. kondens. Materie **15**, 1 (1972)
27. Müller-Vogt, G.: Private Mitteilung

Dr. Dieter Wappler
Physikalisches Institut
der Universität Karlsruhe
D-75 Karlsruhe 1
Engesserstr. 7
Bundesrepublik Deutschland

Phys. cond. Matter 17, 125—134 (1974)

Normal Spectral Emissivity of Platinum in the Red and in the Green in Vacuum and Helium

A. H. Madjid, R. L. Stover, and J. M. Martinez *

Department of Physics, Thermionic Emission Laboratory
The Pennsylvania State University, University Park

Received October 8, 1973

The spectral emissivity of platinum for $\lambda = 0.645$ μm and $\lambda = 0.546$ μm has been measured by using a De Vos blackbody in ultrahigh vacuum and in helium. Changes in emissivity brought about by heat treatment and changing ambiences, are illustrated and discussed and the results are compared with previously published data. The emissivities determined turned out to be appreciably lower than the measurements of previous investigators. The determined values for $1100 < T(\mathrm{K}) < 1800$ are for 0.645 μm between 0.237 and 0.249 in vacuum, and between 0.185 and 0.290 in helium; and for 0.546 μm between 0.250 and 0.312 for vacuum and between 0.219 and 0.321 in helium.

I. Introduction

Considerable discord still exists about the true value of the spectral emissivity of platinum at incandescent temperatures [1—13]. The most recent summaries within this context have been published by Stephen [11] and Worthing [12] and, although these investigators brought a considerable measure of coherency into the field by their own measurements [10, 11], the spread in the results of the last five most recent such determinations [8—11, 13] still is sufficiently large to warrant a reconsideration of this problem within the improved experimental framework which modern techniques, that are available today, afford.

The use of platinum in the laboratory and in technology is widespread because of the chemical inertness and the high melting point of the element. In many such applications sample temperatures have to be known within close limits because the experimental parameters that are to be studied are rather sensitive to relatively small temperature differences. This is true for experiments in which platinum is the primary object under investigation, but it is also true for cases in which this metal plays a secondary role, such as for example, when it is chosen as a substrate to investigate the high temperature properties of other refractory substances. In such experiments, temperatures in the incandescent range are most conveniently determined by using optical pyrometers, but such determinations are contingent on an accurate knowledge of the emissivity of the sample involved and this makes the continuing effort toward more accurate determinations of the emissivity parameters an experimentally meaningful undertaking.

The reason for major discrepancies in different results may be mainly apportioned to differing manufacturing procedures, experimental ambiences, and heat

---

* Present Address: University of Madrid.

treating schedules. Changes in manufacturing processes often change the bulk impurity content of the metal and as manufacturing improves and becomes more standardized, emissivity results by different investigators tend to settle into closer agreement. Impurities and defects which alter the free electron concentration in the metal, or which act as scattering centers, particularly if they tend to aggregate within a few escape depths of the emitted radiation from the surface, will be active in changing the emissivity. Heterovalent impurities that become bound to defects in the surface region will be particularly serious offenders in this respect, but easily diffusing contaminants that are readily dissolved by the metal, such as hydrogen, have the effect of dilating the lattice and should also be guarded against. In short, all processes which alter the resistivity in the pertinent region within a few escape depths of radiation from the surface of the metal will tend to alter the spectral emissivity value. (The escape depth is here defined as the reciprocal of the absorption coefficient.)

The experimental ambience may be aggressive. The resulting etching structure will increase the true surface area over the geometric surface area and this will thus tend to increase the value of the emissivity. It is likely that air or oxygen affect platinum at incandescent temperatures in this way through the formation of a volatile and unstable oxide [14, 15]. The second effect may arise from the absorption of ambient species on the surface. Species that are physically absorbed at the surface are rarely a problem at incandescent temperatures because they are relatively weakly bound and desorb readily even below incandescence. Chemisorption, as a rule, causes pronounced rearrangement in atomic surface structure. Most species nevertheless desorb below 900 K, but hydrogen and carbon must specially be guarded against because they tend to persist on the platinum surface to 1500 K for hydrogen and to at least 1700 K, and very probably even to the melting point, for carbon. Morgan and Somorjai [16], who determined the numerical values listed above, using low energy electron diffraction techniques, report a $Pt(100)-(5 \times 1)$ surface structure for the normal, clean platinum (100) face; a $Pt(100)-(2 \times 2)-H_2$ structure, stable below 1500 K, for hydrogen; and a ring-like diffraction pattern, stable at all temperatures studied by these authors (this is, up to 1700 K) for carbon. The affinity of carbon for platinum surfaces is well known to workers in this field. It is, for example, fairly simple to carry out an evaporation using a platinum wire, or ribbon, heated to near its melting point in a conventional mercury vacuum system; but it is difficult or even impossible to repeat this procedure in organic oil pumped systems. The probable explanation is that appreciable amounts of surface carbon attenuate the evaporation rate so that attempts to compensate by rising the temperature result in the melting of the uncontaminated bulk platinum core and the consequent rupture of the ribbon or wire. Carbon is an impurity that is difficult to avoid in vacuum technology. A low, mass spectrometrically determined background when the ultimate vacuum is attained is not a guard against carbon contamination. The background of CO, CO_2, and residual hydrocarbons may reach appreciable levels during bake-out or outgassing cycles. Some of the easier decomposable compounds may, then, deposite on the cold platinum surfaces and cause carbon contamination by decomposition when these surfaces are in turn outgassed by raising their temperature near the melting point. Fore pumps are particularly serious offenders in this respect. It is impossible to maintain a surface hydrocarbon clean in a system which is fore pumped in a continuous duty cycle, even if multiple liquid N_2 trapping is resorted to. Fore line trapping, short cycle pumping, and using a high quality hydrocarbon diffusion pump oil in the fore pump alleviates this problem. (In order not to damage the pump, close attention must be given to the viscosity and the lubricating properties of the oil in question. Silicon oils are usually ruled out because of their poor lubricating properties which lead to pump damage in a short period of time.) Carbon contamination can usually be removed from platinum surfaces by a high temperature treatment in a low pressure oxygen ambience. Absorption and ion pumped systems are satisfactory, but excessive hydrogen background levels must here be guarded against.

The third and last surface perturbation affecting the emissivity are recrystallization effects due to thermal treatment. Single crystal grains often grow to appreciable size in preferred directions. The preferred grain faces for platinum are the (111) planes [17]. Impurities tend to be swept by the growing grains into the grain boundaries which, as a rule, are prone to grooving. The effect of this grooving on the emissivity is, thus, similar to that of surface damage which tends to raise emissivity values. A discussion of this effect is given by Quinn [18] for tungsten and by Alvares for platinum [19].

II. Experimental

The blackbody tubes were similar in design to those used by De Vos [20, 21] for tungsten and by Martinez and Madjid [22] for tantalum. The tubes were made from 2 mil platinum stock of 99.95% purity supplied by the Engelhard Industries. The dimensions were:

Length of blackbody: 20.3 cm; triangular cross section: $0.64 \times 0.64 \times 0.64$ cm; wall thickness: 0.005 cm; blackbody hole diameter: 0.3 mm.

The tubes were shaped in a triangular die, capped, and spot welded at the ends to corrugated tantalum spring tabs. The tabs were, in turn, spot welded to a truss made from 0.25 cm round tungsten stock which fitted into set-screw sockets drilled into the copper conductors of a commercial stainless steel ultrahigh vacuum feedthrough. The corrugated tabs were carefully dimensioned to balance Joule heat developed within them by the heating current against suspension heat losses with the results that temperature inhomogeneities along the middle 8 cm of the blackbody remained at a virtually undetectable level at all temperatures considered.

The blackbody was mounted in an adsorption and sputter ion pumped stainless steel vacuum chamber and standard bakeout and outgassing techniques were used to attain the ultimate vacuum and to remove surface contaminants from the pertinent surfaces to the extent possible. The vacuum was maintained in the 10^{-9} and 10^{-8} Torr range for low and high temperatures respectively. Measurements in the helium ambient were made using bottle helium (of 99.995% purity and dew point below $-$ 60 C) at near atmospheric pressure.

The optical pyrometer that was used in the experiment was a precision red-green disappearing filament micropyrometer and the readings were made through a sapphire port. The results were based on nine readings at each temperature setting and all data was corrected for window and pyrometer optics absorption before and after the entire measurement series. The pyrometer was compared against a calibrated secondary standard pyrometer lamp prior to the measurement, and the calibration of the latter was checked by a one-point comparison against an institutional standard [23].

Recrystallization effects in reasonably thin and pure platinum sheet stock start to develop while the samples are heated during vacuum processing. Thus, by the time the outgassing and annealing schedules, the surface contamination removal procedures, the temperature uniformity studies, and all the other quality assessments of the blackbody are completed, grain growth in the 2-mil thick platinum is already partially developed. Because of this and because grain growth in platinum has already been adequately studied by McLean and Mykura [17], and Alvares [19], no attempt to determine the "as received" emissivity values was made. We concentrated instead on measuring the emissivity of recrystallized platinum, and on heat treating the metal to a stable surface structure, that is, to a reproducible emissivity characteristic. Towards this purpose, a large grain with a uniform surface was chosen which was located in the immediate vicinity of the blackbody hole, but which was, nevertheless, far enough removed from the emissivity perturbation which the slight indentation around the blackbody hole represents. The pyrometer was carefully focused on an identical spot in the middle of this grain throughout the measurements. A reference temperature of about 1400 K was chosen to monitor the effects of the heat treatment on the emissivity values. This temperature is sufficiently high to allow accurate pyrometer work, yet it is low enough for the platinum not to undergo appreciable structural surface changes even when the blackbody is kept at this temperature for extended periods of time. The stabilizing heat treatment consisted in increasing the temperature by roughly 100 K steps, starting from 1000 K up to 1800 K. The dwell at each intermediate step was one hour. The blackbody, however, was kept at 1800 K for several hours. The temperature was periodically reduced from this temperature to the reference temperature of 1400 K to keep track of any possible changes in the emissivity. It was assumed that the platinum had reached surface structure stability when the difference between two successive emissivity determinations at 1400 K made half an hour apart, remained within the bounds given by the limits of error. This occurred when the platinum had remained between four and five hours at 1800 K. Starting from this temperature, the reported emissivity data was taken in decreasing temperature sequence and all measured temperatures were held for one hour before the respective pyrometer readings were made.

III. Results

Once fully developed, the geometry of the grain structure remaind remarkably stable throughout the entire duration of the experiment. This was confirmed visually and photographically. In Fig. 1 the "as received" surface structure of the rolled sheet stock, Fig. 1 (a), is compared with the grain structure of the fully heat treated platinum, Fig. 1 (b). (The area surrounding the blackbody hole was chosen for display. The slight indentation around the hole, caused by drilling,

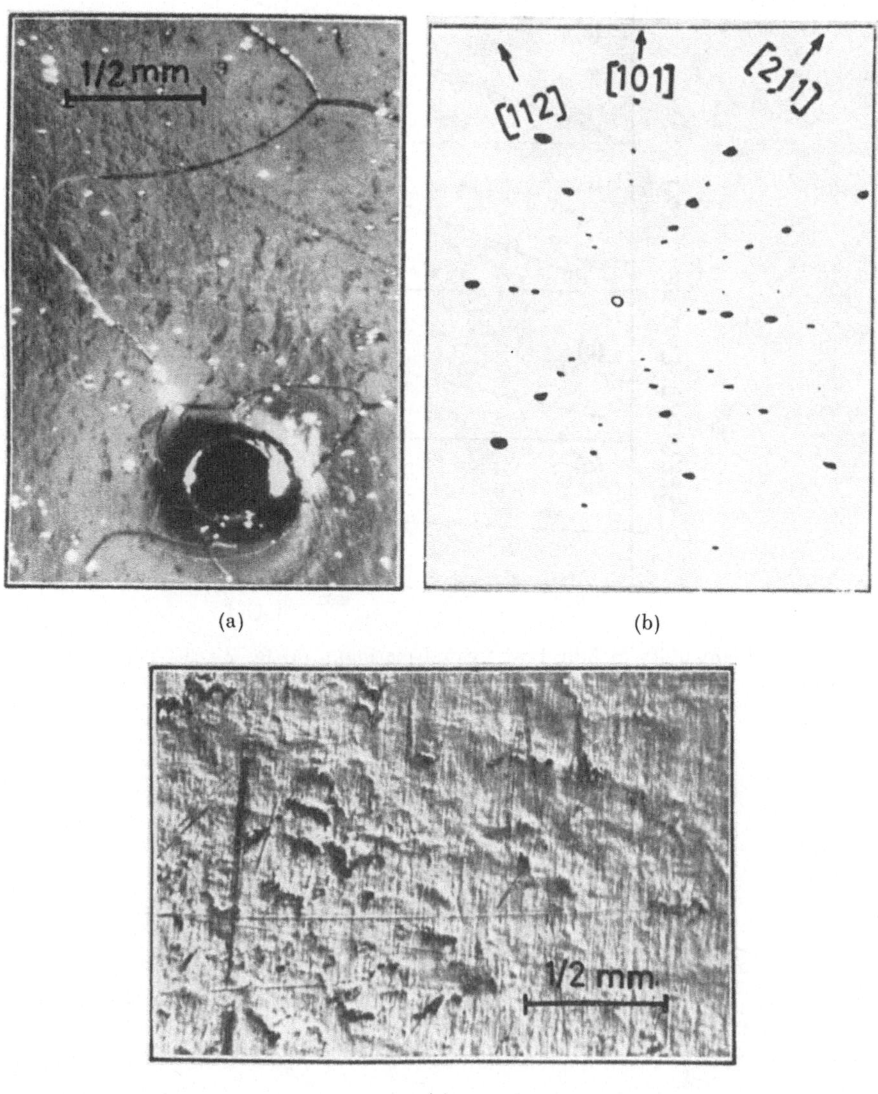

Fig. 1. (a) Micrograph of the region of the platinum blackbody around the blackbody hole after heat treatment. Several large (111) grains are apparent within which a moderate number of ordered triangular etch pits may clearly be discerned. (b) Laue X-ray diffraction pattern establishing that the grains shown above are (111) planes. (c) Micrograph of the "as received" platinum foil. Comparison with (a) shows that heat treatment results in considerable improvement as far as surface smoothness is concerned

appears distinctly blacker than the flat portions of the surface.) The heat treated surface is seen to be considerably smoother than the original "as received" surface. The grain faces are (111) planes. This can be inferred from the Laue reflection pattern shown in Fig. 1 (c). The slight asterism in the pattern indicates

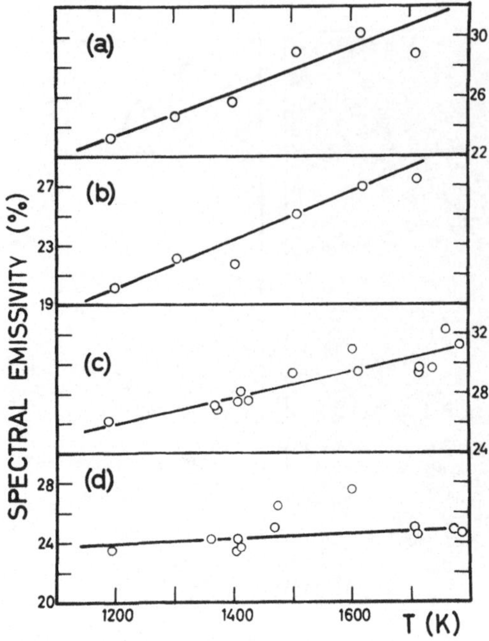

Fig. 2. Spectral emissivity of fully heat treated platinum. (a) for $\lambda = 0.546$ μm and (b) for $\lambda = 0.645$ μm both in helium and (c) for $\lambda = 0.546$ μm and (d) for $\lambda = 0.645$ μm, both in ultra high vacuum

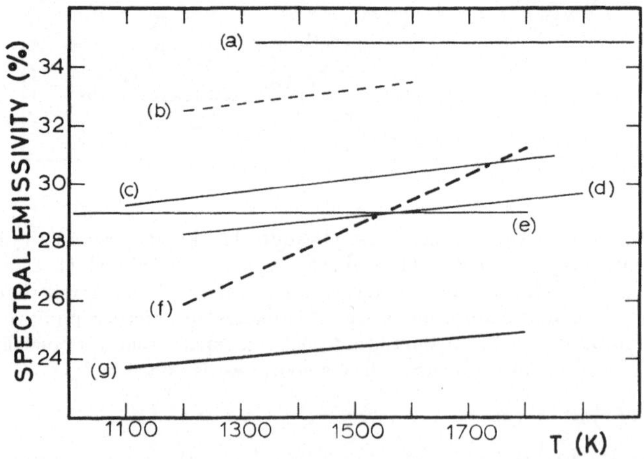

Fig. 3. Comparison of present measurements of ε_λ with previously published results. (a) Henning and Heuse [9], $\lambda = 0.647$ μm; (b) Worthing [10], $\lambda = 0.535$ μm; (c) Worthing [10], $\lambda = 0.665$ μm; (d) Stephens [11], $\lambda = 0.660$ μm; (e) Krishnan and Jain [13], $\lambda = 0.655$ μm; (f) present results for $\lambda = 0.546$ μm; and (g) for $\lambda = 0.645$ μm

strain, but the identity of the plane is unmistakable. The triangular thermal etch pits, distinctly visible in Fig. 1 (b), corroborate this conclusion.

The spectral emissivity was calculated from the measured true or blackbody temperature T, and the brightness temperature S_λ by using the expression,

$$(1/T) - (1/S_\lambda) = (\lambda/C_2) \ln (\varepsilon_\lambda), \tag{1}$$

with λ the wavelength, and $C_2 = 1.43879 \pm 0.00019$ (cm K). The spectral emissivity in the red, 0.645 μm, and in the green, 0.546 μm, for a vacuum, and a helium ambience thus determined are shown in Fig. 2. The lines shown in these graphs are least square fits which were calculated by assuming that within the precision of the measurements, the emissivity varies linearly with temperature over the measured temperature interval of about $1100 < T\,(\mathrm{K}) < 1800$. The corresponding analytical expressions with T in K, are

$$\varepsilon^{\mathrm{vacuum}}_{0.645\mu\mathrm{m}} = 176\,(1) \times 10^{-8}\,T + 0.217\,(7), \tag{2}$$

$$\varepsilon^{\mathrm{vacuum}}_{0.546\mu\mathrm{m}} = 879\,(4) \times 10^{-8}\,T + 0.153\,(5), \tag{3}$$

$$\varepsilon^{\mathrm{He}}_{0.645\mu\mathrm{m}} = 150\,(2) \times 10^{-7}\,T + 0.0197\,(8), \tag{4}$$

$$\varepsilon^{\mathrm{He}}_{0.546\mu\mathrm{m}} = 145\,(3) \times 10^{-7}\,T + 0.0594\,(7), \tag{5}$$

These findings are summarized in the Table for the measurements in the vacuum and the helium ambience respectively and the present measurements are compared with previous results in Fig. 3.

Table. Spectral emissivity of commercial platinum stock in a ultra high vacuum ambience and in a helium ambience, the data pertains to platinum which was heat treated for several hours at 1800 K. T is the true temperature, S_λ is the brightness temperature, and ε_λ is the spectral emissivity. The error in the emissivity was calculated on the basis of an estimated error in the determination of T and S_λ of $\pm$ 1 K, and the use of Eq. (5) for the calculation of ε_λ

T °K	S_λ °K	ε_λ $\lambda = 0.645$ μm	$\pm$ Error	S_λ °K	ε_λ $\lambda = 0.546$ μm	$\pm$ Error
1100	1027	0.237	0.006	1040	0.250	0.008
1200	1114	0.239	0.005	1130	0.259	0.007
1300	1200	0.241	0.005	1221	0.268	0.006
1400	1286	0.242	0.004	1311	0.277	0.005
1500	1370	0.244	0.003	1400	0.285	0.004
1600	1454	0.246	0.003	1489	0.294	0.004
1700	1537	0.248	0.003	1578	0.303	0.004
1800	1618	0.249	0.002	1667	0.312	0.004
Helium Ambience						
1200	1104	0.200	0.004	1125	0.234	0.006
1300	1193	0.215	0.004	1216	0.248	0.005
1400	1282	0.230	0.004	1307	0.263	0.005
1500	1370	0.245	0.003	1398	0.277	0.005
1600	1459	0.260	0.003	1489	0.292	0.004
1700	1548	0.275	0.003	1579	0.306	0.004

132   A. H. Madjid, R. L. Stover, and J. M. Martinez

IV. Discussion

Comparing the four most recent published determinations [9—11, 13] and the present measurements of the spectral emissivity of platinum in Fig. 3 reveals that, almost in every case, each successive measurement of ε_λ as a function of temperature in the red is lower than each previous measurement. An interesting feature, furthermore, is that Worthing (1926) [10], Stephen (1939) [11], and the present authors determined almost identical positive temperature coefficients for the emissivity in the red of about 1.8×10^{-5} (K^{-1}) in a temperature range of roughly $1000 < T < 2000$ (K). Krishnan and Jain's (1954) [13] measurements indicate a constant ε_λ over the interval, but these authors' primary purpose was to measure thermal conductivity at high temperatures, and their use of a composite blackbody is open to some criticism when compared to the more rigorous procedures used by Stephen [13] for example. The comparatively much lower emissivity values determined in this study are probably the result of using an oil free ultrahigh vacuum ambience, superior blackness conditions, and the use of a flat surfaced blackbody big enough to permit large, flat, smooth-faced, single crystal grains to develop and it is, therefore, suggested that the emissivity thus determined is more representative of platinum than previous results.

A comparison of the results displayed in Fig. 2 illustrates that even an inert ambience may have a fundamental effect in the emissivity of a surface. On

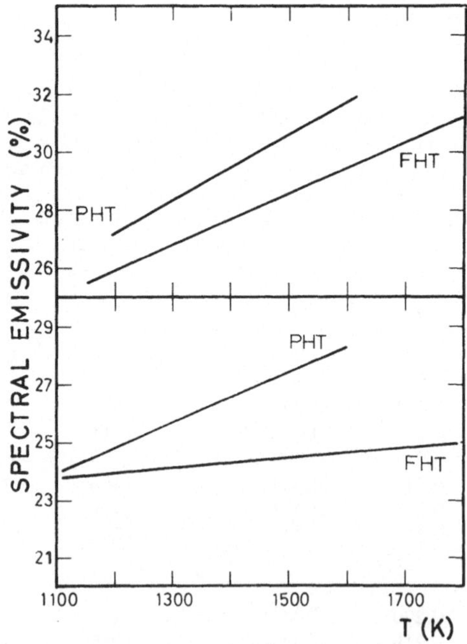

Fig. 4. The effect of heat treatment on the emissivity of platinum. The ambience in both cases is ultra high vacuum. (a) Refers to measurements in the green, $\lambda = 0.546$ μm; and (b) to measurements in the red, $\lambda = 0.645$ μm. The curves marked PHT refer to the emissivity characteristic of an only partially heat treated surface and the curves marked FHT refer to ε_λ of fully heat treated platinum

platinum, the helium, over the ultrahigh vacuum ambience, has, in both cases, the effect of lowering the low temperature end of the emissivity, and of raising the temperature coefficient of ε_λ. The effect of heat treatment is shown in Fig. 4. As has previously been mentioned, partial grain growth in the platinum occurred already during the normal course of vacuum processing and surface cleaning. Shown is, therefore, the characteristic of a surface that is already partially heat treated against that of a fully heat treated surface. The appreciably lower values of ε_λ for the latter surface are probably due to the smoothing of the single crystal grain surface which occurs during this process. Noteworthy, finally, remains the fact that the temperature coefficient of the emissivity with about 8.8×10^{-8} (K^{-1}) is appreciably higher than that in the red which is about 1.8×10^{-8} (K^{-1}). As has been mentioned, the latter value has been verified by at least three independent investigators and considerable confidence may, therefore, be placed in this numerical value. But in the present determination the pyrometer remained focused on the same spot on the platinum grain in question, and the red and the green measurements were made almost concurrently for each separate determination of the respective brightness temperatures. Some confidence may, therefore, be placed in the measured values that show that a marked disparity exists between the temperature coefficients of the emissivity in the green and in the red.

Acknowledgments. We wish to thank Dr. H. J. Kostkowski and the National Bureau of Standards — Gaithersburg, Maryland — for the calibration of our optical pyrometer, and Miss Sandra Wallace for help with the preparation of the manuscript.

References

1. Holborn, L., Kurlbaum, F.: Ann. Phys. **10**, 225 (1903)
2. Waidner, Burgess, G. K.: Bull. Nat. Bur. Stand. **3**, 163 (1907)
3. Laue, Martens: Physik Zeits. **8**, 853 (1907)
4. Féry, Ch., Cheneveau: Comptes rendus **148**, 401 (1909)
5. Henning, F.: Z. Instrumkde **30**, 61 (1910)
6. Mendenhall, C. E.: Astrophys. J. **33**, 91 (1911)
7. McCauley: Astrophys. J. **37**, 164 (1913)
8. Spence, B. J.: Astrophys. J. **37**, 194 (1913)
9. Henning, F., Heuse, W.: Z. Phys. **16**, 63 (1923)
10. Worthing, A. G.: Phys. Rev. **28**, 174 (1926)
11. Stephens, R. E.: J. Opt. Soc. Amer. **29**, 158 (1939)
12. Worthing, A. G.: J. appl. Phys. **11**, 421 (1940)
13. Krishnan, K. S., Jain, S. C.: Brit. J. appl. Phys. **5**, 426 (1954)
14. Corruccini, R. J.: J. Res. nat. Bur. Stand. **47**, 94 (1951)
15. Uhlig, H. H. (ed.): Corrosion Handbook, p. 700. New York: John Wiley & Sons, Inc. 1948
16. Morgan, A. E., Somorjai, G. A.: Surf. Sci. **12**, 405 (1968)
17. McLean, M., Mykura, H.: Acta metallurgica **13**, 1291 (1965)
18. Quinn, T. J.: Brit. J. appl. Phys. **16**, 973 (1965)
19. Alvares, N. J.: Symposium on Thermal Radiation of Solids, ed. S. Katzoff. National Aeronautics and Space Administration, Scientific and Technical Information Division, NASA SP-55 (1965)
20. DeVos, J. C.: Physica **20**, 669 (1954)
21. DeVos, J. C.: Physica **20**, 690 (1954)
22. Martinez, J. M., Madjid, A. H.: J. appl. Phys. **41**, 5322 (1970)

23. We are indebted to Dr. H. J. Kostkowski and the National Bureau of Standards, Gaithersburg, Md., for this calibration. We gratefully acknowledge this help.

Dr. A. H. Madjid
Department of Physics
Thermionic Emission Laboratory
The Pennsylvania State University
University Park
Pennsylvania 16802, USA

Phys. cond. Matter 17, 135—152 (1974)
© by Springer-Verlag 1974

The Electrical Resistivity of Metals at High Temperatures.
An Analysis for Alkali and Noble Metals

Göran Grimvall

Institute of Theoretical Physics, Göteborg, Sweden

Received August 6, 1973

Experimental data for the electrical resistivity of K, Na, Cu, Ag and Au at high temperatures and constant volume have been analysed. In particular, corrections to the most simple theoretical model (often referred to as Ziman's model) arising from lattice anharmonicity, Debye-Waller factors, multi-phonon scattering and terms beyond the Born approximation are discussed. The net effect of all these corrections amounts at most to some 10% decrease in the resistivity close to the melting point.

1. Introduction

The electrical resistivity of metals is well understood in its gross features, but going beyond the most simple approximation in the theory, there are still a number of points to be discussed, such as the influence of lattice anharmonicity, Debye-Waller factors, multi-phonon scattering, thermally created lattice defects and the validity of the Born approximation. It is the purpose of this paper to analyse existing experimental data with respect to the effects mentioned. The elements considered are potassium, sodium, copper, silver and gold for which there are accurate measurements by Dugdale and Gugan [1] and Laubitz [2, 3]. We first briefly discuss the effects involved. In the most simple model, the ideal electrical resistivity (i.e. without impurity scattering) varies linearly with temperature in the high temperature limit. Experiments in general give a more rapid increase. This is mainly due to the fact that the thermal expansion will lower the phonon frequencies. In this paper we are interested in the explicit temperature dependence and therefore consider the resistivity reduced to constant volume. Even so will there be a shift in the phonon frequencies with temperature that affects the resistivity. At high temperatures the phonons will be damped which may be of importance. We should also include a temperature factor of the Debye-Waller type. In addition there will be multi-phonon scattering processes and they will give a contribution to the resistivity opposite in sign to that of the Debye-Waller factor. It is not yet known how effective this cancellation is, and it has even been suggested that the two terms might be paired off exactly [4]. The Born approximation has been the starting point of all published calculations and we discuss its validity. Apart from these major sources of corrections to the high temperature resistivity proper attention must also be given to points like thermally created lattice defects and the dielectric screening and also the validity of the simple transport equation. For some of these effects, e.g. lattice anharmonicity and lattice defects, we can give reasonable numerical estimates. From these values and a comparison with experimental data for the resistivity, we can make statements about the combined effect of essentially Debye-Waller factors, multi-phonon scattering and deviation from the Born approximation. It should be stressed that we are not interested in absolute values of the resistivity but only in its temperature dependence in the high temperature limit. Therefore our analysis makes no assumption about how to calculate numerically the electron-phonon matrix elements.

2. The Resistivity of Metals

a) Some Basic Relations

We follow Green and Kohn [5] and write for the phonon limited electrical resistivity ϱ (also see Baym [6] and Ziman [7])

$$\varrho = \text{const.} \iint (d^2 S_k/v_k)\,(d^2 S_{k'}/v_{k'})\,\{\Phi_k - \Phi_{k'}\}^2 \,|f(k, k')|^2 \,\frac{S'(q\,\omega)\,\beta\,\omega\,d\omega}{1 - e^{-\beta\omega}}. \tag{1}$$

This is the expression one arrives at in a variational solution of the transport equation using the Born approximation. The only important point for our discussion is how the temperature T enters ($\beta = h/k_\mathrm{B}T$). $S(q, \omega)$ is the spectral function for the lattice displacements, the general definition being

$$S(q, \omega) = (1/2\pi N) \int dt\, e^{-i\omega t} \left\langle \sum_{i,j}^{N} e^{-i\,q\cdot R_i(0)}\, e^{i\,q\cdot R_j(t)} \right\rangle_T. \tag{2}$$

The quantity S' in Eq. (1) is equal to S minus the Bragg peak that does not contribute to the resistivity. $R_i(t)$ is the position of the i-th atom at time t and $\langle\ \rangle_T$ denotes a thermal average. $S'(q, \omega)$ can in principle be obtained from neutron scattering experiments [6] and will then include all effects of one- and multi-phonon scattering, Debye-Waller factors etc. but the experimental accuracy is much too low to be of any help in resistivity calculations for solid metals. In practice one has to resort to approximate forms for $S'(q, \omega)$ and it is the validity of these approximations that we are to study in this paper. In the so-called one-phonon approximation and with the neglect of Debye-Waller factors and anharmonicity we have

$$S'(q, \omega) = \sum_p \frac{|\xi_p \cdot q|^2}{e^{\beta\omega}-1}\, A_{qp}(\omega) \tag{3}$$

with

$$A_{qp}(\omega) = (1/2\,M\,\omega_{qp})\,[\delta(\omega - \omega_{qp}) - \delta(\omega + \omega_{qp})], \tag{4}$$

where ξ_p ($p = 1, 2, 3$) are the polarization vectors (longitudinal or transverse vibrations; we have for simplicity assumed one atom per unit cell). The frequency of the mode (q, p) is denoted by $\omega(q, p)$. With Eq. (3) for the spectral function and with $f(k, k')$ calculated in the Born approximation we arrive at a form of Eq. (1) that has been the starting point of almost all numerical calculations for solids and is usually referred to as Ziman's formula [7]. Going to the limit of high temperatures in the expression for ϱ [Eqs. (1) and (3)] we can make an expansion in T^{-2},

$$\varrho/T = \varrho_0 - a\,T^{-2} + b\,T^{-4}. \tag{5}$$

It is seen that the constant ϱ_0 involves an average over ω^{-2} and we therefore write

$$\varrho_0 = R/\Theta_{\mathrm{R}}^2 , \tag{6}$$

where Θ_{R} is a "Debye" temperature relevant for the electrical resistivity and R is a constant. Our interest will be focused on the temperature dependence of the parameter ϱ_0.

b) Validity of the Transport Equation and the Born Approximation

Several comments must be made concerning the validity of Eq. (1). It has been assumed that the ordinary weak-coupling Boltzmann equation is accurate enough. It is well known [8] that this is the case if $(\hbar/\tau E_{\mathrm{F}}) \ll 1$ (or alternatively $1/l k_{\mathrm{F}} \ll 1$). Here E_{F} and k_{F} are the Fermi energy and wave vector respectively and τ and l are the life time and mean free path of the conduction electrons. For the metals under consideration and close to the melting point we have $1/l\, k_{\mathrm{F}} \sim 0.01$ and corrections to the transport equation can not be ruled out. Rubio [9] tried to give a non-perturbative solution in closed form for the strong scattering case which could be used for numerical estimates of the validity of Eq. (1). Unfortunately, Rubio's result has been shown to be correct only to lowest order [10] and the problem treated by him seems to be still unsolved. The accuracy of the transport equation is a more serious problem in liquid metals and has received attention in this context. Faber [11, 12] has discussed in some detail corrections due to the short electron mean free path. In particular one finds that the elementary Ziman formula for the resistivity of liquid metals gives quite reasonable agreement with experiment even for metals where the mean free path is only a few atomic diameters. For liquids the Debye-Waller factor, multi-phonon scattering and anharmonicity are in principle all included in the spectral function and thus require no further correction. It is therefore most unlikely that the simple form of the transport equation should not be accurate enough for the metals we consider.

Another important point is whether the scattering probability is correctly given by the Born approximation. This problem is not quite independent of the previous one as a strong potential may cause a short mean free path and failure of the Born approximation as well. There seems to be no quantitative estimate of the validity of the Born approximation for solids but again the problem has been extensively dealt with for liquids (see e.g. Refs. [11—19]). The least one can say is that the results are as yet inconclusive. In our analysis we make no assumption about how to calculate the transition probability. There may be an average correction to the Born approximation that is temperature independent and in addition there must be terms in a power series of T. It is only the latter terms that affect the temperature dependence of the resistivity and hence show up in our analysis of experimental data. One could refer to the relative success of the Ziman formula even for those liquid metals that have a strong electron-ion interaction and argue that there is no particular reason for a temperature dependent failure of the Born approximation in the metals under consideration. We still include this possibility via a fudge factor $1 + B(T)$ that corrects the resistivity given by Eq. (1).

It has been assumed that the matrix element $f(\boldsymbol{k}, \boldsymbol{k}')$ calculated in the Born approximation is temperature independent. Implicit in this assumption is the

concept of rigid ions, i.e. the interaction between an electron and an individual
ion is not distorted by the thermal vibrations. If it were not so, one would expect
a qualitative difference between the alkali metals and the noble metals, the latter
having larger electron cores. No such difference is observed. In appendix 2 it is
further argued that neither thermal Fermi level smearing nor finite electron life
time effects will affect the dielectric screening constant in the electron-ion potential.

Some more comments can be made. The adiabatic approximation is known to
hold to order $(m_{el}/M_{ion})^{1/2}$ [4] and corrections to this approximation are neglected.
It is also known that the scattering is correctly described to the same order by
the weak coupling approximation i.e. without any many body electron-phonon
renormalisation effects (cf. Prange and Kadanoff [20] and Baym [6]). At the high
temperatures of interest to us, "phonon drag" effects are completely negligible
and the phonon system can be described by the usual Bose-Einstein statistical
factors [21].

c) The Debye-Waller Factor and Multi-Phonon Scattering

It is evident that for plane electron waves the one-phonon scattering processes
in the resistivity should be reduced by a Debye-Waller factor $\exp(-2W(K))$ for
scattering with wave vector transfer K. $W(K)$ has the form

$$W(K) = (\hbar/4MN) \sum_{qp} \frac{|\xi_{qp} \cdot K|^2}{\omega_{qp}} (2n_{qp}^{ph} + 1). \tag{7}$$

Note that $W(K)$ depends quadratically on $|K|$. The scattering is between two
states on the Fermi surface. If we assume that the conduction electrons can be
described by single plane waves, we have $|K| = |k - k'| \leq 2k_F$. In this case,
and for a metal close to the melting point, we can expect a reduction in the
resistivity by some 15% [3]. Bailyn [22] has pointed out that the inclusion of
terms beyond the single plane wave approximation for the electrons can drastically
increase the net effect of the Debye-Waller factor. Let us expand the Bloch
electron wave function in a set of plane waves,

$$\Psi_k(r) = \sum_{G} C_G \exp[i(k + G) \cdot r], \tag{8}$$

where the sum goes over reciprocal lattice vectors G. The matrix elements for
thermal scattering of electrons, $\langle k | H_{el-ph} | k' \rangle$, will be built up of a sum of
matrix elements for plane waves but with a wave vector transfer that may
exceed $2k_F$. Bailyn makes explicit calculations for a particular pair of states
(k, k') for caesium and finds that the effect of Debye-Waller factors may very well
be doubled using the expansion (8) instead of the single plane wave approximation.
The estimate is uncertain, and Bailyn had difficulties to get convergence for the
expansion coefficients C_G, but nevertheless we learn that the Bloch wave character,
which permits large momentum transfers in the scattering, may lead to a much
larger temperature factor than is obtained e.g. in X-ray scattering. At high
temperatures $W(K)$ is linear in T, and therefore the shift in the resistivity due to
the Debye-Waller factors goes as T^2. Up to now we have only considered one-
phonon scattering processes. There must also be multi-phonon scattering, e.g.
three phonon processes where two phonons are created in the modes q and q' and

one is annihilated in the mode q''. At high temperatures, the two-phonon processes give an increase in the resistivity proportional to T^2, the three-phonon processes proportional to T^3 etc. It is very difficult to calculate even the effect of two-phonon scattering. Franzak and Bailyn [23] have made an estimate of this term for the alkali metals at room temperature and they find an enhancement in the resistivity over the one-phonon result by 4%. However, this estimate must be considered to be very uncertain. The terms just referred to correspond to so-called "intrinsic" multi-phonon processes. There are also other multi-phonon processes corresponding to higher order terms in the perturbation expansion. They have been included in the factor $1 + B(T)$ as discussed in the previous section. Sham and Ziman [4] noticed that a series expansion of the Debye-Waller factor and the intrinsic multi-phonon scattering in terms of creation and annihilation operators leads to two expressions that have many terms in common but with opposite signs. Thus the effects of the Debye-Waller factor and the intrinsic multi-phonon scattering tend to cancel, and Sham and Ziman remarked that there might even be exact cancellation. An analytical analysis of the problem seems to be a formidable task and we will use semi-empirical methods.

d) Anharmonicity

The temperature dependence of the electrical resistivity given in the expansion of Eq. (5) comes entirely from the Fermi-Dirac and Bose-Einstein statistical factors that enter the expression for the ideal scattering process. In the last section we saw that an additional temperature dependence arises from Debye-Waller factors and multi-phonon scattering and these two effects can be dealt with assuming perfectly harmonic lattice vibrations. For an anharmonic system an additional temperature dependence may arise from the fact that the peaks in the spectral function are no longer delta functions at temperature independent positions but are broadened, shifted and may even show considerable skewness and structure. In its general form we can write for $A_{qp}(\omega)$ [24, 25] (with Debye-Waller factors and multi-phonon scattering suppressed)

$$A_{qp}(\omega) = \frac{1}{\pi M} \frac{2\,\omega_{qp}\,\Gamma_{qp}(\omega)}{[\omega^2 - \Omega_{qp}^2]^2 + 4\,\omega_{qp}^2\,\Gamma_{qp}^2(\omega)} \tag{9}$$

with

$$\Omega_{qp}^2 = \omega_{qp}^2 + 2\,\omega_{qp}\,\Delta_{qp}(\omega). \tag{10}$$

Here ω_{qp} are the eigen-frequencies of the strictly harmonic Hamiltonian that results from retaining only the first (quadratic) term in an expansion of the crystal energy in the displacements of the lattice sites. The inclusion of anharmonic terms leads to a complex self energy $\Delta_{qp}(\omega) + i\Gamma_{qp}(\omega)$. Note that both the real and imaginary parts of the self energy are frequency dependent and may thus lead to a very complicated spectral function. As a first approximation we could neglect this frequency dependence, i.e. take Δ_{qp} and Γ_{qp} to be constants for a given mode (qp). The resulting $A_{qp}(\omega)$ will have almost Lorentz shaped peaks,

$$A_{qp}(\omega) = \frac{1}{\pi M} \frac{2\omega_{qp}\,\Gamma\,\mathrm{sgn}\,(\omega)}{[\omega^2 - \omega_{qp}^2 - 2\omega_{qp}\,\Delta]^2 + 4\omega_{qp}^2\,\Gamma^2}. \tag{11}$$

If the self energy is small compared to ω_{qp} and only slowly varying with the frequency ω, it can be a good approximation to set

$$\Delta_{qp}(\omega) = \Delta_{qp}(\omega_{qp}); \qquad \Gamma_{qp}(\omega) = \Gamma_{qp}(\omega_{qp}). \tag{12}$$

This gives exactly the form (11), where we note the fact that $\Gamma(\omega)$ is odd in ω while $\Delta(\omega)$ is an even function. Calculations have shown (12) to be a reasonable approximation in many cases, even fairly close to the melting temperature, but we will elaborate more on this point in a later section. Of more importance to us is the temperature dependence of the self energy. In passing, we remind of the fact that we are considering the resistivity at constant volume, and therefore all effects of thermal expansion will be excluded. Theoretical calculations of anharmonicity usually include the cubic and quartic terms of the crystal energy expanded in lattice point displacements. For the quartic contribution to the real part of the self energy we can write in compact notation [26]

$$\Delta_{qp}^4 = \frac{1}{N} \sum_{\text{all modes}} (2n^{\text{ph}} + 1)\, V^4 \tag{13}$$

where V^4 contains the fourth derivatives of the crystal energy. The temperature dependence enters via the Bose-Einstein statistical factor n^{ph} and thus Δ^4 is linear in temperature for $T > \Theta_{\text{D}}$. The expression for the cubic term Δ^3 and the imaginary part of the self energy are somewhat more complicated than Eq. (13) but they both give a contribution that is linear in T for $T > \Theta_{\text{D}}$. Let us now consider how anharmonicity modifies the temperature dependence of the electrical resistivity. It is evident that the dominating effect for $T > \Theta_{\text{D}}$ is a temperature dependence of the asymptotic limit ϱ_0 in the expansion of Eq. (5). When $A_{qp}(\omega)$ is approximated by the almost Lorentz shaped peaks of Eq. (11), the calculation is straight forward (Appendix 1). For each individual phonon mode, the contribution to the resistivity is changed by a factor

$$(1 - 2\Delta_{qp}/\omega_{qp})(1 - 4\Gamma_{qp}^2/\omega_{qp}^2). \tag{14}$$

Calculations have shown [26, 27, 28] that 0.03 is a reasonable order of magnitude estimate for Δ_{qp}/ω_{qp} as well as Γ_{qp}/ω_{qp} in metals at some 0.8 of the melting temperature. Thus the shift Δ_{qp} in the positions of the one-phonon peaks can be expected to be one or two orders of magnitude more important than their broadening as long as the Lorentz form is maintained. The two factors in the expression also differ in their temperature dependence. The shift gives a correction to the resistivity that is linear in T while the broadening leads to a T^2-term. The fact that the shift Δ_{qp} in principle may have either sign will in general not change our qualitative statements. The true peaks in the spectral function may deviate from the Lorentz form, but one easily finds [Appendix Eq. (A.7)] that for peaks that are shifted and then broadened symmetrically around the new positions we have much the same result as for true Lorentz peaks. For skew peaks or for peaks with considerable structure, i.e. more strongly energy dependent self energies, the situation is so complicated that general statements are difficult to make.

The conclusion to be drawn is that the most important effect of anharmonicity on the electrical resistivity is a shift in the position of the one-phonon peaks in the spectral function. Following the notation of Eq. (6) we can write the corresponding correction factor to the resistivity at high temperatures as

$$1 - 2\Delta\Theta_R/\Theta_R, \tag{15}$$

where $\Delta\Theta_R$ can have either sign.

Summing up the contributions from lattice anharmonicity, Debye-Waller factors, multi-phonon scattering and deviations from the Born approximation we find that the resistivity should change by a factor

$$1 + \Delta\varrho/\varrho = (1 - 2\Delta\Theta_R/\Theta_R)\exp(-2W)f(T)h(T)(1 + B(T)). \tag{16}$$

The first two factors on the right hand side refer to the anharmonic shift in the phonon frequencies and the Debye-Waller factor. The fact that anharmonicity also gives a broadening and structure in the one-phonon peaks of the spectral function $S'(q, \omega)$ is taken into account via the factor $f(T)$. This term will also include anharmonic effects that go beyond the third and fourth order terms. The effect of multi-phonon scattering is entirely in $h(T)$. It should be mentioned that for an anharmonic crystal there are interference effects between the one-phonon and multi-phonon scattering [25, 29]. We let the possible influence of this interference be included under our heading multi-phonon scattering. $1 + B(T)$ represents corrections to the Born approximation.

From experiments we will extract the quantity $1 + \Delta\varrho/\varrho$. At low temperatures, $T < \Theta_D$, it is close to unity. If follows from our previous discussion that at higher temperatures, $\Delta\varrho/\varrho$ will first be linear in T and then get a more complicated from as higher order corrections to the various terms in (16) become important. Resistivity measurements alone only give the combined effect (16) and to say anything about the individual factors, e.g. about the cancellation between Debye-Waller factors and multi-phonon scattering, we need further experimental information.

e) Thermal Lattice Defects

Close to the melting point there might be enough thermally created defects to noticeably affect the resistivity. We assume that there is only one type of defects, monovacancies. The extra resistivity $\Delta'\varrho$ is linear in the defect concentration c,

$$\Delta'\varrho = c\,\varrho_{vac}. \tag{17}$$

In equilibrium at temperature T and at zero pressure we have

$$c = n/N = \exp(\Delta S/k)\exp(-Q/kT), \tag{18}$$

where n is the number of defects, N the number of lattice sites, ΔS the entropy change associated with the lattice relaxation around the defect and Q is the formation energy per vacancy. Our experimental data refer to the situation when an external pressure is applied so that the volume of the crystal is kept constant with increased temperature. Thermodynamics easily gives [30]

$$c(P, T) = c(P = 0, T)\exp(-PV_{vac}/kT), \tag{19}$$

where V_{vac} is the formation volume per vacancy and P the applied pressure.

Apart from point defects one could in principle have fluctuating local transformations to other crystal structures. Also, it has been suggested by Kuhlmann-Wilsdorf [31] that melting may be caused by the thermal generation of dislocations when the free energy of dislocations cores becomes zero. However, it was also shown [31] that the thermally generated dislocation density is completely negligible below the melting point. Any excess of dislocations introduced e.g.by cold working of the specimens will be rapidly annealed out at the high temperatures of interest to us. It should also be mentioned that measurements of the electrical resistivity in sodium by Bradshaw and Pearson [32] just around the melting point showed no remarkable temperature behaviour apart from the discontinuity at the phase transformation. We therefore completely neglect the influence of other lattice defects than vacancies for the rest of this paper.

3. Analysis of Experimental Data

a) Potassium

We first discuss the measurements by Dugdale and Gugan [1] for potassium. They measured the electrical resistivity and its pressure dependence from 8 to 295K. From values in the literature on the compressibility and the temperature variation of the lattice parameter they also determined the equation of state i.e. a relation between P, V and T. Thus they could deduce the electrical resistivity at constant volume. The only analysis made in their original paper was the calculation of an effective $\Theta_R(T)$ that fitted the data to the well known Block-Grüneisen formula for the electrical resistivity. When combined with recent measurements of other quantities, the accurate data of Dugdale and Gugan can be used for a detailed analysis of the high temperature resistivity. Following the lines of the preceding discussion (cf. Eq. (5)), we plot ϱ/T as a function of $1/T^2$. If it were not for the corrections $\Delta\varrho/\varrho$ and $\Delta'\varrho$ (Eqs. (16) and (17)), the points would asymptotically approach a straight line for high temperatures. Our analysis requires that asymptote to be drawn. This is not trivial, for as we approach the asymptotic limit in the ideal resistivity, the corrections $\Delta\varrho/\varrho$ start to become important. Our solution is based on a fitting procedure as follows: For temperatures $T < \Theta_R$, the factor $1 + \Delta\varrho/\varrho$ should be fairly close to unity. On the other hand the a/T^2 and higher terms in the expansion of ϱ/T, Eq. (5), are important. Thus the leading term in the difference between the asymptote and the experimental data for $T < \Theta_R$ should be a/T^2. Further, we can get a reasonable order of magnitude estimate of b from the Bloch-Grüneisen formula. These two conditions allow us to draw the asymptote with good accuracy. The slight uncertainty is of no importance for the qualitative results of this paper. Note that although the experimental points for K below 100 K (Fig. 1) seem to lie approximately on a straight line, this is not the asymptote as the two conditions mentioned are strongly violated.

For potassium we find a small increase in ϱ/T at high temperatures ($T > T_m/2$, Figs. 1 and 2) which might be thought to come from vacancies. Unfortunately, there seems to be no published experiments giving ϱ_{vac} for alkali metals and there is also very little information on Q. For sodium, Feder and Charbnau [33] have measured the formation energy 0.42 eV and a defect concentration $0.75 \cdot 10^{-3}$

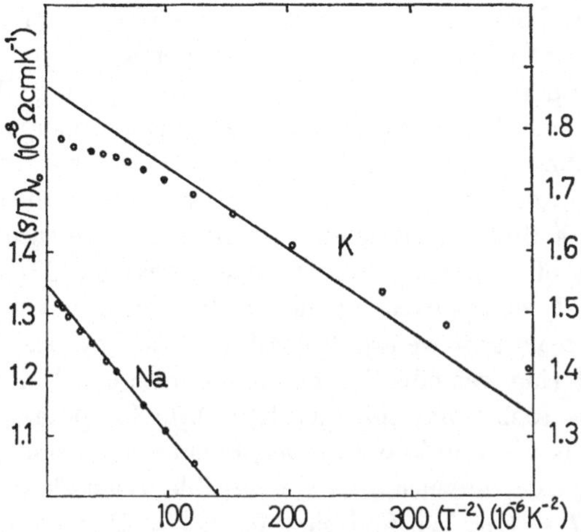

Fig. 1. The electrical resistivity ϱ (given as ϱ/T) as a function of T^{-2} for potassium and sodium at constant volume. Experimental values are due to Dugdale and Gugan [1]. The straight lines are the theoretical asymptotes if anharmonicity, multi-phonon scattering and Debye-Waller factors are neglected and the Born approximation is valid. The experimental accuracy is of the order of the diameter of the circles. The left and right vertical axes refer to sodium and potassium, respectively

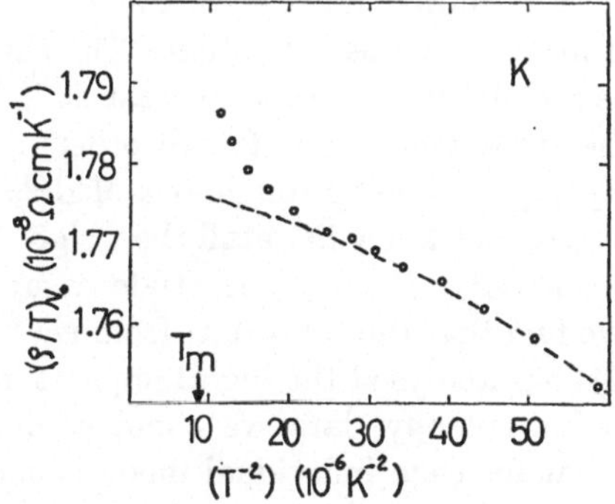

Fig. 2. The resistivity at constant volume for potassium at high temperatures. The dashed curve is a smooth extrapolation from the experimental points at lower temperatures. There is a marked increase in ϱ/T starting around $T = 225\,°K$. The arrow points at the melting point T_m

at the melting point. This formation energy and defect concentration is in good agreement with values obtained by Martin [34] in his analysis of the high temperature thermal expansion of sodium. (It should here be remarked that Martin gives a lattice expansion which is somewhat different from that used by Dugdale and Gugan. The difference is almost constant in the high temperature region and is therefore of no importance for our analysis.) Martin also finds that potassium has almost the same vacancy formation energy and concentration of defects at the melting point as sodium. Assuming the reasonable value $\varrho_\mathrm{vac} = 1\ \mu\,\Omega\,\mathrm{cm/at.\%}$ and V_vac equal to half the atomic volume (cf. data for Au in Ref. [30]) one finds

that the resistivity from vacancies should be two orders of magnitude smaller
than what is necessary to account for the extra increase in ϱ/T (cf. Fig. 2). It also
turns out to be impossible to fit this increase to anything like an exponential
temperature behaviour if one does not take the formation energy Q as low as 0.1 eV,
which is an unrealistic value. We therefore conclude that the scattering from
vacancies can be neglected altogether in our subsequent discussion. For completeness we remark that measurements (at constant pressure) of the high temperature resistivity of sodium by Bradshaw and Pearson [32] showed an upward
curvature when ϱ/T was plotted against T. The volume expansion makes the
situation different from that we are discussing. Bradshaw and Pearson suggested
that vacancy scattering was effective but no definite conclusion could be drawn.

From Fig. 1 we easily find the quantity $\Delta\varrho/\varrho$ for potassium. The result is
plotted in Fig. 3. As we only know the *asymptotic* behaviour of the "ideal" (i.e.
Eq. (5)) temperature dependence, it is not possible to get $\Delta\varrho/\varrho$ with any accuracy
for temperatures $T < \Theta_\mathrm{R}$. We find that for $\Theta_\mathrm{R} < T < 2\Theta_\mathrm{R}$, $\Delta\varrho/\varrho$ is linear in
T as expected. For still higher temperatures we see deviations from this linear
behaviour. For our analysis of $\Delta\varrho/\varrho$ it is essential to have independent information
on the average shift in the phonon frequencies, i.e. $\Delta\Theta_\mathrm{R}/\Theta_\mathrm{R}$. Buyers and Cowley
[26] have calculated the third and fourth order shifts in the phonon frequencies
of potassium at 9, 92, 215 and 299 K using a pseudopotential approach and found
reasonable agreement with their own neutron scattering measurements. The data
are of course restricted to a limited number of modes, most of them being for
the longitudinal branch in the principal directions. The shifts differ considerably
between different modes and are uncertain to at least 20% like in all calculations
of this type. Most of the shifts (third plus fourth order) are negative but a few
are positive. We can only expect to get a rough idea of $\Delta\Theta_\mathrm{R}/\Theta_\mathrm{R}$ from Buyers' and
Cowley's data, because we do not know in detail the weight by which the different
modes enter the total resistivity. Making the crude assumption that all modes
have the same weight, we find that the total shift from Θ_R to 295 K is $\Delta\Theta_\mathrm{R}/\Theta_\mathrm{R} =
-\,(5 \pm 3)\%$. Martin [34] has analysed the high temperature heat capacity of the
alkali metals and from his entropy data we could estimate the average shift
$\Delta_3 + \Delta_4$. It is known that for each individual mode it is exactly the frequency

Table 1. The difference $\Delta\varrho/\varrho$ between the theoretical prediction of the Ziman formula for the
resistivity and the experimental data at constant volume at some 15% from the melting
temperature. In the right column is shown the same difference after a correction of the
theoretical result for the anharmonic frequency shift at constant volume

	$\Delta\varrho/\varrho$ uncorrected (%)	$\Delta\varrho/\varrho$ after correction for lattice anharmonicity (%)
K	$-\,4$	$-\,12$
Na	$-\,1$	$-\,6$
Cu	$-\,3$	$-\,2$
Ag	$-\,10$	$-\,6$
Au	$-\,6$	$+\,9$

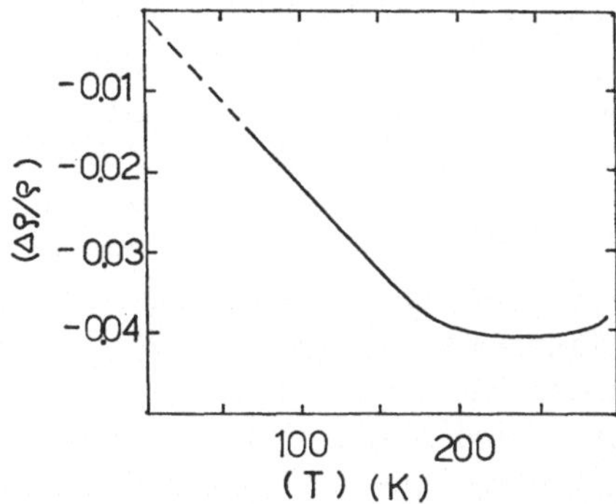

Fig. 3. The quantity $\Delta\varrho/\varrho$ (Eq. (16)) as a function of temperature for potassium

shifts we discuss that go into the mode contribution to the entropy [35]. Neglecting the different weighting of the phonon modes in the resistivity and the entropy, Martins' values would give $\Delta\Theta_R/\Theta_R = - (1.5 \pm 0.5)\%$ for potassium in the interval Θ_R to 295 K. Wallace [36] has also analysed high temperature entropy data and finds $\Delta\Theta_R/\Theta_R = - 4\%$. Finally Kaveh and Wiser [37] in a calculation of the difference between the resistivity at constant pressure and constant volume for potassium and sodium obtained as an estimate for potassium $\Delta\Theta_R/\Theta_R = - 3\%$. This value is based on the temperature and pressure dependence of the elastic constants. Thus there is some uncertainty about the average shift to be used but there is no doubt that it is *negative*, resulting in a positive $\Delta\varrho/\varrho$. Note that the experimental data show unambigously that $\Delta\varrho/\varrho$ is negative at high temperatures. In Table 1 we give the total experimental value for $\Delta\varrho/\varrho$ and also the remaining $\Delta\varrho/\varrho$ when the correction due to $\Delta\Theta_R/\Theta_R$ has been included. The frequency shifts are all from the entropy analysis by Wallace.

We next consider $f(T)$ which has been taken to incorporate the effects of phonon damping, structure in the phonon spectral function and also frequency shifts higher than fourth order. In our discussion of the spectral function it was found that as long as the peaks were Lorentz shaped, their broadening at higher temperatures was expected to be one order of magnitude less important than the effect of the frequency shifts even at high temperatures. A considerable structure or skewness in the spectral function could invalidate this result. However, the calculations of Buyers and Cowley showed that the phonon peaks were very close to Lorentz shaped even at 299 K. Qualitatively the same result has been obtained for aluminium in calculations by Koehler, Gillis and Wallace [27, 28] and also by Sandström (private communication). Further, any symmetric broadening of the peaks in the spectral function would give a T^2-contribution to $\Delta\varrho/\varrho$ whereas the experimental data (Fig. 3) start out linear in T. Thus it seems unlikely that a finite width or structure in the spectral function will be of any real importance for the resistivity. It is more difficult to make any statements about higher order shifts in the phonon frequencies. Theoretical calculations never go beyond the fourth order, but this is due to numerical difficulties. One might expect higher order terms to be important for temperatures $T > 2\Theta_R$. This is equivalent to saying that for these temperatures we expect a non-linear behaviour in $\Delta\varrho/\varrho$.

It now remains to discuss the Debye-Waller factor, multi-phonon scattering and the Born approximation. From Fig. 3 we see that $\Delta\varrho/\varrho$ is negative, $-\,0.04$ at 295 K, whereas the effect of phonon frequency shifts is expected to be a *positive* contribution of some 0.08 at the same temperature. Thus the experiments require some other effects that are linear in T in their leading term. If we dare to trust the Born approximation, the natural explanation is that the Debye-Waller factor does not cancel against the multi-phonon scattering. In view of Bailyn's paper on $\exp(-\,2W)$ [22] and Franzak's and Bailyn's [23] on two-phonon scattering (cf. our previous discussion) this is not unreasonable. Thus we have a simple explanation for the linear behaviour of $\Delta\varrho/\varrho$ up to 200 K. Above this temperature there is a much different T dependence in $\Delta\varrho/\varrho$ (Fig. 3). The origin of this behaviour is very complicated to analyse. A probable cause is still higher order terms in the phonon frequency shifts. Other possibilities are higher order terms in the Debye-Waller factor, the multi-phonon scattering and failure of the Born approximation. It is interesting to note that the non-linear T-dependence in $\Delta\varrho/\varrho$ can not be fitted to a T^2-term (cf. Fig. 3). Therefore it would not suffice in an explanation to include only the next term (in orders of T) in the shifts etc. For the same reason it could not be the effect of anharmonicity in the Debye-Waller factors and multi-phonon scattering. This last conclusion is further supported by the fact that the average frequency shift for potassium was negative. Therefore when this shift is inserted into the Debye-Waller and multi-phonon terms (cf. Eq. (7)) it gives a change in the resistivity which is not only too small compared to the interesting feature in Fig. 3 but also has the wrong sign.

b) Sodium

The resistivity data for sodium [1] confirm the qualitative results obtained for potassium, but the data end at 295 K, i.e. at too low temperatures (compared to the Debye temperature or the melting point) to show the features at high temperatures found for potassium. Glyde and Taylor [38] have recently calculated the anharmonic properties of sodium. Their approach is somewhat different in principle from that used by Cowley *et al.* [29] for potassium but the outcome is much the same. Most of the phonon peaks seem to be reasonably well described with Lorentz peaks at temperature dependent positions. The frequency shifts differ considerably from mode to mode and are not all of the same sign. From Martin's heat capacity data we estimate the average shift from Θ_R to 295 K to be $-\,(2\pm0.5)\%$ and Glyde's and Taylor's calculation is not inconsistent with this value. The shift in Θ_R obtained by Kaveh and Wiser [37] from elastic constants is $-\,4.2\%$. The high temperature entropy analysis by Wallace [36] gives $-\,3.5\%$. Again it is no doubt that the average shift in the phonon frequencies is negative when the temperature rises, thus increasing the difference between the experimental resistivity data and the Ziman formula (see Table 1).

c) Copper, Silver and Gold

Bailyn [22] argued that it was when one abandoned the single plane wave approximation that the Debye-Waller factor became really important. It would therefore be interesting to consider the noble metals where the plane wave

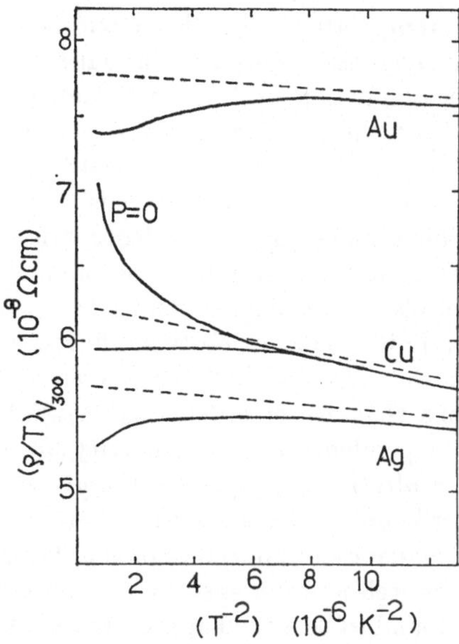

Fig. 4. The electrical resistivity ϱ (given as ϱ/T) as a function of T^{-2} for copper, silver and gold. The experimental results at constant pressure ($P = 0$, only shown for Cu) are due to Laubitz [2, 3]. They have been reduced to constant volume ($V = V_{300}$). The straight lines represent the asymptote in the most simple theory (cf. Fig. 1)

approximation is poor but the conduction bands still are fairly simple so that our basic assumptions are valid. Laubitz [2, 3] has measured the resistivity of the noble metals from 300 to 1200 K (Ag 1100 K) at ordinary pressure. We take his data and reduce them to constant volume using X-ray data for the lattice parameter [39] and the volume dependence of the resistivity given by Dugdale [40]. The result is shown in Fig. 4. Tentative asymptotes are drawn as in Fig. 1. Although the resulting data are not as accurate as for the alkali metals, we immediately see that the general behaviour is the same. The discrepancy between experiment and the standard Ziman formula is very similar for Na, K, Cu, Ag and Au. The interesting point is that although the single plane wave approximation for the conduction electrons in the noble metals is supposed to be less good, this has not resulted in a large $\Delta\varrho/\varrho$. This fact gives partial support to the validity of the "rigid ion" approximation. One may have some doubts about its validity at high temperatures when the amplitude of the lattice vibrations is large. However one also expects a difference in this respect between the alkali metals with small and the noble metals with large ion cores. Data on lattice anharmonicity from Wallace [36] can be used to correct for the shifts in the phonon frequencies just as was made for potassium and sodium. The result is summarized in Table 1.

4. Conclusions

We have analysed high temperature data (reduced to *constant volume*) for the electrical resistivity of the five metals K, Na, Cu, Ag and Au in order to test the validity of the formula used in conventional resistivity calculations. The interest has been focused on the temperature dependence and not on the absolute values of the resistivity. The analysis essentially goes in four steps. First we plot experimental data for ϱ/T as a function of $1/T^2$ and compare with the asymptotic behaviour that results from the standard resistivity formula in which the only temperature dependence enters from one-phonon scattering in a strictly harmonic lattice. The discrepancy between these two curves is what we want to study. We next argue that negligible temperature dependence arises from thermal lattice defects (vacancies, local structure instabilities or premelting generation of dislocations). Also the dielectric screening is shown to be unaffected by thermal Fermi level smearing or finite electron life time effects. The electron mean free path is long enough for the standard Boltzmann equation to be valid. Further we neglect all corrections that are of order $(m_{el}/M_{ion})^{1/2}$. The major sources of additional temperature dependence in the resistivity are lattice anharmonicity, Debye-Waller factors, multi-phonon scattering and corrections to the Born approximation. As a third step we treat lattice anharmonicity. We take a semi-empirical approach and make use of information on anharmonicity obtained from theoretical calculations as well as from specific heat measurements and inelastic neutron scattering. It is found to be a good approximation to neglect finite phonon life time effects and represent the phonons with sharp frequencies that are temperature dependent even at constant volume. After correcting the resistivity for lattice anharmonicity (cf. Table 1), we are left with the fourth and last step, the discussion of Debye-Waller factors, multi-phonon scattering and the Born approximation. The net effect of these terms at some 15% below the melting temperature is summarized in the right column of Table 1. One might have some confidence in the Born approximation at least for the alkali metals Na and K. Assuming its validity, it follows that the cancellation between the Debye-Waller factor and multi-phonon scattering is only approximate. This is in disagreement with a conjecture by Sham and Ziman [4] that the cancellation might be complete. A closer numerical analysis of Debye-Waller factors, multi-phonon scattering and the Born approximation is a formidable task and beyond the scope of this paper. Summing up, the main conclusions are the following:

1) The standard resistivity formula in the one-phonon approximation (often referred to as the Ziman formula) overestimates the temperature dependence of the high temperature electrical resistivity at *constant volume* for alkali (Na, K) and noble (Cu, Ag, Au) metals by up to some 10% (Table 1).

2) The major reasons for the discrepancy between the elementary theory and experiments are the neglect of Debye-Waller factors, multi-phonon scattering, anharmonic effects and the possible failure of the Born approximation. These effects can all be of the same order of magnitude and partially cancelling.

3) The lattice anharmonicity at constant volume is properly taken into account by a shift in the harmonic phonon frequencies but this correction alone does *not* in general give better agreement between theory and experiment (Table 1). This fact should be contrasted to case of the high temperature resistivity at constant *pressure* where phonon frequency shifts due to the thermal expansion account for most of the difference between experiment and the elementary theory [37]. There is clear evidence in potassium for corrections higher than linear in the

temperature (presumably due to anharmonicity) but these corrections are very small even as close as 20% from the melting temperature.

4) It has not been possible to separate the effects of the Debye-Waller factor, multi-phonon scattering and possible failure of the Born approximation. Each of these terms requires very lengthy calculations for numerical estimates. However, if the Born approximation is valid, there can be no complete cancellation between Debye-Waller and multi-phonon scattering corrections as has been suggested by Sham and Ziman [4].

Acknowledgements. This work has been supported by the Swedish Board for Technical Development. Discussions with N. Ashcroft, R. Sandström and A. Sjölander are gratefully acknowledged. I also want to thank S. Dugdale and J. Morgan for reading a preliminary version of the manuscript.

Appendix A

According to Eq. (9) the function $A_{qp}(\omega)$ has the form

$$A_{qp}(\omega) = \frac{1}{\pi M} \frac{2\,\omega_{qp}\,\Gamma_{qp}(\omega)}{[\omega^2 - \Omega_{qp}^2]^2 + 4\,\omega_{qp}^2\,\Gamma_{qp}^2(\omega)}\,. \tag{A.1}$$

An approximation to (A.1) is the Lorentz form

$$A_{qp}(\omega) = (2 M \Omega \pi)^{-1}\left[\frac{\Gamma}{(\omega - \Omega)^2 + \Gamma^2} - \frac{\Gamma}{(\omega + \Omega)^2 + \Gamma^2}\right] \tag{A.2}$$

where Ω and Γ are constants. Any form of $A(\omega)$ must fulfill the sum rule [41] (Debye-Waller factors and multi-phonon scattering suppressed)

$$\int_{-\infty}^{\infty} \omega A(\omega)\,d\omega = M^{-1}\,. \tag{A.3}$$

We are here interested in the average (cf. Eqs. (1) and (3))

$$\int_{-\infty}^{\infty} \omega^{-1} A(\omega)\,d\omega\,. \tag{A.4}$$

Integrals of the type (A.3) and (A.4) can be handled by the method of residues. Let us assume in (A.1) that $\Delta_{qp}(\omega)$ and $\Gamma_{qp}(\omega)$ are slowly varying as a function of ω and that the poles given by this approximation are the only ones present. It is then trivial to check that the sum rule is fulfilled. For the integral (A.4) we get from (A.1)

$$\begin{aligned} M^{-1}\Omega^{-2}&[1 + 4\,\omega_{qp}^2\,\Gamma^2(\omega_{qp})/\Omega^4]^{-1} \\ &\approx M^{-1}\,\omega_{qp}^{-2}[1 - 2\Delta(\omega_{qp})/\omega_{qp}][1 - 4\Gamma^2(\omega_{qp})/\omega_{qp}^2]\,. \end{aligned} \tag{A.5}$$

If we start from the Lorentz function (A.2) instead we find

$$M^{-1}\Omega^{-2}[1 + \Gamma^2/\Omega^2]^{-1} \approx M^{-1}\,\omega_{qp}^{-2}[1 - 2\Delta/\omega_{qp}][1 - \Gamma^2/\omega_{qp}^2]\,. \tag{A.6}$$

Thus there is a slight difference in the effect of broadening between the two approximations for $A_{qp}(\omega)$ but this effect is negligible anyhow compared to that involving Δ. If we now take any form of $A(\omega)$ that consists of two non-overlapping peaks that are symmetric in ω around $\pm\,\Omega$ we find

$$\int\limits_{-\infty}^{\infty}\omega^{-1}A(\omega)\,d\omega = 2\int\limits_{0}^{\infty}\omega\,(\Omega+\omega-\Omega)^{-2}A(\omega)\,d\omega$$

$$= 2\,\Omega^{-2}\int\limits_{0}^{\infty}[1+3(\omega-\Omega)^2/\Omega^2+\cdots]\,\omega\,A(\omega)\,d\omega \qquad (A.7)$$

and with the use of the sum rule (A.3) we are again back to quantities like those in (A.5) and (A.6).

Appendix B

We first estimate the correction in the dielectric screening constant $\varepsilon^0(\boldsymbol{q},\omega)$ due to thermal smearing in the Fermi-Dirac distribution function $n(E_{\boldsymbol{q}},\mu,T)$. Here μ is the chemical potential and $E_{\boldsymbol{q}}$ the electron energy corresponding to wave vector $\boldsymbol{q}$. We can write for $\varepsilon^0(\boldsymbol{q},\omega)$ (see e.g. [42])

$$\varepsilon^0(\boldsymbol{q},\omega) = 1 - (4\pi\,e^2/q^2)\sum_{k}\frac{n(E_{\boldsymbol{k}+\boldsymbol{q}},\mu,T)-n(E_{\boldsymbol{k}},\mu,T)}{E_{\boldsymbol{k}+\boldsymbol{q}}-E_{\boldsymbol{k}}+h(\omega+i\,\delta)}\,. \qquad (B.1)$$

We now observe that

$$n(E_{\boldsymbol{q}},\mu,T) = -\int\frac{\partial n}{\partial E}(E,\mu,T)\,n(E_{\boldsymbol{q}},E,0)\,dE\,. \qquad (B.2)$$

The function $\partial n/\partial E$ is strongly peaked around $E=\mu$ and it is readily seen that once we have the dielectric function at $T=0\,K$ and the temperature dependence of the chemical potential we find the finite temperature dielectric function as a weighted sum over zero temperature functions with varying Fermi levels. Integrals involving the function $\partial n/\partial E$ appear e.g. in the elementary calculation of the heat capacity of an almost degenerate Fermi gas. Exactly the same approach can be used here. Let us insert (B.2) into (B.1) and use the Lindhard or random phase approximation [42] for the zero temperature dielectric function. It is easily found that the finite temperature corrections are of order $(k_{\mathrm{B}}T/\mu)^2$. Special attention must be given to the region around $E_{\boldsymbol{q}}=\mu$ where $\partial\varepsilon/\partial q$ has a logarithmic singularity. However it is easily checked that our qualitative result for the temperature effect still holds.

We now turn to the effect of a finite electron life time τ. Mermin [43] has shown that the dielectric function (at zero temperature) takes the form

$$\varepsilon(\boldsymbol{q},\omega) = 1 + \frac{(1+i/\omega\,\tau)\,[\varepsilon^0(\boldsymbol{q},\omega+i/\tau)-1]}{1+(i/\omega\,\tau)\,[\varepsilon^0(\boldsymbol{q},\omega+i/\tau)-1]/[\varepsilon^0(\boldsymbol{q},0)-1]}\,. \qquad (B.3)$$

In the screened electron-phonon interaction the frequency ω that enters $\varepsilon(\boldsymbol{q},\omega)$ is a phonon frequency. We can also estimate the life time τ from the elementary resistivity formula [44]

$$\varrho = m/(n\,e^2\,\tau)\,. \qquad (B.4)$$

It is found that $1/\omega\tau$ is of the order of unity or larger for metals at high temperatures and consequently can not be neglected. On the other hand $\varepsilon^0(\boldsymbol{q},\omega)$

is correctly given by the static function $\varepsilon^0(q, 0)$ to order $(m_{el}/M_{ion})^{1/2}$ [42]. Therefore to the same accuracy

$$\varepsilon(q, \omega) = \varepsilon^0(q, 0)$$

and finite life time effects can be neglected.

References

1. Dugdale, J. S., Gugan, D.: Proc. Roy. Soc. **270** A, 186 (1962)
2. Laubitz, M. J.: Canad. J. Phys. **45**, 3677 (1967)
3. Laubitz, M. J.: Canad. J. Phys. **47**, 2636 (1969)
4. Sham, L. J., Ziman, J. M.: In: Solid state physics (Edited by Seitz, F., Thurnbull, D.). Vol. 2, p. 221. New York: Academic Press 1963
5. Greene, M. P., Kohn, W.: Phys. Rev. **137**, A 513 (1965)
6. Baym, G.: Phys. Rev. **135**, A 1691 (1964)
7. Ziman, J. M.: Electrons and phonons, p. 360. Oxford: Clarendon Press 1960
8. Chester, G. V.: Rep. progr. Phys. **26**, 411 (1963)
9. Rubio, J.: J. Phys. C **2**, 288 (1969)
10. Ballantine, L.: J. Phys. C **3**, L 16 (1970)
11. Faber, T. E.: Introduction to the theory of liquid metals. Cambridge: University Press 1972
12. Faber, T. E.: In: The physics of metals 1. Electrons (Edited by Ziman, J. M.). p. 282. Cambridge: University Press 1969
13. Springer, B.: Phys. Rev. **136**, A 115 (1964)
14. Greenwood, D. A.: Proc. Phys. Soc. **87**, 775 (1966)
15. Ziman, J. M.: Adv. Phys. **16**, 551 (1967)
16. Greenfield, A. J., Wiser, N.: Adv. Phys. **16**, 591 (1967)
17. Adams, P. D., Ashcroft, N. W.: Adv. Phys. **16**, 597 (1967)
18. Greenfield, A. J., Wiser, N.: Adv. Phys. **16**, 601 (1967)
19. Ashcroft, N. W., Schaich, W.: Phys. Rev. B **1**, 1370 (1970)
20. Prange, R. E., Kadanoff, L. P.: Phys. Rev. **134**, A 566 (1964)
21. Huebener, R. P.: Phys. Rev. **146**, 502 (1966)
22. Bailyn, M.: Phys. Rev. **121**, 1336 (1961)
23. Franzak, E., Bailyn, M.: Phys. Rev. **126**, 2033 (1962)
24. Maradudin, A. A., Fein, A. E.: Phys. Rev. **128**, 2589 (1962)
25. Maradudin, A. A., Ambegaokar, V.: Phys. Rev. **135**, A 1071 (1964)
26. Buyers, W. J. L., Cowley, R. A.: Phys. Rev. **180**, 755 (1969)
27. Koehler, T. R., Gillis, N. S., Wallace, D. C.: Phys. Rev. B **1**, 4521 (1970)
28. Gillis, N. S., Koehler, T. R.: Phys. Rev. B **3**, 3568 (1971)
29. Cowley, R. A., Svensson, E. C., Buyers, W. J. L.: Phys. Rev. Letters **23**, 525 (1969)
30. Huebener, R. P.: In: Lattice defects in quenched materials (Edited by Cotterill, R. M. J., Doyama, M., Jackson, J. J., Meshii, M.), p. 569, New York: Academic Press 1965
31. Kuhlmann-Wilsdorf, D.: Phys. Rev. **140**, A 1599 (1965)
32. Bradshaw, F. J., Pearson, S.: Proc. Phys. Soc. B **69**, 441 (1956)
33. Feder, R., Charbnau, H. P.: Phys. Rev. **149**, 464 (1966)
34. Martin, D. L.: Phys. Rev. **139**, A 150 (1965)
35. Barron, T. H. K.: In: Lattice Dynamics (Edited by Wallis, R. F.), p. 247. New York: Pergamon Press 1965
36. Wallace, D. C.: Thermodynamics of crystals, p. 376. New York: John Wiley and Sons 1972
37. Kaveh, M., Wiser, N.: Phys. Rev. B **6**, 3648 (1972)
38. Glyde, H. R., Taylor, R.: Phys. Rev. B **5**, 1206 (1972)
39. Pearson, W. B.: Handbook of lattice spacings and structures of metals and alloys. Oxford: Pergamon Press 1967
40. Dugdale, J. S.: Science **134**, 77 (1961)
41. Ambegaokar, V., Conway, J. M., Baym, G.: In: Lattice Dynamics (Edited by Wallis, R. F.), p. 261. New York: Pergamon Press 1965

42. Schrieffer, J. R.: Theory of Superconductivity. New York: Benjamin 1964
43. Mermin, M. D.: Phys. Rev. B 1, 2362 (1970)
44. Kittel, C.: Introduction to solid state physics, p. 216. New York: John Wiley and Sons 1967

Dr. Göran Grimvall
Institute of Theoretical Physics
Fack
S-402 20 Göteborg 5
Sweden

Phys. cond. Matter 17, 153—178 (1974)
© by Springer-Verlag 1974

Stability of Lamellar Eutectics

S. Strässler and W. R. Schneider

Brown Boveri Research Center, Baden, Switzerland

Received October 9, 1973

A new theory for the growth of lamellar eutectics is presented. The present analysis is different from previous theories in that the growth problem is formulated in terms of an interface equation. This equation is analysed for steady state solutions and their stability against different perturbations.

The main conclusion from this theoretical analysis is that lamellar eutectics should grow close to the minimum undercooling condition.

The present investigation not only clarifies earlier discussions but it also puts forward a new approach to the complex problem of growth.

1. Introduction

Directionally solidified eutectic alloys may have a structure consisting of parallel plates or rods. This growth has been the subject of many experimental and theoretical studies. The first theoretical work that took into account both the effect of diffusion and the surface energy was done by Zener [1]. He found that the lamellar structure (see Fig. 2) could grow in a range of growth rates at a given undercooling. To remove this ambiguity Zener postulated that the growth rate was the maximum possible at the given undercooling. He predicted from this that the product of the growth velocity v_0 and the square of the lamellar spacing λ should be constant. Tiller [2] proposed a minimum undercooling condition to replace the maximum velocity condition used by Zener. These conditions are formally identical. Hillert [3, 4] extended the work of Zener. He found a solution to the diffusion equation assuming the interface to be plane. Jackson and Hunt [5] generalised Hillers work to a general case, with no restriction on the relative volumes of the two phases, and with the melt on or off eutectic composition. The basic hypothesis in Jackson and Hunt's work is that they base their theory to start with on quantities that are averaged over the lamellas. In particular they equate the average temperature before the two phases α and β, respectively. Lesoult and Turpin [6] in their work on eutectic growth point out that no justification of this equation has been given. Furthermore Hunt and Jackson [5] discuss the stability of the lamellar growth only intuitively, e.g. they do not base it on mathematical equations. Their conclusion is that the growth occurs at or near the minimum undercooling condition. Following the demonstration by Mollard and Fleming [7], that eutectic microstructures could be obtained in Pb-Sn alloys of noneutectic compositions provided that the quotient of temperature gradient to growth velocity was sufficiently large, two stability calculations were proposed to account for this phenomenon [8, 9]. The two perturbation approaches try to generalise the work of Mullins and Sekerka [10] for singlephase alloys. Both

theories take as their starting point the Jackson and Hunt [5] analysis of steady
state lamellar eutectic growth. The two approaches lead to significantly different
predictions of the spatial wavelength of the instability which first occurs. In a
critical discussion of the two approaches Hunt [11] *et al.* conclude that a proper
stability calculation must await a better solution of the steady state problem.

In this work we present a new mathematical method which makes it possible
to obtain a better description of the complex problem of eutectic growth.

We begin in Sect. 2 with a formulation of the problem. A simplified model is
introduced on which the following analysis will be based. In Sect. 3 we introduce
a new mathematical technique which allows us to transform the basic set of
equations (diffusion equations and boundary conditions) into a one-dimensional
integral equation for the interface.

Before applying our interface equation to eutectic growth we shall study the
stability of the uniform advance of a planar interface of a dilute solid alloy into
the molten metal in Sect. 4. In Sects. 5—8 we shall study the steady state and
small perturbations of eutectic growth. We conclude (Sect. 9) with a discussion
of our results.

2. The Model

We consider a system consisting of two phases L and S (liquid and solid)
which are separated by an interface F (Fig. 1). For the sake of simplicity we
assume spatial homogeneity in one direction, reducing the problem from a three
— to a two dimensional one.

Let (ξ, ζ) denote coordinates with respect to a laboratory frame. The interface F_t at time t is the set

$$F_t = \{(\xi, \zeta) \,|\, \zeta = v_0 t + f_t(\xi)\} \tag{2.1}$$

where the function $f_t \colon \mathbb{R} \to \mathbb{R}$ is bounded and has a continous derivative except
at isolated points. Furthermore, we assume that for a steady state

$$\lim_{a \to \infty} \frac{1}{2a} \int_{-a}^{a} f_t(\xi)\, d\xi = \zeta_0 . \tag{2.2}$$

Consequently, the mean position of F_t is given by $\zeta = v_0 t + \zeta_0$, i.e. the solid
phase is growing with the steady state velocity v_0 in the ζ-direction.

We introduce internal coordinates (x, z) by setting

$$x = \xi, \qquad z = \zeta - v_0 t . \tag{2.3}$$

The interface is now given by

$$F_t = \{(x, z) \,|\, z = f_t(x)\} . \tag{2.4}$$

Occasionally we shall drop the index t.

The domains G_L, G_S of the liquid and solid phase respectively, are given by

$$G_{L(S)} = \{(x, z) \,|\, z \underset{(<)}{>} f_t(x)\} . \tag{2.5}$$

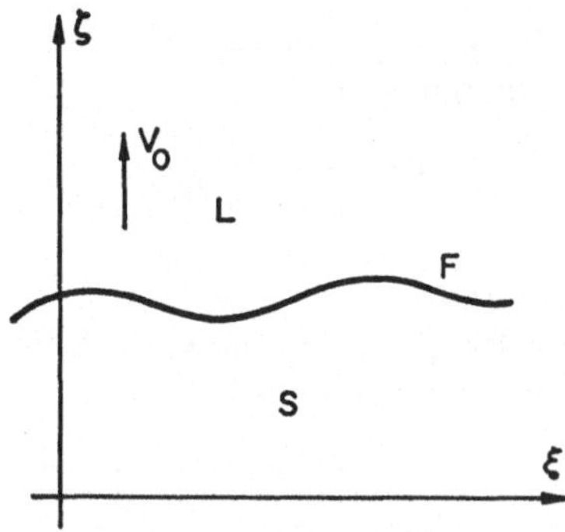

Fig. 1. Schematic representation of the directional liquid (L) solid (S) phasetransformation. F represents the interface

The state of the system is given by a quadruplet $(c_\mathrm{L}, c_\mathrm{S}, T_\mathrm{L}, T_\mathrm{S})$ of functions

$$c_\mathrm{L}, T_\mathrm{L}\colon G_\mathrm{L}\times\mathbb{R}_+ \to \mathbb{R},$$
$$c_\mathrm{S}\, T_\mathrm{S}\colon G_\mathrm{S}\times\mathbb{R}_+ \to \mathbb{R} \tag{2.6}$$

where $c_\mathrm{L(S)}(\boldsymbol{x}, t)$ is the concentration of the solute at time $t\in\mathbb{R}_+$ at the point $\boldsymbol{x}\in G_\mathrm{L(S)}$, $T_\mathrm{L(S)}(\boldsymbol{x}, t)$ is the temperature. These functions have to satisfy certain boundary conditions on the interface F. Material and heat conservation lead to

$$D_\mathrm{S}\left(\frac{\partial c_\mathrm{S}}{\partial n}\right)_F - D_\mathrm{L}\left(\frac{\partial c_\mathrm{L}}{\partial n}\right)_F = \boldsymbol{v}\cdot\boldsymbol{n}\,(c_\mathrm{LF} - c_\mathrm{SF}), \tag{2.7}$$

$$K_\mathrm{S}\left(\frac{\partial T_\mathrm{S}}{\partial n}\right)_F - K_\mathrm{L}\left(\frac{\partial T_\mathrm{L}}{\partial n}\right)_F = \boldsymbol{v}\cdot\boldsymbol{n}\cdot J. \tag{2.8}$$

Here a subscript F denotes boundary values on the interface F (taken as limit of values in G_L, G_S respectively), $D_\mathrm{L(S)}$ diffusion coefficient, $K_\mathrm{L(S)}$ heat conductivity coefficients, J heat of fusion per unit volume, $\boldsymbol{v}$ local velocity of the interface, $\boldsymbol{n}$ unit vector normal to F, given by

$$\boldsymbol{n} = (n_x, n_z) = \frac{(-f', 1)}{(1 + f'^2)^{1/2}}, \tag{2.9}$$

$$\left(' = \frac{\partial}{\partial x},\ \cdot = \frac{\partial}{\partial t}\right).$$

The temperature has to be continuous across F, hence

$$T_\mathrm{LF} = T_\mathrm{SF} \equiv T_F. \tag{2.10}$$

Finally, assuming the principle of local equilibrium to hold, the existence of chemical potentials μ_L, μ_S for the phases L and S is garanteed. If, by a further assumption, the interface kinetics can be neglected, then the continuity condition

$$\mu_\mathrm{LF} = \mu_\mathrm{SF} \tag{2.11}$$

must hold. For a dilute alloy this leads within a certain approximation (Appendix A) to

$$T_F - T_\mathrm{E} = -\,m\,c_\mathrm{LF} + a_0\,K \tag{2.12}$$

where T_E is the melting temperature of the pure component, m is the liquidus slope, a_0 the Thompson constant and

$$K = \frac{f''}{(1 + f'^2)^{3/2}}$$

is the local curvature.

Further boundary conditions at $\zeta = \zeta_S$ and $\zeta = \zeta_L$ have to be imposed, where ζ_L, ζ_S are constants. We postpone the formulation of these conditions.

The full dynamics of our system would be

$$\Delta c_{L(S)} + \frac{v_0}{D_{L(S)}} \frac{\partial c_{L(S)}}{\partial z} = \frac{1}{D_{L(S)}} \frac{\partial c_{L(S)}}{\partial t} \cdot \tag{2.13}$$

and

$$\Delta T_{L(S)} + \frac{v_0}{D'_{L(S)}} \frac{\partial T_{L(S)}}{\partial z} = \frac{1}{D'_{L(S)}} \frac{\partial T_{L(S)}}{\partial t} \tag{2.14}$$

where $D_{L(S)}$, $D'_{L(S)}$ are material and heat diffusion coefficients, respectively.

We replace (2.13), (2.14) by the following equations

$$\Delta c_L + \frac{v_0}{D_L} c_L = 0, \tag{2.15}$$

$$\Delta T_{L(S)} = 0. \tag{2.16}$$

This approximation is valid for perturbations whose shape changes sufficiently slowly. The equation for c_S is dropped altogether assuming that the diffusivity in the solid is small. In order to satisfy (2.7) we assume

$$\left(\frac{\partial c_S}{\partial n} \right)_F = 0 \tag{2.17}$$

and

$$c_{SF} = k c_{LF} \tag{2.18}$$

for a dilute alloy where k is some constant. In case of an eutectic see Appendix A.

We now formulate the boundary conditions at $\zeta = \zeta_{L(S)}$ for the case

$$\zeta_L = + \infty, \quad \zeta_S = - \infty$$

or equivalently for $z_L = + \infty$, $z_S = - \infty$:

$$\lim_{z \to \infty} c_L(x, z) = c^\infty, \tag{2.19}$$

$$\lim_{z \to \infty} \frac{1}{z} T_L(x, z) = g_L, \tag{2.20}$$

$$\lim_{z \to -\infty} \frac{1}{z} T_S(x, z) = g_S. \tag{2.21}$$

3. Mathematical Methods

In this section we solve formally the boundary value problems posed in Sect. 2. This procedure leads to an integro-differential equation for the function f which describes the interface.

The first boundary value problem is to solve

$$\Delta c_{\mathrm{L}} + \frac{v_0}{D_{\mathrm{L}}}\, c_{\mathrm{L}} = 0 \ \text{ in } G_{\mathrm{L}} \tag{3.1}$$

with the boundary conditions

$$\left(\frac{\partial c_{\mathrm{L}}}{\partial n}\right)_F = -\frac{1}{2}\, q\, n_z \ \text{ on } F \tag{3.2}$$

and

$$\lim_{z\to\infty} c_{\mathrm{L}}(x, z) = c^\infty \tag{3.3}$$

where

$$q = \frac{2}{D_{\mathrm{L}}}\, (v_0 + \dot{f})\,(c_{\mathrm{LF}} - c_{\mathrm{SF}}) \tag{3.4}$$

according to (2.7), (2.9).

The following Ansatz

$$c_{\mathrm{L}}(\boldsymbol{x}) = c^\infty + a\, e^{-2\gamma z} + \frac{1}{2\pi} \int dx'\, p(x')\, e^{-\gamma(z - f(x'))} K_0(\gamma\,|\boldsymbol{x} - \boldsymbol{x}_F|)\,, \tag{3.5}$$

$$2\gamma = \frac{v_0}{D_{\mathrm{L}}} \tag{3.6}$$

satisfies (3.1) and (3.3) for arbitrary bounded functions p. Here K_n is the modified Bessel function of order n and the notation $\boldsymbol{x}'_F = (x', f(x'))$ is used in (3.5). Evaluation of the normal derivative gives

$$
\begin{aligned}
\left(\frac{\partial c_{\mathrm{L}}}{\partial n}\right)_F = {}& - 2\gamma a\, e^{-2\gamma f(x)} - \frac{1}{2}\, p(x)\, n_z(x) \\
& - \frac{1}{2\pi} \int dx'\, e^{-\gamma(f(x) - f(x'))}\, \gamma\varrho\, K_1(\gamma\varrho)\, \frac{\boldsymbol{n}(x)\cdot(\boldsymbol{x}_F - \boldsymbol{x}'_F)}{\varrho^2}\, p(x') \\
& - \frac{1}{2\pi}\, n_z(x) \int dx'\, e^{-\gamma(f(x) - f(x'))}\, \gamma\, K_0(\gamma\varrho)\, p(x')
\end{aligned}
\tag{3.7}
$$

(singular integrals are to be understood as principal value integrals) with

$$\varrho = |\boldsymbol{x}_F - \boldsymbol{x}'_F|\,. \tag{3.8}$$

Hence (3.2) leads to the following integral equation for p

$$
\begin{aligned}
p(x) = {}& q(x) - 4a\gamma\, e^{-2\gamma f(x)} \\
& + \frac{1}{\pi} \int dx'\, e^{-\gamma(f(x) - f(x'))}\, \gamma\, K_1(\gamma\varrho)\, \frac{f'(x)(x - x') - (f(x) - f(x'))}{\varrho^2}\, p(x') \\
& - \frac{1}{\pi} \int dx'\, e^{-\gamma(f(x) - f(x'))}\, \gamma\, K_0(\gamma\varrho)\, p(x')\,.
\end{aligned}
\tag{3.9}
$$

The second boundary value problem is to solve

$$\Delta T_{\mathrm{L(S)}} = 0 \quad \text{ in } G_{\mathrm{L(S)}} \tag{3.10}$$

with the boundary conditions

$$T_{\mathrm{LF}} = T_{\mathrm{SF}}, \tag{3.11}$$

$$K_{\mathrm{S}}\left(\frac{\partial T_{\mathrm{S}}}{\partial n}\right)_{F} - K_{\mathrm{L}}\left(\frac{\partial T_{\mathrm{L}}}{\partial n}\right)_{F} = -Q\,n_z\,, \tag{3.12}$$

$$\lim_{z\to\infty}\frac{1}{z}\,T_{\mathrm{L}}(x, z) = g_{\mathrm{L}}\,, \tag{3.13}$$

$$\lim_{z\to\infty}\frac{1}{z}\,T_{\mathrm{S}}(x, z) = g_{\mathrm{S}}\,, \tag{3.14}$$

where Q is assumed to be given. It will be identified with

$$Q = -J(v_0 + \dot{f}) \tag{3.15}$$

according to (2.8), (2.9).

The following Ansatz

$$T(x) = T_0 + \tfrac{1}{2}(g_{\mathrm{L}} + g_{\mathrm{S}})\,z + (g_{\mathrm{L}} - g_{\mathrm{S}})\,\Psi(x) + \Phi(x) \tag{3.16}$$

with

$$\Psi(x) = \frac{1}{4\pi}\int dx'\,\ln\frac{(x - x')^2 + (z - f(x'))^2}{(x - x')^2} \tag{3.17}$$

and

$$\Phi(x) = -\frac{1}{2\pi}\int dx'\,h_1(x')\,\frac{\partial}{\partial x'}\ln|x - x'_F| \tag{3.18}$$

satisfies (3.11), (3.13), (3.14) (for details see Appendix B) if we set

$$\begin{aligned} T_{\mathrm{L}}(x) &= T(x), & x \in G_{\mathrm{L}}, \\ T_{\mathrm{S}}(x) &= T(x), & x \in G_{\mathrm{S}}. \end{aligned} \tag{3.19}$$

The boundary condition (3.12) leads to the following integral equation

$$\begin{aligned} h(x) = \frac{2}{K_{\mathrm{L}} + K_{\mathrm{S}}}\Bigg\{ &Q(x) - \frac{1}{2}(g_{\mathrm{L}} + g_{\mathrm{S}})\,K_{\mathrm{L}} - K_{\mathrm{S}}) \\ &+ \frac{K_{\mathrm{L}} - K_{\mathrm{S}}}{2\pi}\int dx'\,h(x')\,\frac{f'(x)(x - x') - (f(x) - f(x'))}{(x - x')^2 + (f(x) - f(x'))^2}\Bigg\} \end{aligned} \tag{3.20}$$

for the function

$$h(x) = \frac{dh_1}{dx} + (g_{\mathrm{L}} - g_{\mathrm{S}})\,. \tag{3.21}$$

4. Stability of a Planar Interface

In this section we shall show that the equations derived in Sect. 3 applied to the freezing of a dilute binary alloy admit a planar interface as solution. The stability of this solution against small perturbations of a certain form is investigated.

The interface equation is

$$T_F - T_{\mathrm{E}} = -m\,c_{\mathrm{LF}} + a_0\,K \tag{4.1}$$

according to (2.12). The boundary values T_F, c_{LF} are given by

$$T_F(x) = T_0 + \tfrac{1}{2}(g_{\mathrm{L}} + g_{\mathrm{S}})\,f(x) + (g_{\mathrm{L}} - g_{\mathrm{S}})\,\Psi(x_F) + \Phi(x_F)\,, \tag{4.2}$$

$$c_{LF}(x) = c^{\infty} + a\,e^{-2\gamma f(x)} + \frac{1}{2\pi} \int dx'\, e^{-\gamma(f(x)-f(x'))} K_0(\gamma\,|\,x_F - x'_F|)\,p(x') \tag{4.3}$$

where Ψ and Φ are given by (3.17), (3.18), respectively. Obviously, T_F depends on the unknown functions f, h, similarly c_{LF} on f, p, where h and p have to satisfy the integral equations (3.20), (3.9), respectively.

Setting $f = 0$, $p = 0$, $h_1 = 0$ we obtain from (3.9), (3.4) and (2.18)

$$a = (1 - k)\,c_{LF} \tag{4.4}$$

and from (3.21), (3.15)

$$g_{\mathrm{L}} - g_{\mathrm{S}} = \frac{2}{K_{\mathrm{L}} + K_{\mathrm{S}}} \left\{ - J v_0 - \frac{1}{2}\,(g_{\mathrm{L}} + g_{\mathrm{S}})\,(K_{\mathrm{L}} - K_{\mathrm{S}}) \right\}. \tag{4.5}$$

From (4.2) we get

$$T_F = T_0 \tag{4.6}$$

and from (4.3)

$$c_{LF} = c^{\infty} + a\,. \tag{4.7}$$

Combining (4.4) and (4.7) yields

$$a = c^{\infty}\!\left(\frac{1}{k} - 1\right) \tag{4.8}$$

and

$$c_{LF} = \frac{c^{\infty}}{k}\,. \tag{4.9}$$

Finally, we obtain from (4.1) by inserting (4.9)

$$T_F = T_{\mathrm{E}} - m\,\frac{c^{\infty}}{k}\,. \tag{4.10}$$

The complete solution of the boundary value problems for c_{L} and T in the case of a planar interface is now given by

$$T(x) = T_{\mathrm{E}} - m\,\frac{c^{\infty}}{k} + \frac{1}{2}\,(g_{\mathrm{L}} + g_{\mathrm{S}})\,z + \frac{1}{2}\,(g_{\mathrm{L}} - g_{\mathrm{S}})\,|z|, \tag{4.11}$$

$$c_{\mathrm{L}}(x) = c^{\infty} + c^{\infty}\!\left(\frac{1}{k} - 1\right) e^{-2\gamma z}. \tag{4.12}$$

We now investigate the stability of this solution against small perturbations of a form described below. Setting $A = A^{(0)} + A^{(1)}$ for any quantity A, where $A^{(0)}$ is the unperturbed value of A and $A^{(1)}$ the deviation caused by the perturbation, we have $f^{(0)} = 0$, $p^{(0)} = 0$, $h_1^{(0)} = 0$ and

$$T_F^{(0)} = T_0 = T_{\mathrm{E}} - m\,\frac{c^{\infty}}{k}\,, \tag{4.13}$$

$$c_{LF}^{(0)} = c^{\infty} + a = \frac{c^{\infty}}{k}\,. \tag{4.14}$$

We consider now a perturbation of the interface of the form

$$f^{(1)}(x, t) = \hat{f}^{(1)}\, e^{-\varkappa t} \sin \lambda x\,. \tag{4.15}$$

160   S. Strässler and W. R. Schneider

In accordance with the assumption of the perturbation being small, we linearise the equations which determine $h_1^{(1)}$, $p^{(1)}$, $c_{LF}^{(1)}$, $T_F^{(1)}$. These equations are reduced to algebraic equations by the Ansatz

$$A^{(1)} = \hat{A}^{(1)} e^{-\varkappa t} \begin{cases} \sin \lambda x \\ -\lambda^{-1} \cos \lambda x \end{cases} \tag{4.16}$$

for any of the above quantities:

$$\hat{h}_1^{(1)} = \frac{2}{K_L + K_S} \left\{ J\varkappa - \frac{1}{2}(K_L - K_S)(g_L - g_S) \right\} \hat{f}^{(1)}, \tag{4.17}$$

$$\hat{p}^{(1)} = 4\gamma(1-k)\hat{c}_{LF}^{(1)} - \frac{2}{D}\varkappa(1-k)c_{LF}^{(0)}\hat{f}^{(1)} + 8a\gamma^2\hat{f}^{(1)} - \frac{\gamma}{(\lambda^2 + \gamma^2)^{1/2}}\hat{p}^{(1)}, \tag{4.18}$$

$$\hat{c}_{LF}^{(1)} = -2a\gamma\hat{f}^{(1)} + \frac{1}{2}\frac{1}{(\lambda^2 + \gamma^2)^{1/2}}\hat{p}^{(1)}, \tag{4.19}$$

$$\hat{T}_F^{(1)} = \left(\bar{g} - \frac{J\varkappa}{\lambda}\frac{1}{K_L + K_S} \right)\hat{f}^{(1)} \tag{4.20}$$

with

$$\bar{g} = \frac{K_L g_L + K_S g_S}{K_L + K_S}. \tag{4.21}$$

Solving (4.18), (4.19) for $\hat{p}^{(1)}$, $\hat{c}_{LF}^{(1)}$ we obtain

$$\hat{p}^{(1)} = 4a\gamma \frac{\left(2k\gamma - \dfrac{\varkappa}{v_0}\right)(\lambda^2 + \gamma^2)^{1/2}}{(\lambda^2 + \gamma^2)^{1/2} + (2k-1)\gamma}\hat{f}^{(1)}, \tag{4.22}$$

$$\hat{c}_{LF}^{(1)} = 2a\gamma \frac{\gamma - \dfrac{\varkappa}{v_0} - (\lambda^2 + \gamma^2)^{1/2}}{(\lambda^2 + \gamma^2)^{1/2} + (2k-1)\gamma}\hat{f}^{(1)}. \tag{4.23}$$

The interface equation yields

$$\hat{T}_F^{(1)} = -mc_{LF}^{(1)} - \lambda^2 a_0 \hat{f}^{(1)}. \tag{4.24}$$

Inserting (4.23) into (4.24) and comparing with (4.20) leads to the following equation for $\varkappa$

$$\left(\frac{J}{\lambda}\frac{1}{K_L + K_S} + \frac{2a\gamma\dfrac{m}{v_0}}{(\lambda^2 + \gamma^2)^{1/2} + (2k-1)\gamma} \right)\varkappa$$

$$= \bar{g} + a_0\lambda^2 + 2a\gamma m \frac{\gamma - (\lambda^2 + \gamma^2)^{1/2}}{(\lambda^2 + \gamma^2)^{1/2} + (2k-1)\gamma}. \tag{4.25}$$

This equation is equivalent to Eq. (4.20) in Ref. [10].

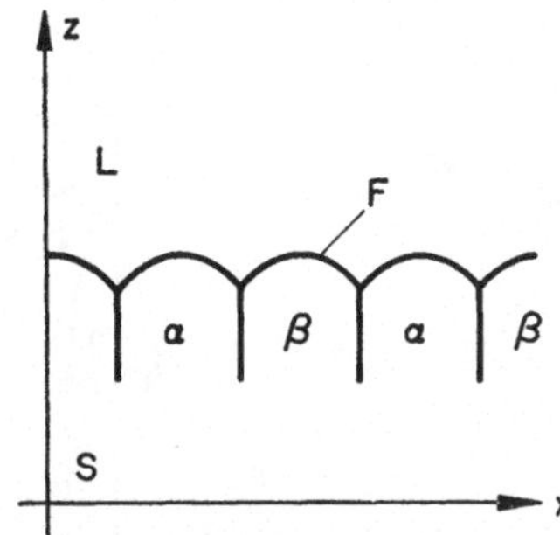

Fig. 2. Schematic representation of lamellar growth

The general stability condition $\mathrm{Re}\,\varkappa > 0$ for perturbations with time dependence $\exp(-\varkappa t)$ reduces here to $\varkappa > 0$ as (4.25) has only real solutions. This condition leads to

$$C < \frac{1 + B/\Lambda^2}{1 - \Lambda/\{[1 + (\Lambda/2k)^2]^{1/2} + (1 - 1/2k)\,\Lambda\}} \tag{4.26}$$

where

$$C = c^\infty\left(\frac{1}{k} - 1\right)\frac{m}{\bar{g}}\frac{v_0}{D_{\mathrm{L}}}, \qquad B = \frac{a_0}{\bar{g}}\left(k\frac{v_0}{D_{\mathrm{L}}}\right)^2.$$

The stability equations (4.25) and (4.26) have been obtained previously in Ref. [10] using a different method. A detailed discussion can be found in Ref. [12].

5. Lamellar Growth of Eutectics

In this section we apply the model and the methods introduced in Sect. 2 and 3 to the lamellar growth of a solid consisting of two phases α and β (Fig. 2). Let $\{x_n\}$ be a set of R, indexed by $\mathbb{Z}$, such that

$$d_1 < |x_{n+1} - x_n| < d_2, \quad n \in \mathbb{Z} \tag{5.1}$$

for some positive constants d_1, d_2. The interface F splits into pieces F_n defined by

$$F_n = \{(x, z)\,|\,x_n \leq x \leq x_{n+1},\; z = f(x)\}. \tag{5.2}$$

We assume F_n to be the boundary of the phase n of the solid with

$$n = \begin{cases} \alpha, n \text{ even} \\ \beta, n \text{ odd}. \end{cases} \tag{5.3}$$

In general the points x_n will depend on the time t. At the points x_n the left and right derivatives of f with respect to x are different and satisfy (see Fig. 3)

$$f'(x_n + 0, t) = \tan(\theta_n - \varphi_n), \tag{5.4}$$

$$f'(x_n - 0, t) = \tan(\theta_{n-1} + \varphi_n) \tag{5.5}$$

where

$$\theta_n = \begin{cases} \theta_\alpha\ n \text{ even} \\ \theta_\beta\ n \text{ odd} \end{cases} \tag{5.6}$$

162   S. Strässler and W. R. Schneider

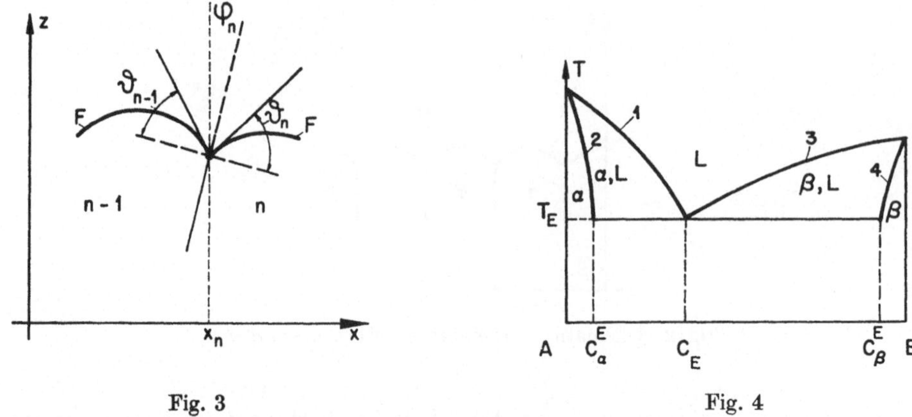

Fig. 3   Fig. 4

Fig. 3. Definition of the angles θ_n, θ_{n-1} and φ_n which occur in Eqs. (5.4) and (5.5)

Fig. 4. Phase diagram showing definition of c_α^{E}, c_β^{E}, c_{E}, T_{E}

and are determined by the surface tensions $\sigma_{\alpha\beta}$, $\sigma_{\mathrm{L}\alpha}$, $\sigma_{\mathrm{L}\beta}$ according to

$$\sigma_{\mathrm{L}\alpha}\sin\theta_\alpha + \sigma_{\mathrm{L}\beta}\sin\theta_\beta = \sigma_{\alpha\beta}, \tag{5.7}$$

$$\sigma_{\mathrm{L}\alpha}\cos\theta_\alpha = \sigma_{\mathrm{L}\beta}\cos\theta_\beta. \tag{5.8}$$

The function $c_{SF}(x)$ entering the boundary condition (2.7) is given by

$$c_{SF}(x) = c_n, \qquad x_n \leqq x \leqq x_{n-1} \tag{5.9}$$

where

$$c_n = \begin{cases} c_\alpha^{\mathrm{E}}, & n \text{ even} \\ c_\beta^{\mathrm{E}}, & n \text{ odd} \end{cases} \tag{5.10}$$

where c_α^{E}, c_β^{E} are given constants (see Fig. 4). In case of an eutectic the interface equation (2.11) becomes (see Appendix A)

$$T_F(x) - T_{\mathrm{E}} = m_n(c_{\mathrm{LF}}(x) - c_{\mathrm{E}}) + a_n K(x), \qquad x_n \leqq x \leqq x_{n+1} \tag{5.11}$$

where

$$m_n = \begin{cases} -m_\alpha, & n \text{ even} \\ m_\beta, & n \text{ odd} \end{cases} \tag{5.12}$$

and

$$a_n = \begin{cases} a_\alpha, & n \text{ even} \\ a_\beta, & n \text{ odd} \end{cases} \tag{5.13}$$

The constants T_{E}, c_{E} are given (their physical meaning is eutectic temperature and concentration, respectively, Fig. 4).

In the following we shall linearise the equations for T_F, c_{LF}, p and h with respect to f. The angles φ_n, $n \in \mathbb{Z}$, are then given by

$$\tan\varphi_n = \frac{\dot{x}_n}{v_0}. \tag{5.14}$$

Furthermore we assume the diffusion length $l_D = D_L/v_0$ to be large compared to a characteristic length of the interface, e.g. d_2. Hence we retain only terms up to first power in γ.

In this way we obtain from (3.9)

$$p(x) = q(x) - 4a\gamma + \frac{1}{\pi} \int dx'\, p(x') \frac{f'(x)(x-x') - (f(x) - f(x'))}{(x-x')^2} \qquad (5.15)$$

and from (3.5)

$$c_L(x) = c^\infty + a - 2a\gamma f + \frac{1}{2\pi} \int dx'\, p_1(x') \frac{x'-x}{|x_F' - x|^2} \qquad (5.16)$$

where $x_F' = (x', 0)$ in view of the linearisation with respect to f. p_1 is related to p via

$$p = \frac{d}{dx} p_1. \qquad (5.17)$$

(The conditions on p_1 in order to satisfy the asymptotic behavior (3.3) are discussed in Appendix B.) From (5.16) we obtain

$$c_{LF}(x) = c^\infty + a - 2a\gamma f + \frac{1}{2\pi} \int dx'\, p_1(x') \frac{1}{x'-x} \qquad (5.18)$$

or symbolically

$$c_{LF} = c^\infty + a - 2a\gamma f + \tfrac{1}{2} H p_1 \qquad (5.19)$$

where Hg denotes the Hilbert transform of the function g:

$$Hg(x) = \frac{1}{\pi} \int dx'\, g(x') \frac{1}{x'-x} \qquad (5.20)$$

(remind that singular integrals are defined as principal value integrals).

Linearizing (3.16) and (3.20) leads to

$$T(x) = T_0 + \tfrac{1}{2}(g_L + g_S) z + \tfrac{1}{2}(g_L - g_S)|z|$$
$$- \frac{1}{2\pi} \int dx'\, h_1(x') \frac{x'-x}{|x_F' - x|^2} \qquad (5.21)$$

and

$$h(x) = \frac{2}{K_L + K_S}\Bigg\{ Q(x) - \frac{1}{2}(g_L + g_S)(K_L - K_S)$$
$$+ \frac{K_L - K_S}{2\pi} \int dx'\, h(x') \frac{f'(x)(x-x') - (f(x) - f(x'))}{(x-x')^2}\Bigg\}. \qquad (5.22)$$

Finally, we get from (5.21)

$$T_F = T_0 + \tfrac{1}{2}(g_L + g_S) f + \tfrac{1}{2}(g_L - g_S)|f| - \tfrac{1}{2} H h_1. \qquad (5.24)$$

The interface equation (5.11) is taken into full account, i.e. the curvature $K(x)$ is not linearized with respect to f. Integrating (5.11) from x_n to x_{n+1} gives

$$-a_n\{\sin(\theta_n - \varphi_n) + \sin(\theta_n + \varphi_{n+1})\} = \int_{x_n}^{x_{n+1}} \{T_F - T_E - m_n(c_{LF} - c_E)\}\, dx. \qquad (5.25)$$

Similarly, integrating from x_n to x leads to

$$a_n\left\{\frac{f'(x)}{(1 + f'(x)^2)^{1/2}} - \sin(\theta_n - \varphi_n)\right\} = \int\limits_{x_n}^{x} \{T_F - T_E - m_n(c_{LF} - c_E)\}\, dx'.$$

(5.26)

Combining (5.25), (5.26) we arrive at

$$a_n\frac{f'(x)}{(1 + f'(x)^2)^{1/2}} = \int\limits_{x_n}^{x} \{T_F - T_E - m_n(c_{LF} - c_E)\}\, dx'$$

$$- \frac{x - x_n}{x_{n+1} - x_n}\int\limits_{x_n}^{x_{n+1}} \{T_F - T_E - m_n(c_{LF} - c_E)\}\, dx' \quad (5.27)$$

$$+ a_n\sin(\theta_n - \varphi_n) - a_n\frac{x - x_n}{x_{n+1} - x_n}\{\sin(\theta_n - \varphi_n) + \sin(\theta_n - \varphi_{n+1})\}.$$

This equation offers itself for an iterative method to determine f: Suppose an approximation f^a given. Using Eqs. (5.15), (5.19), (5.22), (5.24) we obtain the corresponding approximations p^a, h^a, c_{LF}^a, T_F^a. Hence the right hand side of (5.27) may be evaluated (the points x_n are to be known for all t, allowing to calculate the angles φ_n via (5.14)). Solving for f' and integrating we get a new approximation f^b for f. Even if the first approximation f^a does not satisfy (5.5), (5.6) the new approximation f^b will, as is easily seem from (5.27). We shall apply this procedure in the next section to the steady state situation.

6. Steady State of an Eutectic

In this section we shall apply the iterative procedure indicated at the end of Sect. 5 to the steady state of a lamellar eutectic. As starting approximation we take $f^{(0)} = 0$. Hence (5.15) and (3.4) reduce to

$$p = q - 4a\gamma \tag{6.1}$$

and

$$q = 4\gamma(c_{LF} - c_n) \tag{6.2}$$

where (5.9) has been taken into account. Hence

$$p = 4\gamma(c_{LF} - c_n - a). \tag{6.3}$$

Now c_{LF} is given by (5.19) and depends itself on p, but if one keeps only terms up to first order in γ in (6.3) only the first two terms of (5.19) contribute, thus leading to

$$p^{(0)} = 4\gamma(c^\infty - c_n). \tag{6.4}$$

The conditions on p_1 (related to p by (5.17)) as discussed in Appendix B are satisfied if

$$\int\limits_{x_n}^{x_{n+2}} p(x)\, dx = 0. \tag{6.5}$$

Inserting (6.4) leads to

$$(c^\infty - c_n)(x_{n+1} - x_n) + (c^\infty - c_{n+1})(x_{n+2} - x_{n+1}) = 0. \tag{6.6}$$

Hence up to an irrelevant translation the points x_n are given by

$$x_n = \begin{cases} \frac{1}{2} n\lambda & n \text{ even} \\ \frac{1}{2}(n-1)\lambda + \eta\lambda, & n \text{ odd}, \end{cases} \tag{6.7}$$

i.e. a periodic structure is imposed on the lamellar spacing, a period of length λ consisting of a fraction η of phase α, a fraction $1-\eta$ of phase β, where η is given by

$$(c^\infty - c_\alpha^{\mathrm{E}})\eta + (c^\infty - c_\beta^{\mathrm{E}})(1-\eta) = 0 \tag{6.8}$$

or

$$\eta = \frac{c_\beta^{\mathrm{E}} - c^\infty}{c_\beta^{\mathrm{E}} - c_\alpha^{\mathrm{E}}}. \tag{6.9}$$

Obviously, $p^{(0)}$ according to (6.4) is a periodic function and may be written as a Fourier series:

$$p^{(0)}(x) = 8\gamma\,\Delta c\,\frac{1}{\pi}\sum_{n=1}^\infty \frac{1}{n}\sin\left(n\pi\eta\right)\cos\left\{2\pi n\left(\frac{x}{\lambda} - \frac{\eta}{2}\right)\right\} \tag{6.10}$$

where

$$\Delta c = c_\beta^{\mathrm{E}} - c_\alpha^{\mathrm{E}}. \tag{6.11}$$

Inserting (6.10) into (5.19) yields

$$c_{\mathrm{LF}}^{(0)}(x) = c^\infty + a + c_1(x) \tag{6.12}$$

with

$$c_1(x) = 2\gamma\,\Delta c\,\frac{1}{\pi^2}\sum_{n=1}^\infty \frac{1}{n^2}\sin\left(n\pi\eta\right)\cos\left\{2\pi n\left(\frac{x}{\lambda} - \frac{\eta}{2}\right)\right\}. \tag{6.13}$$

With $h_1^{(0)} = 0$ we obtain from (5.24)

$$T_F^{(0)} = T_0 \tag{6.14}$$

whereas (5.22) leads to (4.5).

Inserting (6.12), (6.14) into (5.25) we get

$$-2a_\alpha \sin\theta_\alpha = \lambda\eta[T_0 - T_{\mathrm{E}} + m_\alpha(c^\infty + a - c_{\mathrm{E}})] + m_\alpha\int_0^{\lambda\eta} dx\,c_1(x), \tag{6.15}$$

$$-2a_\beta \sin\theta_\beta = \lambda(1-\eta)[T_0 - T_{\mathrm{E}} - m_\beta(c^\infty + a - c_{\mathrm{E}})] - m_\beta\int_{\lambda\eta}^\lambda dx\,c_1(x). \tag{6.16}$$

Solving (6.15), (6.16) for $T_{\mathrm{E}} - T_0 \equiv \Delta T$ yields

$$\Delta T = \frac{2m}{\lambda}\left[\frac{1}{\eta}\frac{a_\alpha}{m_\alpha}\sin\theta_\alpha + \frac{1}{1-\eta}\frac{a_\beta}{m_\beta}\sin\theta_\beta\right] + \lambda\cdot\frac{mv_0\Delta c}{D_{\mathrm{L}}}\frac{1}{\eta(1-\eta)}\Psi(\eta) \tag{6.17}$$

where

$$\frac{1}{m} = \frac{1}{m_\alpha} + \frac{1}{m_\beta}$$

and

$$\Psi(\eta) = \frac{1}{\pi^3}\sum_{n=1}^\infty \frac{1}{n^3}\sin^2(\pi n\eta). \tag{6.18}$$

The quantity ΔT is called average undercooling and attains its minimum with respect to a variation in λ for

$$v_0\,\lambda_{\min}^2 = 2\,\frac{D_{\mathrm{L}}}{\Delta c}\left\{(1-\eta)\,\frac{a_\alpha}{m_\alpha}\sin\theta_\alpha + \eta\,\frac{a_\beta}{m_\beta}\sin\theta_\beta\right\}\cdot\frac{1}{\Psi(\eta)}\,. \qquad (6.19)$$

Eq. (6.17) is identical to the average undercooling equation derived by Jackson and Hunt. To apply the minimum undercooling principle to find a unique value for λ (see Eq. (6.19)) is only an heuristic argument. It will be the purpose of Sect. 7 and 8 to illuminate the validity of this principle. In the following we shall show how one can improve on the zeroth order approximation $f \equiv 0$ used in deriving Eq. (6.17). Solving (6.15), (6.16) for $(c^\infty + a - c_{\mathrm{E}})$ yields

$$-(m_\alpha + m_\beta)(c^\infty + a - c_{\mathrm{E}})$$
$$= \frac{2}{\lambda}\left(\frac{a_\alpha\sin\theta_\alpha}{\eta} - \frac{a_\beta\sin\theta_\beta}{1-\eta}\right) + 2\,\lambda\,\Delta c\,\gamma\left(\frac{m_\alpha}{\eta} - \frac{m_\beta}{1-\eta}\right)\Psi(\eta)\,. \qquad (6.20)$$

Eq. (6.20) determines the constant a. Keeping only zero order terms in γ we obtain

$$a = c_{\mathrm{E}} - c^\infty - \frac{2}{\lambda(m_\alpha + m_\beta)}\left(\frac{a_\alpha\sin\theta_\alpha}{\eta} - \frac{a_\beta\sin\theta_\beta}{1-\eta}\right)\,. \qquad (6.21)$$

Finally a new approximation for f is obtained by inserting (6.12), (6.14) into (5.27): In the interval $0 \leq x \leq \lambda\eta$ we have

$$\frac{f'}{(1+f'^2)^{1/2}} = \frac{m_\alpha}{a_\alpha}\frac{v_0\,\Delta c}{D_{\mathrm{L}}}\lambda^2\frac{1}{2\pi^3}\sum_{n=1}^{\infty}\frac{1}{n^3}\sin(n\pi\eta)$$
$$\cdot\left\{\sin\left[2\pi n\left(\frac{x}{\lambda} - \frac{\eta}{2}\right)\right] + \sin(\pi n\eta)\right\}$$
$$- \frac{x}{\lambda\eta}\frac{m_\alpha}{a_\alpha}\frac{v_0\,\Delta c}{D_{\mathrm{L}}}\lambda^2\Psi(\eta) + \sin\theta_\alpha - 2\frac{x}{\lambda\eta}\sin\theta_\alpha \qquad (6.22)$$

and in the interval $x\eta \leq x < \lambda$

$$\frac{f'}{(1+f'^2)^{1/2}} = -\frac{m_\beta}{a_\beta}\frac{v_0\,\Delta c}{D_{\mathrm{L}}}\lambda^2\frac{1}{2\pi^3}\sum_{1=1}^{\infty}\frac{1}{n^3}\sin(n\pi\eta)$$
$$\cdot\left\{\sin 2\pi n\left(\frac{x}{\lambda} - \frac{\eta}{2}\right) - \sin n\pi\eta\right\}$$
$$+ \frac{1}{\lambda(1-\eta)}(x - \lambda\eta)\frac{m_\beta}{a_\beta}\frac{v_0\,\Delta c}{D_{\mathrm{L}}}\lambda^2\Psi(\eta)$$
$$+ \sin\theta_\beta - 2\frac{1}{\lambda(1-\eta)}(x - \lambda\eta)\sin\theta_\beta\,. \qquad (6.23)$$

Solving (6.22) and (6.23) for f' and integrating (numerically on a computer) yields the new approximation for f.

Alternatively we may proceed as follows: We insert into the last term of (5.15) the approximative value $p^{(0)}$ thus obtaining

$$p = p^{(0)} + p^{(1)} \qquad (6.24)$$

where

$$p^{(1)}(x) = \frac{1}{\pi} \int dx' \, p^{(0)}|(x') \frac{f'(x)(x-x') - (f(x) - f(x'))}{(x-x')^2} \qquad (6.25)$$

which may be written symbolically (after some manipulation) as

$$p^{(1)} = HD(p^{(0)}f) - D(fHp^{(0)}) \qquad (6.26)$$

where H denotes Hilbert transform as above and D stands for differentiation. Correspondingly we obtain from (5.18)

$$c_{LF} = c_{LF}^{(0)} + c_{LF}^{(1)} \qquad (6.27)$$

where $c_{LF}^{(0)}$ is given by (6.12) and

$$c_{LF}^{(1)} = \tfrac{1}{2} HD^{-1} p^{(1)} \qquad (6.28)$$

(D^{-1} is defined by $DD^{-1} = identity$, it is unique up to an arbitrary constant which however does not enter because the Hilbert transform of a constant is zero).

Using (6.26) and taking $H^2 = - identity$ into account we get

$$c_{LF}^{(1)} = - \tfrac{1}{2}[p^{(0)}f + H(fHp^{(0)})]. \qquad (6.29)$$

Similarly we derive from (5.22)

$$h = h^{(0)} + h^{(1)} \qquad (6.30)$$

where $h^{(0)} = (g_L - g_S)$ and

$$h^{(1)} = (g_L - g_S) \frac{K_L - K_S}{K_L + K_S} HDf. \qquad (6.31)$$

Correspondingly we have

$$T_F = T_F^{(0)} + T^{(1)} \qquad (6.32)$$

with $T_F^{(0)} = T_0$ and

$$T^{(1)} = - \tfrac{1}{2} HD^{-1} h^{(1)} + \tfrac{1}{2}(g_L + g_S)f \qquad (6.33)$$

or with (6.31)

$$T^{(1)} = \bar{g}f \qquad (6.34)$$

where

$$\bar{g} = \frac{g_L K_L + g_S K_S}{K_L + K_S}. \qquad (6.35)$$

Inserting all results into the interface equation (5.11) the following integro-differential equation for f results:

$$a_n \frac{f''}{(1+f'^2)^{3/2}} = T_0 - T_E + \bar{g}f$$
$$- m_n\{a + c^\infty - c_E - 2a\gamma f + c_1 - \tfrac{1}{2} p^{(0)}f - \tfrac{1}{2} H(fHp^{(0)})\}. \qquad (6.36)$$

Here c_1 is given by Eq. (6.13). To solve Eq. (6.36) is obviously much harder than Eqs. (6.22), (6.23) (even numerically) and we shall not use it for this purpose but postpone its application to Sect. 8 where stability against deformation of the steady state solution of f is discussed.

7. Perturbation of the Steady State (I)

Within the approximation $f = 0$ discussed in Sect. 6 the only possible perturbation consists of varying the points x_n separating the phases α and β. These points will be denoted by x_n^ε to exhibit explicitly their dependence on the perturbation parameter ε. We assume x_n^ε to be of the form

$$x_n^\varepsilon = x_n^\varepsilon + \varepsilon y_n \tag{7.1}$$

where
$$y_n = y_n(t) = y_n(0)\, e^{-\varkappa t}. \tag{7.2}$$

The following periodicity condition

$$y_{n+2N} = y_n, \quad n \in \mathbb{Z} \tag{7.3}$$

is imposed for some integer N. The unperturbed points x_n^0 are given by Eq. (6.7).

We are going to prove that for $\lambda < \lambda_{\min}$ where $\lambda_{\min}$ is the lamellar spacing obtained by the minimum undercooling principle, (see Eq. 6.19) there exists always a perturbation for which $\varkappa < 0$. This implies that an eutectic cannot grow for $\lambda < \lambda_{\min}$. Thus we see that the principle of minimum undercooling does not determine the value of λ uniquely but determines a lower bound.

In the following, terms of order higher than one in ε are neglected. Thus by Eq. (5.4) we have

$$\varphi_n = -\,\varepsilon y_n \frac{\varkappa}{v_0} \tag{7.4}$$

and

$$b_n^\varepsilon \equiv a_n \{\sin(\theta_n - \varphi_n) + (\sin\theta_n + \varphi_{n+1})\}$$
$$= 2\, a_n \sin\theta_n + \varepsilon\, a_n \cos\theta_n (y_n - y_{n+1}) \frac{\varkappa}{v_0}. \tag{7.5}$$

The perturbation induces a change in the functions p, and c_1 which therefore will be denoted by p^ε and c_1^ε.

No change will occur in h_1 (which therefore remains zero) and in T_F (which keeps its value T_0 according to (6.14)). Eq. (5.25) may be written as

$$\frac{1}{x_{n+1}^\varepsilon - x_n^\varepsilon} \int_{x_n^\varepsilon}^{x_{n+1}^\varepsilon} \{T_F - T_E - m_n(c^\infty + a - c_E)\}\, dx$$
$$= \frac{1}{x_{n+1}^\varepsilon - x_n^\varepsilon} \left\{ m_n \int_{x_n^\varepsilon}^{x_{n+1}^\varepsilon} c_1^\varepsilon\, dx - b_n^\varepsilon \right\}. \tag{7.6}$$

Now the left hand side of (7.6) does not depend on ε. Hence the functions

$$F_n(\varepsilon) = \frac{1}{x_{n+1}^\varepsilon - x_n^\varepsilon} \left\{ m_n \int_{x_n^\varepsilon}^{x_{n+1}^\varepsilon} c_1^\varepsilon\, dx - b_n^\varepsilon \right\} \tag{7.7}$$

have to satisfy

$$\left. \frac{dF_n}{d\varepsilon} \right|_{\varepsilon=0} = 0. \tag{7.8}$$

Setting

$$c_1^\varepsilon = c_1^0 + \varepsilon\, C \tag{7.9}$$

(7.8) leads to

$$(y_{n+1} - y_n)\left\{m_n \int_{x_n^0}^{x_{n+1}^0} c_1^0\, dx - 2\, a_n \sin\theta_n - a_n \cos\theta_n \frac{\varkappa}{v_0}(x_{n+1}^0 - x_n^0)\right\}$$

$$= m_n(x_{n+1}^0 - x_n^0)\left\{y_{n+1}\, c_1^0(x_{n+1}^0) - y_n\, c_1^0(x_n^0) + \int_{x_n^0}^{x_{n+1}^0} C\, dx\right\}. \tag{7.10}$$

Now the perturbed function p^ε is given by

$$p^\varepsilon(x) = 4\,\gamma\,(c^\infty - c_n), \qquad x_n^\varepsilon \leqq x \leqq x_{n+1}^\varepsilon. \tag{7.11}$$

Setting

$$p^\varepsilon = p^0 + \varepsilon\, P \tag{7.12}$$

we obtain

$$P(x) = -\,4\,\gamma\,\Delta c \sum_{m=0}^{2N-1}(-1)^m\, y_m\, \delta(x - x_m^0), \qquad 0 \leqq x \leqq N\lambda \tag{7.13}$$

and

$$P(x + N\lambda) = P(x) \tag{7.14}$$

as p^ε is periodic with period $N\lambda$ according to (7.3). In addition P has to satisfy

$$\int_0^{N\lambda} P(x)\, dx = 0. \tag{7.15}$$

From (7.13)−(7.15) the following Fourier series representation is obtained

$$P(x) = -\,4\,\gamma\,\Delta c\,\frac{1}{N\lambda}\sum_{m=0}^{2N-1}(-1)^m\, y_m \sum_{k=1}^{\infty}\cos\left(\frac{2\pi k}{N\lambda}(x - x_m^0)\right). \tag{7.16}$$

The connection between c_1^ε and p^ε is $c_1^\varepsilon = \frac{1}{2}HD^{-1}p^\varepsilon$, hence

$$C = \tfrac{1}{2}HD^{-1}P \tag{7.17}$$

leading with (7.16) to

$$C(x) = -\,2\,\gamma\,\Delta c\,\frac{1}{\pi}\sum_{m=0}^{2N-1}(-1)^m\, y_m \sum_{k=1}^{\infty}\frac{1}{k}\cos\left(\frac{2\pi k}{N\lambda}(x - x_m^0)\right). \tag{7.18}$$

Inserting (7.18) and (6.13) for c_1^0 into (7.10) a system of equations of the form

$$\sum_{n=0}^{N-1} A_{m-n} Y_n = 0, \qquad m = 0, \ldots, N-1 \tag{7.19}$$

is obtained where

$$Y_n = \begin{pmatrix} y_{2n} \\ y_{2n+1} \end{pmatrix} n = 0, 1, \ldots, N-1 \tag{7.20}$$

and A_k, $k \in \mathscr{Z}$, are 2×2 matrices, satisfying

$$A_{k+N} = A_k, \qquad k \in \mathbb{Z}. \tag{7.21}$$

The system (7.19) may be decoupled by Fourier transform into

$$\tilde{A}_m \tilde{Y}_m = 0, \quad m = 0, 1, \ldots, N - 1 \tag{7.22}$$

where

$$\tilde{Y}_m = \sum_{k=0}^{N-1} e^{(2\pi i/N)mk} Y_k \tag{7.23}$$

and similarly for $\tilde{A}_m$.

The condition for a non trivial solution is

$$\det \tilde{A}_m = 0 \tag{7.24}$$

where $m \in \{0, 1, \ldots, N - 1\}$. Setting $\tilde{Y}_{m'} = 0$ for $m' \neq m$ the corresponding perturbation mode $(y_0^{(m)}, y_1^{(m)}, \ldots, y_{2N-1}^{(m)})$ is given by

$$Y_k^{(m)} = \frac{1}{N} e^{-(2\pi i/N)km} \tilde{Y}_m, \quad k = 0, 1, \ldots, N - 1. \tag{7.25}$$

Eq. (7.24) leads after a rather lengthy calculation to the following quadratic equation

$$\begin{aligned}
&Q + \hat{\tau}\{\beta_1 - Q\beta_0 + \Sigma(Q\eta + 1 - \eta)\} + \hat{\tau}^2\{\Delta\eta(1 - \eta) - \beta_0\beta_1 + \Psi(\eta)\Sigma\} \\
&+ \hat{\varkappa}^2 R\eta(1 - \eta) + \hat{\varkappa}\{Q\eta + R(1 - \eta)\} \\
&+ \hat{\tau}\hat{\varkappa}\{\eta\beta_1 - R\beta_0(1 - \eta) + \eta(1 - \eta)\Sigma(1 + R)\} = 0
\end{aligned} \tag{7.26}$$

where

$$\hat{\varkappa} = \varkappa \frac{\lambda}{v_0} \frac{1}{2\,\mathrm{tg}\,\theta_\alpha} \tag{7.27}$$

and

$$\hat{\tau} = \lambda^2 v_0 \frac{\Delta c}{D_\mathrm{L}} \frac{m_\alpha}{2\,a_\alpha \sin\theta_\alpha}. \tag{7.28}$$

The quantities in the coefficients of (7.26) are as follows:

$$\Psi(\eta) = \frac{1}{\pi^3} \sum_{n=1}^{\infty} \frac{1}{n^3} \sin^2(\pi n \eta),$$

$$\Phi(\eta) = \frac{1}{2\pi^2} \sum_{n=1}^{\infty} \frac{1}{n^2} \sin(2\pi n \eta),$$

$$\beta_0 = \eta\,\Phi(\eta) - \Psi(\eta),$$

$$\beta_1 = (1 - \eta)\,\Phi(\eta) + \Psi(\eta),$$

$$\varphi_1(\eta) = \frac{N^2}{4\pi^2} \sum_{k=1}^{\infty} \frac{1}{(Nk - m)^2} \sin\left(\frac{2\pi}{N}(Nk - m)\eta\right),$$

$$\varphi_2(\eta) = \frac{N^2}{4\pi^2} \sum_{k=1}^{\infty} \frac{1}{(Nk - m)^2} \cos\left(\frac{2\pi}{N}(Nk - m)\eta\right),$$

$$\psi_1(\eta) = \frac{N^2}{4\pi^2} \sum_{k=1}^{\infty} \frac{1}{(Nk + m)^2} \sin\left(\frac{2\pi}{N}(Nk + m)\eta\right),$$

$$\psi_2(\eta) = \frac{N^2}{4\pi^2} \sum_{k=1}^{\infty} \frac{1}{(Nk + m)^2} \cos\left(\frac{2\pi}{N}(Nk + m)\eta\right),$$

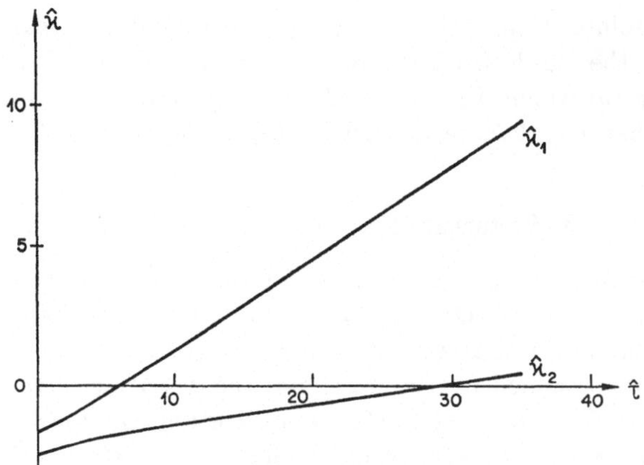

Fig. 5. The time constants $\hat{\varkappa}_{1,2}$ (Eq. 7.30) as a function of the parameter $\hat{\tau}$ which according to Eq. (7.28) is the normalised lamellar spacing

$$\Delta = (\varphi_1(\eta) + \psi_1(\eta))^2 + (\psi_2(\eta) - \varphi_2(\eta))^2 - (\varphi_2(0) - \psi_2(0))^2,$$

$$\Sigma = (\psi_2(0) - \varphi_2(0) + \varphi_2(\eta) - \psi_2(\eta))\,\mathrm{ctg}\,\frac{\pi\,m}{N} - \varphi_1(\eta) - \psi_1(\eta),$$

$$Q = \frac{a_\beta}{a_\alpha}\,\frac{m_\alpha}{m_\beta}\,\frac{\sin\theta_\beta}{\sin\theta_\alpha},$$

$$R = \frac{a_\beta}{a_\alpha}\,\frac{m_\alpha}{m_\beta}\,\frac{\cos\theta_\beta}{\cos\theta_\alpha}. \tag{7.29}$$

From (7.26) we obtain two solutions $\hat{\varkappa}_i(\hat{\tau}; Q, R, N, m, \eta)$, $i = 1, 2$. They are plotted in Fig. 5 for $Q = 1$, $R = 1$, $N = 5$, $m = 1$, $\eta = .4$. For $\hat{\tau} = 0$

$$\hat{\varkappa}_i = \frac{1}{2\,R\,\eta\,(1 - \eta)}\,\{-[R(1 - \eta) + Q\,\eta] \pm |R(1 - \eta) - Q\,\eta|\} \tag{7.30}$$

holds.

With increasing N the curve $\hat{\varkappa}_1$ is becoming steeper and its intersection with the $\hat{\tau}$-axis moves to the left, whereas the intersection of the curve $\hat{\varkappa}_2$ with the $\hat{\tau}$-axis moves to the right. In the limit $N \to \infty$ Eq. (7.26) becomes

$$\hat{\varkappa}\hat{\tau}\eta(1 - \eta)(1 + R) - \hat{\tau}^2\Psi(\eta) + \tau(Q\,\eta + 1 - \eta) = 0 \tag{7.31}$$

with the two solutions

$$\hat{\tau} = 0, \qquad \hat{\varkappa}\ arbitrary \tag{7.32}$$

and

$$\hat{\varkappa} = \frac{\tau\,\Psi(\eta) - (Q\,\eta + 1 - \eta)}{\eta(1 - \eta)(1 + R)} \tag{7.33}$$

(the limit curves do not satisfy (7.30) because of the non uniformity of the limit).

The intersection of the curve $\hat{\varkappa}$ given by (7.33) with the $\hat{\tau}$-axis is given by

$$\hat{\tau}_{\min} = \frac{Q\,\eta + 1 - \eta}{\Psi(\eta)} \tag{7.34}$$

and we have absolute stability for $\hat{\tau} > \hat{\tau}_{\min}$. It is remarkable that the point $\hat{\tau}_{\min}$ which separates the stable from the unstable $\hat{\tau}$ values coincides with the value of $\hat{\tau}$ given by (6.19) which was obtained from the principle of minimum undercooling (recall that $\hat{\tau}$ and λ are related by Eq. (7.28)).

8. Perturbation of the Steady State (II)

In the preceeding section we have found that the lamellar eutectic cannot grow for $\lambda < \lambda_{\min}$. In this section we show that there exist perturbations which grow exponentially also if $\lambda > \lambda_{\max}$. Contrary to $\lambda_{\min}$, $\lambda_{\max}$ depends on the temperature gradient $\bar{g}$ (defined by Eq. 4.21) and the deviation of the concentration in the liquid $c_{\mathrm{E}} - c^{\infty}$ from the eutectic concentration. Obviously if $\lambda_{\max} < \lambda_{\min}$ there exists no value of λ for which the lamellar eutectic structure can grow. The condition $\lambda_{\max} > \lambda_{\min}$ therefore will imply conditions on the value of $\bar{g}$ and $c_{\mathrm{E}} - c^{\infty}$. A particular example will be evaluated at the end of this section.

To derive an equation for $\lambda_{\max}$ we must go beyond the simple approximation $f \equiv 0$ which was sufficient for Sect. 6.

We set $f = f_0 + f_1$ where f_0 is the steady state interface and f_1 a perturbation. The time dependence of f_1 is assumed to be of the form $\exp(-\varkappa t)$. From (6.36) the following linear integro-differential equation is derived:

$$(1 + f_0'^2)^{-3/2} f_1'' - 3(1 + f_0')^{-5/2} f_0' f_0'' f_1'$$

$$= \frac{\bar{g}}{a_n} f_1 + \frac{m_n}{a_n} \left\{ 2(c_{\mathrm{E}} - c^{\infty})\gamma f_1 + \frac{1}{2} p^{(0)} f_1 + \frac{1}{2} H(f_1 H p^{(0)}) \right.$$

$$\left. + \frac{\varkappa}{D_{\mathrm{L}}} K[(c_{\mathrm{E}} - c_n)f_1] \right\}. \quad (8.1)$$

We have neglected a term proportional to the heat of fusion J arising from the second term in (3.15). The last term in (8.1) stems from the second term in (3.4) In addition we made the approximation

$$a = c_{\mathrm{E}} - c^{\infty} \tag{8.2}$$

which is obtained by setting the rhs of (6.20) equal to zero. The linear operator K is defined by

$$(Kg)(x) = (1/2\pi) \int dx' \, K_0(\gamma \, |x - x'|) g(x'). \tag{8.3}$$

For constant g it reduces to $Kg = g/(2\gamma)$ whereas for periodic g with average zero it takes the form $K = \frac{1}{2} H D^{-1}$ in the limit $\gamma \to 0$. For f_0' we use the approximation defined by Eqs. (6.22) and (6.23). We are looking for solutions of (8.1) of the form

$$f_1 = \sum_{n=0}^{\infty} b_n \cos\left\{ 2\pi n \left(\frac{x}{\lambda} - \frac{\eta}{2} \right) \right\}. \tag{8.4}$$

Inserting (8.4) into (8.1) an infinite set of linear equations is obtained:

$$\sum_{n=0}^{\infty} A_{mn} b_n = 0, \quad m = 0, 1, 2, \ldots. \tag{8.5}$$

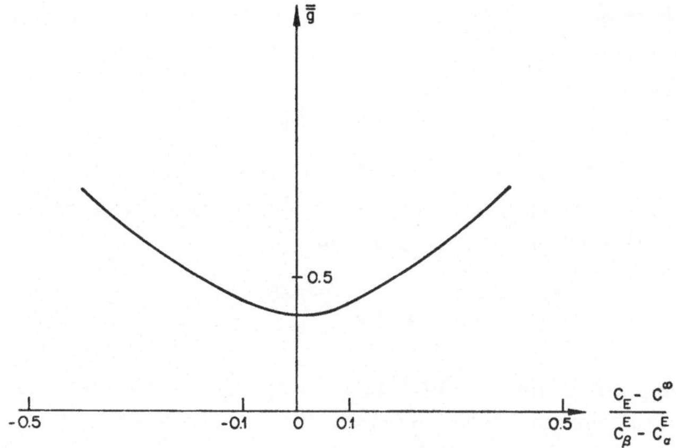

Fig. 6. Normalised temperature gradient $\bar{\bar{g}}$ above which stable lamellar growth occurs as a
function of the off eutectic composition

If we truncate (8.4) to

$$\sum_{n=0}^{N} A_{mn} b_n = 0, \qquad m = 0, 1, 2, \ldots, N \tag{8.6}$$

a non-trivial solution exists if

$$\Delta_N = 0 \tag{8.7}$$

where Δ_N is the determinant of the matrix (A_{mn}) of (8.6). Now Δ_N is a function
of the time constant $\varkappa$, the temperature gradient $\bar{g}$, the off eutectic concentration
difference $c_E - c^\infty$, the lamellar spacing λ and the parameters (see Sect. 5) ΔC,
v_0, D_L, m_α, m_β, a_α, a_β, θ_α and θ_β. (8.7) can be looked at as an equation for the
time constant $\varkappa$. The $N + 1$ solutions $\varkappa_i$ will correspond to $N + 1$ different
perturbations f_1 for the interface. Stability of growth is determined by the time
constant $\varkappa_c$, defined as the minimum of the set $\{\varkappa_i\}$. Numerical computations
have shown that $\varkappa_c$ decreases for increasing values of λ and passes through zero
at $\lambda_{\max}$. For values $\lambda > \lambda_{\max}$ the lamellar structure is unstable with respect to
the corresponding perturbation. If $\lambda_{\min} < \lambda_{\max}$, the question arises what the
precise value of λ is. Obviously to answer this question a new concept must be
introduced. In the range $\lambda_{\min} < \lambda < \lambda_{\max}$ all perturbations are stable. For each
λ in this range there will exist a particular perturbation with smallest time con-
stant $\varkappa_s(\lambda)$. By definition, we have $\varkappa_s(\lambda_{\min}) = \varkappa_s(\lambda_{\max}) = 0$. The absolute maximum of the function $\varkappa_s$ defines the most stable state. It will occur at some point
$\lambda = \lambda_0$. We have performed calculations using the perturbations of Sect. 7 and
this section. We find for reasonable values of the parameters that λ_0 is always
very close to $\lambda_{\min}$. For the example $a_\beta = a_\alpha$, $m_\alpha = m_\beta$, $\theta_\alpha = \theta_\beta = 70^0$, $c^\infty = c_E$,
$c_\beta^E - c_E = (c_\beta^E - c_\alpha^E)/2$ we find

$$\lambda_0 = \lambda_{\min} + (\lambda_{\max} - \lambda_{\min}) .005 . \tag{8.8}$$

To say it simply: lamellar eutectic growth occurs close to the minimum undercooling condition.

Whether lamellar growth can occur at all, is beyond the minimum undercooling principle. According to our theory it is possible only if $\lambda_{\max} > \lambda_{\min}$. From the critical growth condition

$$\lambda_{\max} = \lambda_{\min} \tag{8.9}$$

we can obtain conditions for the temperature gradient and the off eutectic composition. If we use the same parameters as above except that $c_{\mathrm{E}} \neq c^{\infty}$ and introduce the normalised temperature gradient

$$\bar{\bar{g}} = \bar{g}\,\frac{2 D_{\mathrm{L}}}{\Delta c \cdot v_0\, m_\alpha} \tag{8.10}$$

we obtain the growth diagram of Fig. 6. In agreement with experiments [7], the temperature gradient must be increased for increasing off eutectic composition to obtain stable lamellar growth.

9. Discussion and Summary

First we wish to state the main approximations of the present work. The two dimensional model we have introduced in Sect. 2 is adequate to describe the steady state of lamellar growth. But in principle there may exist perturbations perpendicular to the $x - z$-plane (Fig. 2) that grow exponentially in time, even if all the perturbations we considered are stable. A further approximation in this paper was, that we have used the steady state equations for the thermal and diffusion fields. This approximation is adequate to find the stability condition $\varkappa = 0$, but is only a approximation for $\varkappa \neq 0$. In order to simplify the equations in Sect. 5 we have assumed that $\lambda v_0/D_{\mathrm{L}} \ll 1$. Here λ is either the lamellar spacing or a characteristic length of a perturbation. A further approximation was, that we have used an iterative approach to find solutions for the interface, starting with a flat interface.

In Sect. 4 we applied our interface equations derived in Sect. 3 to investigate the stability of the planar interface of a dilute binary alloy. The stability condition (4.26) is identical with the one of Mullins and Sekerka [10]. In Sects. 5 and 6 we applied our method to study the properties of the steady state lamellar growth. Our result (6.17) for the average undercolling agrees with the result of Hunt and Jackson [5]. (6.17) corresponds to a completely flat interface. In the remaining part of Sect. 6 we show how one can obtain a first correction for f using the general interface equation (5.27). In addition we define an explicit linear integro-differential equation for f (Eq. 6.36). In Sects. 7 and 8 new and important results are developed. The calculation in Sect. 7 shows that the minimum undercooling condition defines a lower bound $\lambda_{\min}$ for the lamellar spacing. In Sect. 8 we show that there exists also an upper bound $\lambda_{\max}$. Because lamellar growth is only possible for $\lambda_{\max} > \lambda_{\min}$ we can derive conditions for the temperature gradient and the off eutectic concentration. An example is worked out in Fig. 6. Our calculation seems to solve the controversy between Refs. [8] and [9] in that we show that both long-wavelength and short-wavelength perturbations together determine the growth conditions. Finally we give an argument at the end of Sect. 8 why growth should always occur close to minimum undercooling.

Some attempts have been made to derive a thermodynamic stability criterium from the principle of minimum entropy production [6, 13]. But because the boundary conditions make growth problems intrinsically non-linear these attempts had to fail.

Appendix A

In this section a derivation of the interface equations

$$T_F - T_{\mathrm{E}} = \begin{cases} - m_\alpha(c_{\mathrm{LF}} - c_{\mathrm{E}}) + a_\alpha K \\ m_\beta(c_{\mathrm{LF}} - c_{\mathrm{E}}) + a_\beta K \end{cases} \tag{A.1}$$

is given. The upper (lower) line in (A.1) holds on the interface between the solid phase $\alpha\,(\beta)$ and the liquid L.

The composite whose phase diagram is given in Fig. 4 consists of two components A and B, concentrations refer to B. The chemical potentials of A in the phases α, β, L are denoted by μ_{A}^α, μ_{A}^β, $\mu_{\mathrm{A}}^{\mathrm{L}}$, similarly for B. The variables on which the chemical potential depends are temperature T, concentration c and pressure p. The respective partial derivatives are denoted as follows

$$\frac{\partial \mu}{\partial T} = - s, \qquad \frac{\partial \mu}{\partial p} = v, \qquad \frac{\partial \mu}{\partial c} = r. \tag{A.2}$$

Local equilibrium at the interface between α and L demands

$$\mu_x^\alpha(T_F, c_{\alpha F}, p_{\alpha F}) = \mu_x^{\mathrm{L}}(T_F, c_{\mathrm{LF}}, p_{\mathrm{LF}}) \tag{A.3}$$

for $x = \mathrm{A}$, B.

The pressure in the liquid is supposed to be uniform with some prescribed value p_0, hence $p_{\mathrm{LF}} = p_0$. The local pressure $p_{\alpha F}$ on the interface in the solid phase α is given by

$$p_{\alpha F} = p_0 + \sigma_{\mathrm{L}\alpha} K \tag{A.4}$$

where $\sigma_{\mathrm{L}\alpha}$ denotes the surface tension between the phases α and L and K is the local curvature of the interface.

Linearising the lhs of (A.3) in the point $(T_{\mathrm{E}}, c_\alpha^{\mathrm{E}}, p_0)$, the rhs in the point $(T_{\mathrm{E}}, c_{\mathrm{E}}, p_0)$ we obtain

$$\begin{aligned} - s_x^\alpha(T_F - T_{\mathrm{E}}) &+ r_x^\alpha(c_{\alpha F} - c_\alpha^{\mathrm{E}}) + v_x^\alpha \sigma_{\mathrm{L}\alpha} K \\ &= - s_x^{\mathrm{L}}(T_F - T_{\mathrm{E}}) + r_x^{\mathrm{L}}(c_{\mathrm{LF}} - c_{\mathrm{E}}) \end{aligned} \tag{A.5}$$

for $x = \mathrm{A}$, B.

In the special case of a plane interface we have

$$T_F - T_{\mathrm{E}} = - m_\alpha(c_{\mathrm{LF}} - c_{\mathrm{E}}) \tag{A.6}$$

and

$$T_F - T_{\mathrm{E}} = - n_\alpha(c_{\alpha F} - c_\alpha^{\mathrm{E}}) \tag{A.7}$$

by linearising the equation of state $T = T(c)$. Here $- m_\alpha$ and $- n_\alpha$ are the slopes of the curves 1 and 2 of Fig. 4 at the points c_{E} and c_α^{E}, respectively. For this case (A.5) leads to

$$n_\alpha s_x^\alpha + r_x^\alpha = m_\alpha s_{x_i}^{\mathrm{L}} + r_x^{\mathrm{L}} \tag{A.8}$$

176   S. Strässler and W. R. Schneider

for $x = $ A, B. Furthermore, we have from the Gibbs-Duhem-relation for fixed temperature and pressure

$$r_A^\alpha (1 - c_\alpha^E) + r_B^\alpha c_\alpha^E = 0, \tag{A.9}$$

$$r_A^L (1 - c_E) + r_B^L c_E = 0. \tag{A.10}$$

Eqs. (A.8)—(A.10) allow to eliminate the four quantities $r_A^\alpha, r_B^\alpha, r_A^L, r_B^L$ from (A.5). Solving (A.5) for $(T_F - T_E)$ and $(c_{\alpha F} - c_\alpha^E)$ leads to

$$T_F - T_E = - m_\alpha (c_{LF} - c_E) + a_\alpha K \tag{A.11}$$

and

$$c_{\alpha F} - c_\alpha^E = (m_\alpha / n_\alpha)(c_{LF} - c_E) + b_\alpha K. \tag{A.12}$$

Here a_α and b_α are given by

$$a_\alpha = \sigma_{L\alpha} \frac{v_A^\alpha (1 - c_\alpha^E) + v_B^\alpha c_\alpha^E}{(s_A^\alpha - s_A^L)(1 - c_\alpha^E) + (s_B^\alpha - s_B^L) c_\alpha^E}, \tag{A.13}$$

$$b_\alpha = \frac{\sigma_{L\alpha}}{n_\alpha} \left\{ \frac{v_A^\alpha (1 - c_E) + v_B^\alpha c_E}{(s_A^\alpha - s_A^L)(1 - c_E) + (s_B^\alpha - s_B^L) c_E} \right.$$
$$\left. - \frac{v_A^\alpha (1 - c_\alpha^E) + v_B^\alpha c_\alpha^E}{(s_A^\alpha - s_A^L)(1 - c_\alpha^E) + (s_B^\alpha - s_B^L) c_\alpha^E} \right\} \tag{A.14}$$

By replacing $- m_\alpha$ by m_β, $- n_\alpha$ by n_β (m_β, n_β being the slopes of the curves 3 and 4 at c_E and c_β^E respectively) and changing α into β for the rest of the quantities the corresponding equations for the interface between β and L is obtained.

Remark 1: For $n_\alpha \to \infty$, $n_\beta \to \infty$ we get

$$c_{\alpha F} = c_\alpha^E, \qquad c_{\beta F} = c_\beta^E \tag{A.15}$$

which is the approximation used in the text.

Remark 2: In a similar way the interface equation for a dilute composite is derived:

$$T_F - T_A = - m c_{LF} + a_0 K \tag{A.16}$$

where

$$a_0 = \sigma_{LS} \frac{v_A^S}{s_A^S - s_A^L} = \sigma_{LS} \frac{T_A}{J_A}. \tag{A.17}$$

Here T_A is the melting temperature, J_A the heat of fusion of component A. For c_{SF} we get

$$c_{SF} = \frac{m}{n} c_{LF} \equiv k c_{LF} \tag{A.18}$$

where m and n are the slopes of the curves 1 and 2 of Fig. 4 at the point $c = 0$.

Appendix B

In this section we shall discuss several mathematical details not treated in the text. We start with the investigation of the properties of the function

$$\Psi(x) = \frac{1}{4\pi} \int dx' \ln \frac{(x - x')^2 + (z - f(x'))^2}{(x - x')^2} \tag{B.1}$$

where f is uniformly continuous and bounded. If f is constant, say z_0, we denote by $\Psi_0(\boldsymbol{x})$ the corresponding function given by (B.1). An easy calculation gives

$$\Psi_0(\boldsymbol{x}) = \tfrac{1}{2}|z - z_0|. \tag{B.2}$$

Obviously Ψ_0 satisfies

$$\varDelta\Psi_0 = 0 \tag{B.3}$$

in the domains

$$G^0_{\mathrm{L(S)}} = \{\boldsymbol{x} = (x, z)\,|\,z \underset{(<)}{\gtrless} z_0\}. \tag{B.4}$$

Our aim is to show that Ψ has analogous properties, i.e.

$$\varDelta\Psi = 0 \tag{B.5}$$

in the domains

$$G_{\mathrm{L(S)}} = \{\boldsymbol{x} = (x, z)\,|\,z \underset{(<)}{\gtrless} f(x)\}. \tag{B.6}$$

As $|f(x)| < c$ for some constant c the following inequalities hold

$$\tfrac{1}{2}|z - c| \leq \Psi(\boldsymbol{x}) \leq \tfrac{1}{2}|z + c| \tag{B.7}$$

hence

$$\lim_{z \to \pm\infty} (1/z)\,\Psi(\boldsymbol{x}) = \pm\tfrac{1}{2}. \tag{B.8}$$

From

$$\Psi(x + a, z + b) - \Psi(x, z) = \frac{1}{4\pi} \int dx' \ln \frac{(x - x')^2 + (z + b - f(x' + a))^2}{(x - x')^2 + (z - f(x'))^2} \tag{B.9}$$

the following inequality can be derived

$$|\Psi(x + a, z + b) - \Psi(x, z)|^2 < \sup_{y \in \mathbb{R}} |(z + b - f(y + x + a))^2 - (z - f(y + x))^2|. \tag{B.10}$$

From (B.10) and the uniform continuity of f follows the continuity of Ψ.

If we define

$$\Psi^{(n)}(x) = \frac{1}{4\pi} \int\limits_{-n}^{n} dx' \ln \frac{(x - x')^2 + (z - f(x'))^2}{(x - x')^2} \tag{B.11}$$

for integer n the sequence $\{\Psi^{(n)}\}$ tends to Ψ pointwise as $n \to \infty$. Let $g \in \mathscr{S}(\mathbb{R}^2)$ (Schwartz test function space) with support in G_{L} or G_{S}. An easy calculation gives

$$(\Psi^{(n)}, \varDelta g) = \int d^2x\, \Psi^{(n)}(\boldsymbol{x})\, \varDelta g(\boldsymbol{x}) = -2 f\!\int dx \left(\frac{1}{n - x} + \frac{1}{n + x}\right) \bar{g}(x) \tag{B.12}$$

where

$$\bar{g}(x) = \int dz\, g(x, z) \tag{B.13}$$

and f denotes the principal value integral.

The right side of (3.12) tends to zero as $n \to \infty$, hence

$$(\Psi, \varDelta g) = 0 \tag{B.14}$$

for all $g \in \mathscr{S}(\mathbb{R}^2)$ with support in G_{L} or G_{S}. It follows from Weyl's lemma that

$$\varDelta\Psi = 0 \quad \text{in } G_{\mathrm{L}}, G_{\mathrm{S}}. \tag{B.15}$$

Next we consider

$$\Phi(\boldsymbol{x}) = -\frac{1}{2\pi} \int dx'\, h_1(x')\, \frac{\partial}{\partial x'} \ln|\boldsymbol{x} - \boldsymbol{x}'_F| \tag{B.16}$$

where $\boldsymbol{x}'_F = (x', f(x'))$. We assume f to be bounded and continuous and having continuous and bounded derivatives f', f'' in the complement of a set S, which is finite or countable without accumulation point. At the points $x \in S$ the left and right side limits of f' and f'' may be different. Furthermore we assume h_1 to be piecewise continuous and $h_1 = h'_2$, where h_2 is continuous and bounded and vanishes at the points of S. A partial integration leads then from (B.16) to

$$\Phi(\boldsymbol{x}) = \frac{1}{2\pi} \int dx'\, h_2(x')\, \frac{\partial^2}{\partial x'^2} \ln|\boldsymbol{x} - \boldsymbol{x}'_F| \tag{B.17}$$

or explicitly

$$\Phi(\boldsymbol{x}) = \frac{1}{2\pi} \int dx'\, h_2(x') \left| \frac{A}{\varrho^4} - \frac{z f''(x')}{\varrho^2} \right| \tag{B.18}$$

with

$$\varrho^2 = (x - x')^2 + (z - f(x'))^2 \tag{B.19}$$

and

$$A = \varrho^2 \{1 + f'^2(x') + f(x') f''(x')\} - 2 B^2, \tag{B.20}$$

$$B = (x - x') + f'(x')(z - f(x')). \tag{B.21}$$

For the first part in Eq. (B.18) one easily derives the asymptotic behavior $O(|z|^{-1})$ as $|z| \to \infty$ uniformly in x. The same holds for the second part only if an additional assumption on the function $h_2 f''$ (which is continuous and bounded by the above assumptions) is made, e.g. $h_2 f'' = g'$, where g is continuous and bounded. For the case $f = 0$ this additional condition becomes redundant.

References

1. Zener, C.: Trans. AIME **167**, 550 (1946)
2. Tiller, W. A.: A.S.M. Liquid Metals and Solidification, p. 276. Cleveland Ohio 1958
3. Hillert, M.: Jernkont. Annl. **141**, 757 (1957)
4. Hillert, M.: The Mechanism of Phase Transformations in Crystalline Solids, p. 231, Monograph and Report Series No. 33, Institute of Metals (1969)
5. Jackson, K. A., Hunt, J. D.: Trans. AIME **236**, 1129 (1966)
6. Lesoult, G., Turpin, M.: Mem. Sci. Rev. Metal. **66**, 619 (1969)
7. Mollard, F. R., Flemings, M. C.: Trans. TMS AIME **239**, 1526 (1967)
8. Hurle, D. T. J., Jakeman, E.: Int. Conf. on Crystal Growth, Birmingham U.K. July, 1968, J. cryst. Growth **3**, 574 (1968)
9. Cline, H. E.: Trans. AIME **242**, 1613 (1968)
10. Mullins, W. W., Sekerka, R. F.: J. appl. Phys. **34**, 323 (1963)
11. Hunt, J. D., Hurle, D. T. J., Jackson, K. A., Jakeman, E. J.: Trans AIME **1**, 381 (1970)
12. Sekerka, R. F.: J. cryst. Growth **10**, 239 (1971)
13. Tiller, W. A.: J. appl. Phys. **34**, 3615 (1963)

Dr. S. Strässler
Brown Boveri Research Center
CH-5401 Baden, Switzerland

Phys. cond. Matter 17, 179—181 (1974)

Exchange Effect on the Electronic Specific Heat

David Y. Kojima and A. Isihara

Department of Physics, State University of New York at Buffalo, Buffalo, New York

Received October 8, 1973

By improving the evaluation of the contribution from the first order exchange graphs reported recently by Tsai and Isihara, the electronic specific heat of an electron gas is evaluated to order e^2. It is shown that the specific heat contains a logarithmic term.

In a recent article (hereafter to be called I), Tsai and one of the present authors [1] discussed a statistical mechanical approach to the electronic specific heat. In that paper, the grand partition function is evaluated for a low but finite temperature so that the specific heat is obtained by a temperature differentiation of the internal energy. In contrast, Gell-Mann and Pines [2] used an approximate formula in which a momentum derivative of the energy at 0 °K appears. It is the purpose of the present paper to remark on the contribution of the first order exchange graphs, improving the calculation in I.

The contribution to the grand partition function from the first order exchange graphs is given by (cf. Eqs. (2.8) and (3.12) of I)

$$\ln \Xi_x = \frac{V\beta}{64\,\pi^6} \int d\boldsymbol{q}\,d\boldsymbol{p}\,u(q)\,f(p)\,f(\boldsymbol{p}+\boldsymbol{q}) \tag{1}$$

$$= \frac{V e^2}{8\,\pi^3 \beta}\,J \tag{2}$$

where $u(q) = 4\pi e^2/q^2$, $f(p)$ is the Fermi distribution and

$$J = \pi \int\limits_{-\infty}^{\eta} d\eta\,\{F_{-1/2}(\eta)\}^2 , \tag{3}$$

$$F_{-1/2}(\eta) = \frac{1}{\Gamma(-\tfrac{1}{2}+1)} \int\limits_{0}^{\infty} dx\,\frac{x^{-1/2}}{e^{x-\eta}+1}. \tag{4}$$

$$\eta = \beta\,p_{\mathrm{F}}^2. \tag{5}$$

The integral J can be rewritten as

$$J = J_0 + J_1 \tag{6}$$

where

$$J_0 = 4 \int\limits_{0}^{\infty} x\,f(x)\,dx , \tag{7}$$

$$J_1 = \int\limits_{0}^{\infty} g(x)\,f(x)\,dx , \tag{8}$$

180   D. Y. Kojima and A. Isihara

$$g(x) = 2 \int_0^\infty u^{-1/2}(1-u)^{-1} \ln(1 - e^{-x|1-u|}) \, du \, . \tag{9}$$

The integral J_0 is simply

$$J_0 = 2\,\eta^2 + (2\,\pi^2/3) \tag{10}$$

in agreement with Eq. (2.10) of I. In I, this result was obtained from Eq. (1) based on a consideration of the common volume of the two Fermi spheres corresponding to $f(p)$ and $f(\boldsymbol{p}+\boldsymbol{q})$. In this consideration, it is important to express the integral correctly in conformity with the geometrical shape of the common volume. We remark that Eq. (2.9) of I must be replaced by

$$J_0 = \frac{V\beta}{64\,\pi^6} \int d\boldsymbol{q}\, u(q) \left\{ 2 \int_{p/2}^{p_{\mathrm{F}}} dz \int_0^{(p_{\mathrm{F}}^2-z^2)^{1/2}} 2\,\pi\,\varrho\,d\varrho \right.$$
$$\left. - \frac{\pi^2}{3\,\beta^2} \int d\boldsymbol{p}\, \theta\,(p_{\mathrm{F}}^2 - p^2)\, \delta'(p_{\mathrm{F}}^2 - (\boldsymbol{p}+\boldsymbol{q})^2) \right\} \tag{11}$$

where θ is a step function and δ is the Dirac delta function.

The integral J_1 is our main concern in this paper. One can write

$$g(x) = g_{00}(x) + \sum_j g_j(x) \tag{12}$$

where $j = 0, 2, 4, \ldots$ and

$$g_{00}(x) = -2 \int_1^\infty v^{-1}(1+v)^{-1/2} \ln(1 - e^{-xv}) \, dv \, , \tag{13}$$

$$g_j(x) = -\frac{1}{x}\,\frac{(2j+1)!!}{2^{j-1}(j+1)\,x^j} \sum_{n=1}^\infty n^{-j-2} \left\{ 1 - e^{-nx} \sum_{m=0}^j \frac{(n\,x)^m}{m!} \right\} . \tag{14}$$

We have conducted our calculation in such a way that only one Fermi distribution function appears in the integrand of J. Applying the Sommerfeld method and neglecting terms with $\exp(-\eta)$, we find

$$\int_0^\eta g(x)\,dx = A - \frac{\pi^2}{3}\,\ln\eta + \sum_{j=2,4,\ldots} \frac{(2j+1)!!}{j\,2^{j-1}(j+1)} \left(\sum_{n=1}^\infty n^{-j-2} \right) \eta^{-j}$$
$$- \frac{2\,\pi^2}{9\,\eta^{3/2}} \left\{ 1 - \frac{3}{10\,\eta} - \frac{3}{8\,\eta^2} - \cdots \right\} .$$

With C for the Euler constant, A is given by

$$A = -\frac{\pi^2}{3}\,C - 2 \sum_{n=1}^\infty n^{-2} \ln n - \frac{\pi^2}{6} \sum_{j=2,4,\ldots} \frac{(2j+1)!!}{j(j+1)!\,2^{j-1}}$$
$$+ \frac{\pi^2}{3} \left(2^{1/2} - \frac{1}{2} \ln \left| \frac{2^{1/2}+1}{2^{1/2}-1} \right| \right) . \tag{15}$$

The contribution from the derivatives of $g(x)$ is easy to evaluate. We finally arrive at

$$J = 2\,\eta^2 \left\{ 1 + \frac{\pi^2}{3\,\eta^2} \left(1 + \frac{3\,A}{2\,\pi^2} - \frac{1}{2}\ln\eta \right) + \frac{10\,\pi^4}{2.12^2\,\eta^4} \right.$$
$$\left. - \frac{\pi^2}{9\,\eta^{7/2}} \left(1 - \frac{3}{10\,\eta} - \cdots \right) + \cdots \right\} . \tag{16}$$

Following the golden rule of the grand ensemble method, one can express the Fermi momentum as a function of the density n. The evaluation of the internal energy and the specific heat is straightforward. We find

$$c_V = c_V^0 \left\{ 1 + \left(\frac{4}{9\pi^4} \right)^{1/3} r_s \left[\frac{5}{2} + \frac{3A}{\pi^2} - \ln\left(\frac{p_0^2}{kT} \right) \right] + \cdots \right\} \qquad (17)$$

where

$$c_V^0 = p_0\, k^2\, T/6 \qquad (18)$$

is the ideal gas term and $p_0^2 = 3\pi^2 n$.

In contrast to I, the result given by Eq. (17) contains a logarithmic term which becomes dominant towards the lowest temperature. However, because a perturbation method has been employed caution is necessary about the term. The constant terms other than the logarithmic term in Eq. (17) are important for r_s of order 3 to 5. Taking into consideration higher order graphs, we estimate the lowest temperature up to which Eq. (17) is meaningful to be around 0.5 °K. We shall report on the contribution of higher order graphs in the near future.

Acknowledgement. While the publication of this note has been postponed due to the leave taken by one of the authors (A. I.), the authors learned that a similar result without the constant term has been obtained independently by B. Horovitz and R. Thieberger.

References

1. Tsai, J. T., Isihara, A.: Phys. kondens. Materie **14**, 200 (1972)
2. Pines, David: Phys. Rev. **92**, 620 (1953), Murray Gell-Mann, Phys. Rev. **106**, 369 (1957)

Dr. D. Y. Kojima
Dr. A. Isihara
Department of Physics
State University of New York at Buffalo
Buffalo, N.Y. 14214, USA

Phys. cond. Matter 17, 183—187 (1974)

Note on the Hall Effect in Ice

P. Gosar *

Laboratoire de Spectrométrie Physique, Université I de Grenoble, B.P. 53, 38041 Grenoble,
France

Received June 18, 1973 / In Revised Form July 24, 1973

The magnetic field dependence of the polarization correlation function is investigated. It
is shown that the Hall effect is extremely small and probably inobservable because of the
lattice anti-polarization produced by the motion of ionic defects and because of the incoherent
motion of Bjerrum defects.

1. Introduction

Our present understanding of the protonic conduction in ice is based on the
concepts developped by Bjerrum [1], Onsager [2, 3], Gränicher [4], Eigen [5, 6],
Jaccard [7], and others. Orientational L, D and ionic H_3O^+, HO^- defects parti-
cipate in the conduction process. Thermally activated interbond jumps and intra-
bond tunnelings of protons are responsible for the diffusion of orientational and
ionic defects, respectively. Onsager and Dupuis [2] pointed out that the successive
proton tunnelings should be correlated in order to account for the high dc proton
mobility. The theory of the correlated tunneling has been worked out by Gosar
[8, 9] and Hofacker et al. [10, 11]. These theories show that the formation of a
ionic energy band which is very narrow but otherwise similar to the excitonic or
electronic bands in semiconductors is likely. The motion of ionic defects is therefore
considered as a coherent wave motion in the fundamental band. The coherence
length should be of the order of a few intermolecular distances.

The high mobility of ionic defects in ice suggested a search for the experimental
detection of the Hall current. Stationary experiments by Bullemer and Riehl [12,
13] seemed to confirm by a positive sign Hall mobility the presence of protons as
charge carriers in ice. However, recent pulse experiments by Kern and Bullemer
[14] only establish an upper limit 10^{-2} cm² V⁻¹ sec⁻¹ for the Hall mobility.

We show in this theoretical note that the Hall field should be negligible if the
present picture of the conduction processes in ice is correct. Some doubts on the
possibility of the Hall effect in ice have already been expressed by Eigen et al. [6],
but no formal proof has been given.

The motion of ions in ice cannot be considered as a simple motion of free
carriers. When a ion moves through the crystal, it leaves behind a trace of turned
molecules. This polarization prevents the passage of another ion along the same
path and would eventually stop the dc conduction if orientational or Bjerrum
defects were not present. The diffusion of orientational defects destroys the polari-

---

* On leave of absence from Physics Department and Institute "J. Stefan", University
of Ljubljana, Ljubljana, Yugoslavia.

zation by reorienting the molecules. The effect of the internal polarization on the structure of the ionic bands and the mobility has been recently investigated by Minagawa [15]. In his calculation the protonic disorder of the ice crystal is taken into account. Minagawa showed that the energy bands in the three-dimensional ice crystal retain some characteristics of the one-dimensional bands. The band width is smaller than the band width of the corresponding three-dimensional tight-binding band and the density of states has maxima near the band edges. This resembles very much the case of one-dimensional bands and should have a decisive effect on the Hall mobility of carriers.

2. Theory of the Hall Effect

The Hall conductivity is equal to the anti-symmetric part $\sigma_{\mu\nu}^{(a)}(\omega)$ of the conductivity tensor $\sigma_{\mu\nu}(\omega)$ of the crystal in an external magnetic field $\boldsymbol{B}$:

$$\sigma_{\mu\nu}^{(a)}(\omega) = \tfrac{1}{2}\left[\sigma_{\mu\nu}(\omega) - \sigma_{\nu\mu}(\omega)\right]; \quad \mu,\nu = x,y,z,\tag{1}$$

where ω is the frequency of the applied electric field. It is well known that the conductivity may be calculated from the correlation functions for the current density or the polarization $\boldsymbol{P}$ [16, 17]. A particularly useful form of the correlation function for our study represents the imaginary-time thermodynamic Green's function

$$D_{\mu\nu}(\tau) = -\langle T\, P_\mu(\tau)\, P_\nu(0)\rangle;$$
$$P_\mu(\tau) = \exp(H\tau)\, P_\mu \exp(-H\tau); \quad \hbar = 1,\tag{2}$$

where τ is a real variable in the interval $-\beta < \tau < \beta$. Further, $\langle\ \rangle$ represents the thermodynamic average and T is a time ordering operator. H is the Hamiltonian of the crystal in the magnetic field. $\sigma_{\mu\nu}(\omega)$ may be expressed in terms of the analytic continuation of the Fourier transform $D_{\mu\nu}(z_n)$, $z_n = 2\pi i n/\beta$, n being integer, as

$$\sigma_{\mu\nu}(\omega) = -i\omega\, V D_{\mu\nu}(\omega + i\varepsilon),\tag{3}$$

where V is the volume of the crystal and ε is an infinitesimal positive parameter. $D_{\mu\nu}(z)$ is an analytic function with a branch cut along the real axis and allows a systematic perturbation expansion [18]. Only a linear dependence of $\sigma_{\mu\nu}(\omega)$ on $\boldsymbol{B}$ should be calculated.

First we shall investigate if the motion of ions alone could give rise to the Hall effect. Let us consider a perfect ice crystal with only one H_3O^+ ion. All the hydrogen bonds in the crystal are occupied by one proton in a way which conserves the entity of water molecules and the ion intact. Many such protonic configurations are possible (Bernal-Fowler configurations). As the motion of the ion is a slow tunneling process, we shall use the tight-binding approach. The starting point for this approach is a representation of localized ionic states. A particular localized state of the crystal is specified by the protonic configuration i and by the vibrational excitations of oxygens and protons. The tunneling takes place in the vibrational ground state of the proton in the corresponding hydrogen bond. At this stage of the calculation the lattice vibrations may be disregarded. For this reason we shall first consider only the vibrational ground states and we shall denote the localized states of the crystal simply by $|i\rangle$. The Hamiltonian H may be split into a diagonal H_0 and off-diagonal H_1 part with respect to the basis $|i\rangle$. The energy $\langle i| H |i\rangle$ is practically the same for all configurations. Off-diagonal elements $\langle j| H |i\rangle$ describe the tunneling processes and are different from zero when one can transform the configuration i into the configuration j by the translation of one proton toward a neighbouring oxygen atom. Such translations are possible only in the three bonds which connect the H_3O^+ ion with the neighbouring oxygens.

Let us now consider the effect of the homogeneous magnetic field B on the wave functions and the matrix elements of H. Two effects of the magnetic field must be distinguished. First, if $\psi_R(r)$ represents a localized wave function of the proton in the potential well around the point R in the presence of a magnetic field the wave function $\psi_{R'}(r)$ for a proton in an identical well situated at R' is [19, 20]

$$\psi_{R'}(r) = \exp\{- i(e/2) [B \times (R' - R)] \cdot r\} \, \psi_{R'}(r) \,, \tag{4}$$

where e is the proton charge. The phase factor in (4) describes the effect of the magnetic field on the translational motion of the charge carrier and is essential for the understanding of the Hall effect. Second, the magnetic field also modifies to some extent the localized wave functions because of the linear coupling of the orbital angular momentum of the proton to the field B. If the ground state of the proton in the potential well is an s state, the protonic wave function remains almost unchanged. In any case the modifications of the localized wave functions cannot alone produce the Hall effect.

Because of (4) the tunneling matrix element for the transition from $|i\rangle$ to $|j\rangle$ is approximately equal to [19]

$$\langle j| H_1 |i\rangle = \exp\{i(e/2) B \cdot [R_j \times R_i]\} J_{ji} \,, \tag{5}$$

where R_j and R_i are the position vectors of the centers of two potential wells in the hydrogen bond where the transition took place and J_{ji} is the tunneling matrix element at zero magnetic field. J_{ji} is real. The diagonal matrix elements of H are not affected by the magnetic field in the first approximation.

The perturbation expansion of $D_{\mu\nu}(\tau)$ with respect to H_1 is now possible. Note that the polarization operator P is diagonal in the localized representation. The individual terms of the expansion contain the factors of the type

$$\langle i| H_1(\tau_n) |j\rangle \langle j| H_1(\tau_{n-1}) |k\rangle \cdots \langle l| H_1(\tau_m) |p\rangle$$
$$\times \langle p| P_\mu(\tau) |p\rangle \langle p| H_1(\tau_{m-1}) |r\rangle \cdots \langle s| H_1(\tau_2) |t\rangle$$
$$\times \langle t| H_1(\tau_1) |i\rangle \langle i| P_\nu(0) |i\rangle \,, \tag{6}$$

where τ_n are the intermediate times. The expression (6) describes the process in which the H_3O^+ ion moves from the initial state $|i\rangle$ through the intermediate states $|j\rangle$, $|k\rangle$, $|l\rangle$, ... back to the original state. The configuration of all protons in the final state must be exactly the same as in the initial state. This is possible only if all hydrogen bonds along the path of the ion are traversed by the ion equal times in the opposite directions. The transfer of the proton in the hydrogen bond produces the polarization which can be completely destroyed only by another passage in the opposite direction if the rotations of water molecules are excluded. The area of the surface which could be spread over the ion path is therefore zero. This is a peculiarity of the ice model. If the H_3O^+ ion would be a simple charged particle like an charged interstitial, one could imagine quite easy a path with non-zero surface. The motion of an interstitial does not produce any specific polarization of the lattice. Therefore, no restrictions on the form of the closed path are imposed in this case. Because of the path restrictions the tunneling matrix elements in (6) appear always in pairs

$$\langle j | H_1(\tau_m) | k \rangle \langle k | H_1(\tau_n) | j \rangle . \tag{7}$$

The inspection of (5) shows that (7) does not depend on the magnetic field. The conductivity tensor is not changed by the presence of the magnetic field. No Hall effect can exist. The situation is exactly the same as in the case of the one-dimensional motion of charge carriers where the surface of the closed path is also zero and the Hall effect evidently can not be observed although the dc conductivity is finite.

It rests to investigate if the motion of ions combined with the diffusion of Bjerrum defects could produce the Hall effect. The restoration of the original protonic configuration after the passage of the H_3O^+ ion along the closed path with non-zero surface is possible if an L or D defect diffuses to the path, makes a tour of it in the allowed direction, and afterwards returns to its original position in the crystal. Such a closed path requires at least six hydrogen bonds. We see that the elementary process which could contribute to the Hall effect is extremely complicated also if one completely disregards the fact, that the motion of an orientational defect is possible only by the thermal activation of the proton in the corresponding hydrogen bond. Already the complexity of these processes indicates their low probability. A thermal activation requires the cooperative action of many lattice modes which are coupled to the proton. After the excitation the water molecule may rotate producing the motion of the defect. It is very unlikely that the local excitation of the crystal could persist for time intervals which are long in the atomic scale. Because of the coupling to lattice vibrations and internal electromagnetic fields a fast de-excitation takes place. In such circumstances there are so many possible ways of the evolution of the system that the amplitude of individual intermediate states is small and the coherence is lost. The evolution paths via activation processes therefore cannot contribute to $-\langle T P_\mu(\tau) P_\nu(0) \rangle$. A quantitative estimate of the above statements is clearly difficult. Nevertheless, it is generally accepted that the thermal activation or de-excitation is not a coherent or phase conserving process. We conclude that the participation of the orientational defects in the conduction mechanism cannot give rise to the Hall effect. Also the fact that the concentration of Bjerrum defects near the melting point is much higher than for the ions cannot change this conclusion. The charge carrying capacity of defects is high and they easily depolarize the lattice. But this depolarization is a result of many incoherent processes and does not allow the appearance of the Hall current which can be produced only by the coherent motion of charge carriers in the magnetic field.

For the sake of completeness let us note that the matrix elements for the rotation of excited water molecules depend on the magnetic field in a similar way as tunneling matrix elements (5) as the rotation corresponds to the motion of the proton from the potential well in the occupied bond to the unoccupied potential well in the adjacent bond. This however can not contribute to the Hall effect because of the reasons already mentioned.

In conclusion, we see that, according to the present concepts of the protonic conduction, the Hall effect in ice is practically inobservable. The coherent motion of ionic defects is in fact one-dimensional. The Hall effect is possible only if the conduction mechanism could be described by a single particle motion which in addition is not thermally activated, i.e. the consecutive particle jumps must be mutually correlated or coherent.

Acknowledgement: The author wishes to thank Prof. B. Bullemer for most valuable discussions.

References

1. Bjerrum, N.: Kgl. Danske Videnskab. Selskab. Mat.-Fys. Medd. **27**, 32, 41 (1951)
2. Onsager, L., Dupuis, M.: Rendiconti S.I.F. — Corso X, Bologna 294 (1960)
3. Onsager, L., Dupuis, M.: In: Electrolytes, ed. B. Pesce, p. 27. New York: Pergamon Press 1962
4. Gränicher, H.: Phys. kondens. Materie **1**, 1 (1963) — review article
5. Eigen, M., De Maeyer, L.: Proc. Roy. Soc. (London) A **247**, 505 (1958)
6. Eigen, M., De Maeyer, L., Spatz, H. Ch.: Ber. Bunsenges. Phys. Chem. **68**, 19 (1964)
7. Jaccard, C.: Helv. Phys. Acta **32**, 89 (1959)
8. Gosar, P.: Nuovo Cimento **30**, 931 (1963)
9. Gosar, P., Pintar, M.: Phys. Status Solidi **4**, 675 (1964)
10. Fischer, S. F., Hofacker, G. L.: In: Physics of ice. Eds. N. Riehl, B. Bullemer, and H. Engelhardt, p. 369. New York: Plenum Press 1969
11. Fischer, S. F., Hofacker, G. L., Ratner, M. A.: J. Chem. Phys. **52**, 1934 (1970)
12. Bullemer, B., Riehl, N.: Phys. Letters **22**, 411 (1966)
13. Bullemer, B., Riehl, N.: Phys. kondens. Materie **7**, 248 (1968)
14. Kern, W., Bullemer, B.: to be published
15. Minagawa, I.: preprint
16. Kubo, R.: J. Phys. Soc. Japan **12**, 570 (1957)
17. Gosar, P.: Phys. Status Solidi **10**, 91 (1965)
18. Abrikosov, A. A., Gorkov, L. P., Dzyaloshinskii, I. E.: Methods of quantum field theory in statistical physics. Englewood Cliffs: Prentice-Hall 1963
19. Holstein, T., Friedman, L.: Phys. Rev. **163**, 1019 (1968)
20. Landau, L. D., Lifshitz, E. M.: Quantum Mechanics. London: Pergamon Press 1958

Prof. Dr. P. Gosar
Physics Department and Institute "J. Stefan",
University of Ljubljana
19 Jadranska
61000 Ljubljana, Yugoslavia

Phys. cond. Matter 17, 189—214 (1974)
© by Springer-Verlag 1974

Spinpolarized OPW-Band Structure of Magnetic Semiconductors with Application to EuO. Part I

Karl Lendi*

Institut für Theoretische Physik
Eidg. Technische Hochschule, Zürich, Switzerland

Received July 17, 1973

The OPW-method has been generalized for the calculation of spinpolarized energy bands in magnetic semiconductors and applied to the ferromagnetic semiconductor EuO.

Part I of this paper contains the general formulation of the method, a grouptheoretical analysis and qualitative predictions for the band structure. It is shown why the application of the OPW-method to heavy substances, as for instance to magnetic semiconductors, gives rise to difficulties due to instabilities in the related eigenvalue problem.

Introduction

A lot of attempts have been made towards an understanding of the physical properties of rare-earth compounds [1]. In particular optical experiments have shown up an interesting connection between magnetism and electronic structure in the series of the Europium-chalcogenides [2, 3, 4]. However, any interpretation of optical spectra on the basis of experimental results only is almost impossible. A detailed classification of the nature and structure of interband transitions as observed in reflectivity-measurements for instance can be given only in connection with an electronic band structure calculation of the solid. Therefore we have set up a generalized OPW-treatment for the particular case of heavy ferromagnetic semiconductors and finally have applied it to EuO.

Since the rocksalt-structure of EuO suggests an ionic configuration, whereas optical measurements show a semiconductor behavior, the choice of the method for the determination of the band structure seems to be a risk. It is in this respect that the OPW-method should be best suited because of its ability to treat both cases [5, 6]. In any case difficulties are expected to arise from the heaviness of the elements involved and from the special role evidently played by the electrons of the magnetic $4f$-shell. These are thought to be well localized inside the atomic core, but nevertheless highlying in energy because of the strong repulsive contribution of the centrifugal potential. One therefore cannot exclude some kind of overlapping or mixing between band states and localized orbital states, which should not give rise to difficulties similar to those met in the application of the OPW-method to the transition-metals [7, 8] because of the drastic difference between d- and f-states. Including the f-electrons as localized core states the present calculation gives for the first time an application of the OPW-method to

---

* *Present address:* Centro de Investigación del Instituto Politécnico Nacional, Departamento de Física, Apartado Postal 14—740; México 14, D.F. México.

a situation where the highest core level energy falls into the region of band energies, actually separating valence- and conduction bands from each other. It must be emphasized that there is no adjustment leading to this extraordinary position of the $4f$-level. The zeroth Fourier coefficient as the only adjustable parameter is used to fit the optical gap.

In a first chapter a general spin polarized version of the OPW-method is given from a point of view mainly determined by the difficulties arising in its application to heavy substances. The consequences are most seriously reflected in instabilities in the solutions of the OPW-eigenvalue problem, as discussed in chapter 2.

In view of all uncertainties related to necessary approximations and extended computer calculations it seems to be most desirable to know as much as possible about a band structure in advance. Here group theory provides a powerful tool in an exploratory investigation and allows for quite reliable predictions of energy bands rather independent of crystal potentials. Such an analysis is carried through in chapter 3, based on a complete symmetry classification of all possible allowed types of plane waves in the empty fcc-lattice for several symmetry-directions and -points in the Brillouin zone.

1. Spinpolarized OPW-Formalism

1.1. General Formulation

The OPW-method is based on the idea that the inner electron shells of an atom do not depend on whether it is free or bound in a crystal, because the outermost valence electrons are supposed to be able to screen completely all crystal field effects. A rough estimate of average crystal field energies supports this assumption and gives an idea of which electrons are to be treated as well-localized core electrons and which form valence- or conduction bands.

The scope of the OPW-method exclusively consists in the calculation of valence- and conduction band states assuming the core states to be known already from some type of selfconsistent field calculation for the free atom. This approximation, however, can be unsatisfactory. A detailed discussion of related problems and possibilities of improvements is given by F. Herman *et al.* [9]. In principle it should be possible to describe valence- and conduction states by superpositions of plane waves, because these functions form a complete set in Hilbert space. In practice a relatively small number of waves is expected to give already a reasonable convergence in the band energies. The central and non-trivial question in this procedure is the way the plane wave functions are to be enumerated. As discussed later the physical idea of counting these functions with respect to increasing kinetic energy may not be satisfactory from the mathematical point of view, because in this way almost linearly dependent OPW-functions might occur. Disregarding these difficulties for the moment, we first have to formulate the model mathematically.

Under certain conditions [10] the crystal-potential can be restricted to the unit cell and all wave functions are normalized in the volume Ω_0 of this cell. We assume that we know the solutions

$$H_a^\sigma \psi_\alpha^\sigma(x) = (T + v_a^\sigma(x))\, \psi_\alpha^\sigma(x) = \varepsilon_\alpha^\sigma \psi_\alpha^\sigma(x)\,.$$

$$(\psi_\alpha^\nu, \psi_\beta^\sigma) = \int_{\Omega_0} d^3x\, \overline{\psi}_\alpha^\nu(x)\, \psi_\beta^\sigma(x) = \delta_{\alpha\beta}\, \delta_{\nu\sigma} \qquad \nu, \sigma = \pm 1\,;\ \alpha, \beta = 1, 2, \ldots, f$$

where $H_{\rm a}^\sigma$ is the Hamiltonian belonging to the free atom or ion, f the number of occupied electron states, and α is representative of the set of quantum numbers characterizing these states completely (excluding spin). Neglecting any coupling between spin and orbital motion the spin is considered to be a good quantum number and is exhibited explicitly as the upper index.

According to the introductory assumptions any perturbation of ψ_α^σ through the lattice-contribution $(v_{\rm c}^\sigma - v_{\rm a}^\sigma)$ will be neglected. The functions $\{\psi_\alpha^\sigma\}$ span a $2f$-dimensional subspace $\mathscr{H}_1$ of the total Hilbert space $\mathscr{H}$ of one-particle eigenstates of the operator $H_{\rm c}^\sigma = T + v_{\rm c}^\sigma$ $(\sigma = \pm 1)$. Denoting the orthogonal complement of $\mathscr{H}_1$ by $\mathscr{H}_2$, the space $\mathscr{H}$ is decomposed in

$$\mathscr{H} = \mathscr{H}_1 \oplus \mathscr{H}_2$$

and one has to find a basis in $\mathscr{H}_2$, starting with plane waves

$$\varphi_j^\sigma(\boldsymbol{k}, \boldsymbol{x}) = \zeta^\sigma \frac{\exp\{i(\boldsymbol{k} + \boldsymbol{K}_j)\cdot\boldsymbol{x}\}}{\sqrt{\Omega_0}}$$

$\boldsymbol{K}_j$: reciprocal lattice vector

$(j = 1, 2, \ldots)$

ζ^σ: Pauli-spinor

$$(\zeta^\nu, \zeta^\sigma) = \delta_{\nu\sigma}, \qquad (\varphi_i^\nu, \varphi_j^\sigma) = \delta_{ij}\,\delta_{\nu\sigma} \qquad \nu, \sigma = \pm 1\,.$$

If P_1 and $P_2 = 1 - P_1$ are projectors on $\mathscr{H}_1$ and $\mathscr{H}_2$ respectively, with the properties

$$\begin{aligned} P_1\,\psi_\alpha^\sigma &= \psi_\alpha^\sigma, & P_2\,\psi_\alpha^\sigma &= 0 & &\forall\,\alpha, \sigma \\ P_1\,\varphi_j^\sigma &\in \mathscr{H}_1, & P_2\,\varphi_j^\sigma &\in \mathscr{H}_2 & &\text{for any } \varphi \in \mathscr{H}\,, \end{aligned}$$

this basis is constructed through the relation

$$\hat{\varphi}_j^\sigma(\boldsymbol{k}, \boldsymbol{x}) = P_2\,\varphi_j^\sigma(\boldsymbol{k}, \boldsymbol{x}) \tag{1.1}$$

$\{\hat{\varphi}_j^\sigma\}$ is called the set of orthogonalized plane waves (OPW).

It should be realized that this procedure transforms the originally orthonormalized set $\{\varphi_j^\sigma\}$ into a set of functions $\{\hat{\varphi}_j^\sigma\}$, which are not orthogonal any more between each other and non-normalized, leading therefore to a somewhat unusual basis in a Hilbert space. This situation always gives rise to complications in the related eigenvalue problem. Furthermore the basis set could be "overcomplete", i.e. linearly dependent functions are constructed in this way.

The solutions of the Schrödinger equation, restricted to $\mathscr{H}_2$ read

$$H_{\rm c}^\sigma\,\Phi_i^\sigma(\boldsymbol{k}, \boldsymbol{x}) = E_i^\sigma(\boldsymbol{k})\,\Phi_i^\sigma(\boldsymbol{k}, \boldsymbol{x}) \tag{1.2}$$

with

$$\Phi_i^\sigma(\boldsymbol{k}, \boldsymbol{x}) = \sum_{l=1}^\infty s_{li}^\sigma(\boldsymbol{k})\,\hat{\varphi}_l^\sigma(\boldsymbol{k}, \boldsymbol{x}) \tag{1.3}$$

where the sum is to be taken over linearly independent functions only. i is a general band index. In a practical treatment it is hoped that $\mathscr{H}_2$ can be successfully replaced by $\mathscr{H}_2^{(n)}$, a finite-dimensional subspace of dimension n, where n is determined by the condition, that the corresponding approximated spectrum $E_i^{(n)}(\boldsymbol{k})$ differs from the exact one $E_i(\boldsymbol{k})$ only within a certain prescribed error limit δ, at least for the lowest, most interesting bands, say $1 \leq i \leq 5$. This implies the existence of a convergence property of the form

$$\left| E_i^{(n)}(\boldsymbol{k}) - E_i^{(m)}(\boldsymbol{k}) \right| < \delta_i(n) \qquad m > n;\; 1 \leq i \leq 5\,. \tag{1.4}$$

Unfortunately it is impossible to give a general proof of a similar relation within the framework of the OPW-method for reasons which will become clear later. If there is convergence it will depend on the way the functions $\{\hat{\varphi}_j^\sigma\}$ have been enumerated. Furthermore one cannot expect uniform convergence in k because of the strong dispersion in the empty lattice (see also chapter 3).

In the following it is useful and physically intuitive to give the formulation of (1.2) in terms of simple plane waves $\{\varphi_j^\sigma\}$. The relation

$$P_1 \varphi_j^\sigma = \sum_{\substack{\alpha=1 \\ \nu=\pm 1}}^{f} \psi_\alpha^\nu (\psi_\alpha^\nu, \varphi_j^\sigma) \tag{1.5a}$$

and the abbreviation

$$\chi_i^\sigma(k, x) = \sum_{l=1}^{n} s_{li}^\sigma(k)\, \varphi_l^\sigma(k, x) \tag{1.5b}$$

lead to

$$(T + v_c^\sigma)\, \chi_i^\sigma + \sum_{\alpha, \nu}(E_i^\sigma(k) - \varepsilon_\alpha^\nu)\, \psi_\alpha^\nu(\psi_\alpha^\nu, \chi_i^\alpha) = E_i^\sigma(k)\, \chi_i^\sigma . \tag{1.6}$$

The second term on the left side of this equation can be interpreted as a repulsive potential v_R^σ, compensating the attractive part v_c^σ to a certain degree. This at least is true as long as the spectrum of core levels does not overlap with the band spectrum. In contrast to v_c^σ, v_R^σ is a non-local operator. With the definitions

$$\hat{A}^\sigma \chi_i^\sigma = (T + v_c^\sigma)\, \chi_i^\sigma - \sum_{\alpha, \nu} \varepsilon_\alpha^\nu\, \psi_\alpha^\nu(\psi_\alpha^\nu, \chi_i^\sigma) , \tag{1.7}$$

$$\hat{B}^\sigma \chi_i^\sigma = \chi_i^\sigma - \sum_{\alpha, \nu} \psi_\alpha^\nu(\psi_\alpha^\nu, \chi_i^\sigma) \tag{1.8}$$

for the operators $\hat{A}^\sigma$ and $\hat{B}^\sigma$, one gets for (1.6) the generalized eigenvalue equation

$$\hat{A}^\sigma \chi_i^\sigma = E_i^\sigma(k)\, \hat{B}^\sigma \chi_i^\sigma . \tag{1.9}$$

After calculating the matrix elements of $\hat{A}^\sigma$ and $\hat{B}^\sigma$ with simple plane waves, one is left with the algebraic general eigenvalue problem

$$(A^\sigma - \lambda B^\sigma)\, x^\sigma = 0 \tag{1.10}$$

where A^σ and B^σ are $(n \times n)$-matrices. This shows that the OPW-method does not lead to the usual special eigenvalue problem of the type $(A - \lambda 1)\, x = 0$; there is, however, the advantage, that the matrices A^σ and B^σ do not depend on the eigenvalues λ, which is not the case in other well-known methods of band theory (APW; KKR; [11]). For this reason the OPW-formalism allows for a convenient and effective numerical treatment, as discussed in chapter 1 of Part II.

1.2. Matrix Elements of the Operators $\hat{A}^\sigma$ and $\hat{B}^\sigma$

The atomic wave functions ψ_α^σ, the corresponding eigenvalues $\varepsilon_\alpha^\sigma$ and the crystal potential $v_c^\sigma(x)$ together with the relations (1.7) and (1.8) define the operators $\hat{A}^\sigma$ and $\hat{B}^\sigma$ explicitly. Since the pioneering work of Herman and Skillman (HS) [12], the atomic data can be considered known to a good degree of accuracy at least for non-magnetic atoms. In the case of magnetic shells, the (HS)-methods have to be generalized (see chapter 1 of Part II) to calculate a spin dependent radial part $P_{nl}^{\sigma}(r)$ of the total wave function

$$\psi_{nlm}^{\sigma}(\boldsymbol{x}) = \frac{P_{nl}^{\sigma}(r)}{r}\, y_l^m(\vartheta, \varphi)\,. \tag{1.11}$$

$\sigma = +1$ denotes a spin parallel to that of the magnetic shell. For convenience we introduce the following abbreviations for the matrixelements:

$$(\varphi_i^{\mu}, \hat{A}^{\,\sigma} \varphi_j^{\sigma}) = a_1 + a_2 + a_3\,, \qquad (\varphi_i^{\mu}, \hat{B}^{\sigma} \varphi_j^{\sigma}) = b\,, \tag{1.12}$$

$$a_1 = (\varphi_i^{\mu}, T\, \varphi_j^{\sigma})\,, \tag{1.13a}$$

$$a_2 = (\varphi_i^{\mu}, v_c^{\sigma}\, \varphi_j^{\sigma})\,, \tag{1.13b}$$

$$a_3 = -\sum_{n,l,m,\nu} \varepsilon_{nl}^{\nu}(\varphi_i^{\mu}, \psi_{nlm}^{\nu})\,(\psi_{nlm}^{\nu}, \varphi_j^{\sigma})\,, \tag{1.13c}$$

$$b = (\varphi_i^{\mu}, \varphi_j^{\sigma}) - \sum_{n,l,m,\nu}(\varphi_i^{\mu}, \psi_{nlm}^{\nu})\,(\psi_{nlm}^{\nu}, \varphi_j^{\sigma})\,. \tag{1.13d}$$

These expressions still can be brought into a form directly accessible to numerical computations (for a derivation see [10, 13]):

$$a_1 = (\boldsymbol{k} + \boldsymbol{K}_j)^2\, \delta_{ij}\, \delta_{\mu\sigma}\,, \tag{1.14a}$$

$$a_2 = \frac{\delta_{\mu\sigma}}{\Omega_0} \int_{\Omega_0} d^3x\, v_c^{\sigma}(\boldsymbol{x})\, \exp\{i\,(\boldsymbol{K}_j - \boldsymbol{K}_i)\cdot\boldsymbol{x}\}\,, \tag{1.14b}$$

$$a_3 = -\sum_{n,l} \varepsilon_{nl}^{\sigma} R_{nlj}^{\sigma}(\boldsymbol{k})\, R_{nli}^{\sigma}(\boldsymbol{k})\, P_l(\cos\theta_{ij})\, \delta_{\mu\nu}\, \delta_{\nu\sigma}\,, \tag{1.14c}$$

$$b = \delta_{ij}\, \delta_{\mu\sigma} - \sum_{n,l} R_{nlj}^{\sigma}(\boldsymbol{k})\, R_{nli}^{\sigma}(\boldsymbol{k})\, P_l(\cos\theta_{ij})\, \delta_{\mu\nu}\, \delta_{\nu\sigma}\,. \tag{1.14d}$$

The orthogonality-coefficients $R_{nlj}^{\sigma}(\boldsymbol{k})$ are defined by

$$R_{nlj}^{\sigma}(\boldsymbol{k}) = \sqrt{\frac{4\pi}{\Omega_0}\,(2l+1)} \int_0^{r_0} dr\, r\, j_l\,(|\,\boldsymbol{k} + \boldsymbol{K}_j\,|\, r)\, P_{nl}^{\sigma}(r) \left(\frac{4\pi}{3}\, r_0^3 = \Omega_0\right). \tag{1.15}$$

The P_l's are Legendre-polynomials, the j_l's spherical Bessel functions, and

$$\cos\theta_{ij} = \frac{(\boldsymbol{k} + \boldsymbol{K}_i)\cdot(\boldsymbol{k} + \boldsymbol{K}_j)}{|\,\boldsymbol{k} + \boldsymbol{K}_i\,|\,|\,\boldsymbol{k} + \boldsymbol{K}_j\,|}\,.$$

1.3. The Crystal Potential v_c^{σ}

It can be shown (see chapter 3) that the qualitative structure of an energy band scheme is determined by the symmetry of the lattice already, and any analysis and comparison of published band calculations remarkably underlines this fact. The crudely approximated crystal potential then determines some quantitative details in the band structure, but refinements in the crystal potential usually have little influence on it. Furthermore the principal difficulties in the application of the OPW-method particularly to heavy substances lies in the construction of the basis in $\mathscr{H}_2^{(n)}$ which is almost independent of crystal field effects. For these reasons we only describe the construction of a crude potential, particularly related to EuO.

Assuming an ionic configuration $Eu^{2+}-O^{2-}$, all charge densities of core electrons are spherically symmetric, in particular the density of the half-filled $4f$-shell representing a multiplet groundstate $^8S_{7/2}$. The Fourier coefficients $\tilde{v}(q)$ of the pure Coulomb potentials of electrons and protons are calculated from the charge densities $\tilde{\varrho}(q)$ through Poisson's equation

$$\tilde{v}(q) = \frac{8\pi}{q^2} \tilde{\varrho}(q) \qquad (q = |\boldsymbol{q}|) . \tag{1.16}$$

Atomic units are used throughout:

$$\hbar = 1 , \qquad m_e = \tfrac{1}{2} , \qquad a_0 = 1 \qquad (a_0 : \text{Bohr-radius}) .$$

The electronic charge density

$$\varrho_e(\boldsymbol{x}) = \sum_{n, l, m, \nu} |\psi_{nlm}^\nu(\boldsymbol{x})|^2 \qquad (= \varrho_e(r), \ r = |\boldsymbol{x}|)$$

is most conveniently expressed in terms of a surface charge density $\tau(r)$ directly obtainable from the tables for the radial functions:

$$\tau(r) = 4\pi r^2 \varrho_e(r) = \sum_{n,l,\nu} (2l+1) [P_{nl}^\nu(r)]^2 . \tag{1.17}$$

The Fourier transform $\tilde{\varrho}_e(q)$ reads then

$$\varrho_e(q) = \frac{1}{\Omega_0} \int_{\Omega_0} d^3x \, \varrho_e(\boldsymbol{x}) \exp\{- i\, \boldsymbol{q} \cdot \boldsymbol{x}\} = \frac{1}{\Omega_0} \int_0^{r_0} dr \, \tau(r)\, j_0(q r) , \tag{1.18}$$

(j_0 being the zeroth Bessel function). The total potential $\tilde{v}^\sigma(q)$ (neglecting the index e from now on) is made up of three contributions from the nuclear charge Z, the electrons and the exchange, respectively:

$$\tilde{v}^\sigma(q) = \tilde{v}_p(q) + \tilde{v}_e(q) + \tilde{v}_x^\sigma(q) ,$$

$$\tilde{v}_p(q) = - \frac{8\pi Z}{\Omega_0} \frac{1}{q^2} , \tag{1.19}$$

$$\tilde{v}_e(q) = \frac{8\pi}{\Omega_0} \frac{1}{q^2} \int_0^{r_0} dr \, \tau(r)\, j_0(q r) , \tag{1.20}$$

$$\tilde{v}_x^\sigma(q) = - \frac{8\pi\beta}{\Omega_0} \int_0^{r_0} dr\, r\, j_0(q r) [r\, s^\sigma(r)]^{1/3} . \tag{1.21}$$

$s^\sigma(r)$ denotes the "spin-density" occurring in the Slater exchange

$$v_x^\sigma(r) = - \beta \left[\frac{s^\sigma(r)}{r^2} \right]^{1/3} , \qquad \beta = 3 \left(\frac{3}{4\pi^2} \right)^{1/3} \tag{1.22}$$

and in the case of Eu it is related to $\tau(r)$ through

$$s^{+1}(r) = \tau(r) + 7\,[P_{43}^{+1}(r)]^2 \,, \tag{1.23a}$$

$$s^{-1}(r) = \tau(r) - 7\,[P_{43}^{-1}(r)]^2 \,. \tag{1.23b}$$

This difference in $s^\sigma(r)$ together with the spin dependent expressions for a_3 and b give rise to the splitting into spinpolarized energy bands. For the band calculation an exchange parameter of $\frac{2}{3}\,\beta$ has been used without any attempt to introduce a screening of the long, overestimated tail of the exchange potential (see e.g. [14]). The main influence of screening appears in the zeroth Fourier coefficient, which is taken to be the only adjustable parameter of this calculation. The atomic calculations have been performed using the full exchange parameter β, but it must be emphasized that screening is automatically accounted for through a cut-off as described by Herman and Skillman [12].

In an ionic configuration the long range part of the potential produced by the ions surrounding the considered cell has to be included. This potential, $v_i(\boldsymbol{x})$ say, is not spherically symmetric and so strong that the original assumptions for the OPW-method have to be refined somewhat.

An expansion of $v_i(\boldsymbol{x})$ in terms of spherical harmonics, however, shows the spherically symmetric part $v_i(r)$ to give by far the dominant contribution [15, 16, 17]. Actually $v_i(r)$ is the potential of ionic charges smeared out uniformly over the surfaces of spheres whose radius corresponds to the distance of nearest, second-nearest neighbours etc. Inside the first sphere therefore $v_i(r)$ is a constant [18, 19], namely the Madelung potential

$$v_{\mathrm{M}} = \pm\,\frac{8\,\alpha_{\mathrm{M}}}{a} \tag{1.24}$$

where a is the lattice constant and α_{M} the Madelung constant $(= 1.74756$ for NaCl-lattices). v_{M} cannot modify the wave functions ψ_α^σ, but all eigenvalues $\varepsilon_\alpha^\sigma$ in (1.13c) and (1.14c) have to be replaced by the shifted values $\varepsilon_\alpha^\sigma \pm v_{\mathrm{M}}$, the sign depending on the sign of the charge of the ion in question.

Finally the generalization for a_l and b in (1.13) in case of two different atoms A and B in the unit cell, separated by a distance $|\boldsymbol{D}|$, is given to be

$$a_l = a_l^{\mathrm{A}} + \exp\{-\,i\,\boldsymbol{q}\cdot\boldsymbol{D}\}\,a_l^{\mathrm{B}} \qquad l = 2, 3 \,, \tag{1.25a}$$

$$b = b^{\mathrm{A}} + \exp\{-\,i\,\boldsymbol{q}\cdot\boldsymbol{D}\}\,(b^{\mathrm{B}} - \delta_{ij}\,\delta_{\mu\sigma}) \,. \tag{1.25b}$$

Remark on Relativistic Corrections

The intention of this work is to give a simple approach to the band structure of heavy magnetic semiconductors disregarding relativistic effects first. In principle one could think of adding relativistic corrections on the basis of perturbation theory. However, the amount of work would equal almost that of a full relativistic calculation, because then all three terms occurring in second order of the fine-structure constant [12] have to be considered, and important contributions would originate from the orthogonalization to the inner shells. For the present purpose, and particularly in view of the current experimental situation (see chapter 2, Part II), we find the nonrelativistic approach to give quite satisfactory results.

2. Instability of the OPW-Eigenvalue Problem

As mentioned in the preceding chapter, the construction of the basis in $\mathcal{H}_2$, as prescribed by the conventional OPW-method, may be troublesome and give rise to serious instabilities in at least some of the solutions of the eigenvalue problem $(A - \lambda B)\,x = 0$. It will be shown that this fact is reflected in the properties of the matrix B (the spin index σ is omitted in this discussion). From the definitions (1.5a) and (1.8) it follows that

$$\hat{B} = P_2 = 1 - P_1 \tag{2.1}$$

i.e. $\hat{B}$ is an orthogonal projector with the usual properties

$$\hat{B}^2 = \hat{B}, \qquad \hat{B}^* = \hat{B}$$

and therefore its spectrum is limited by

$$0 \leq \hat{B} \leq 1. \tag{2.2}$$

After restriction of $\mathcal{H}_2$ to $\mathcal{H}_2^{(n)}$ the above condition also holds for the spectrum of the $(n \times n)$-matrix B, whose elements are given according to (1.13d) by the overlap integrals between OPW-functions:

$$b_{ij} = (\hat{\varphi}_i, \hat{\varphi}_j) = \delta_{ij} - (\varphi_i, P_1 \varphi_j). \tag{2.3}$$

This suggests that quasi-linear dependences in the system will lead to eigenvalues of B close to zero. For illustration let us recall the decomposition

$$\mathcal{H} = \mathcal{H}_1 \oplus \mathcal{H}_2^{(n)} \oplus \mathcal{H}_3$$

where $\mathcal{H}_3$ is a complement neglected in a numerical treatment. Now, according to the OPW-construction,

$$\varphi_l \in \mathcal{H}_1 \oplus \mathcal{H}_2^{(n)} \qquad l \leq n$$

and with

$$|\alpha|^2 + |\beta|^2 = 1,$$

$$\varphi_l = \alpha f_l + \beta g_l, \qquad f_l \in \mathcal{H}_1, \qquad g_l \in \mathcal{H}_2^{(n)}.$$

Starting with the set $\{\varphi_j; j = 1, 2, \ldots, m\}$ of simple plane waves, the procedure tries to construct a new set $\{\hat{\varphi}_j; j = 1, 2, \ldots, n\}$ of functions spanning the space $\mathcal{H}_2^{(n)}$. The central question remains of whether $n = \dim \mathcal{H}_2^{(n)}$ is equal to or different from m. If for a pair of functions $\{\varphi_l, \varphi_k\}$ a relation of the form

$$\varphi_l = \alpha f_l + \bar{\beta} g_l, \qquad \varphi_k = -\beta f_l + \bar{\alpha} g_l \qquad l, k \leq m,\ l \neq k \tag{2.4}$$

happens to hold, then clearly

$$n = m - 1$$

and that in turn means, that the rank of the matrix B is smaller than its dimension. Although in practice no exact linear dependence is likely to occur, already an approach to this situation may be extremely cumbersome for any numerical treatment. Rather unexpectedly the instabilities will affect the lowest, namely the valence bands mainly, and one should not hope that these are reproduced very reliably.

It can be seen from (2.3) that the matrix B differs from the unit matrix 1 the more the bigger the matrix elements of the projector P_1 are. These are increasing with increasing number of occupied core states as seen from (1.14d) and the corresponding numerical values given in the Appendix of Part II. The breakdown of the procedure mainly depends therefore on the heaviness of the elements involved in the calculation. Further consequences and details will be discussed in connection with a numerical example in chapter 1 of Part II.

3. Group Theoretical Approach to the Band Structure

Besides giving a quite reliable prediction of the qualitative features of a band structure a complete group theoretical classification of energy bands allows for a deduction of useful selection rules for optical interband transitions and furthermore for an effective reduction of high-dimensional matrices with multiply degenerate spectrum to those of lower dimension with simple spectrum [20, 21]. This last point is of essential importance for any numerical treatment (see chapter 1 of Part II).

3.1. General Definitions

To a symmetry group G of the Hamiltonian belongs a decomposition of Hilbert space

$$\mathcal{H}_2^{(n)} = \bigoplus_{i=1}^{r} \mathfrak{H}_i$$

i.e. into r irreducible subspaces $\mathfrak{H}_i$. The basis functions in these spaces are found from any arbitrary function $F \in \mathcal{H}_2^{(n)}$ by means of projection operators $P_{\nu\nu}^j$:

$$P_{\nu\nu}^j = \frac{d_j}{h} \sum_{s=1}^{h} \bar{U}_{\nu\nu}^j(g_s)\, \hat{U}(g_s) \qquad \begin{matrix} j = 1, 2, \ldots, r \\ \nu = 1, 2, \ldots, d_j \end{matrix} \tag{3.1}$$

$\{g_s\}$ denotes the set of group elements, h the group order, $\bar{U}_{\nu\nu}^j$ a complex conjugate diagonal element of the j-th irreducible representation of dimension d_j, and $\hat{U}(g_s)$ is a corresponding unitary operator defined by

$$(\hat{U}(g_s)\, F)\,(\boldsymbol{x}) = F(g_s^{-1}\boldsymbol{x}) \tag{3.2}$$

taking $\{g_s\}$ to be orthogonal transformations in $\mathbf{R}^3$. A function f_ν^j transforms according to the ν-th row of the j-th irreducible representation, if it is an eigenfunction of $P_{\nu\nu}^j$ with eigenvalue 1:

$$P_{\nu\nu}^j f_\nu^j = f_\nu^j. \tag{3.3}$$

Knowing one single f_ν^j, the other partner functions are obtained from

$$f_\lambda^j = P_{\lambda\nu}^j f_\nu^j \qquad \lambda \neq \nu \tag{3.4}$$

and the system $\{f_\nu^j;\ \nu = 1, 2, \ldots, d_j\}$ spans an orthogonal basis in $\mathfrak{H}_j$. Because of the proportionality

$$P_{\nu\nu}^j F \sim f_\nu^j$$

all basis functions, starting with arbitrary F, can be constructed, if all projectors $P_{\nu\nu}^j$ are known. As a generalization it can be said that any arbitrary set $\{F_1, F_2, \ldots, F_m\}$ will give rise to m or fewer functions transforming like f_ν^j. These, however, must not be identical with f_ν^j nor orthogonal between each other.

The projector properties

$$P^j_{\nu\nu} P^l_{\mu\mu} = \delta_{\nu\mu}\,\delta_{jl}\,P^l_{\mu\mu}\,, \qquad P^j_{\mu\lambda} = P^{*j}_{\mu\lambda}$$

guarantee the orthogonality between two functions f^j_ν and g^l_μ transforming according to inequivalent irreducible representations or according to different rows of one and the same irreducible representation. This in general is not true for different but equivalent representations, and therefore the functions obtained from the application of the projector technique often have to be orthonormalized afterwards by a Schmidt-procedure.

3.2. Special Groups

The energy bands $E_i(\mathbf{k})$ are analyzed in special symmetry directions and — points of the Brillouin zone where the band extrema, which are most important for optical transitions, will be preferably found. The labeling is given in Fig. 1 which shows the Brillouin zone for the fcc- or rocksalt lattice respectively. All corresponding groups have to be found together with their irreducible representations. For illustration let us consider in detail the group G_Δ of order $h_\Delta = 8$, consisting of all operations which leave the reduced vector $\mathbf{k}$ $(\mathbf{k} = 2\pi/a\,(0, 0, \lambda)$, $0 < \lambda < 1)$ in a (001)-direction invariant. Table 1 contains all information needed

Table 1. Irreducible representations of G_Δ

Class	Element	Operation	Δ_1	$\Delta_{1'}$	Δ_2	$\Delta_{2'}$	Δ_5
E	g_1	xyz	1	1	1	1	$\begin{pmatrix} 1 & 0 \\ 0 & 1 \end{pmatrix}$
$C^2_{4\perp}$	g_2	$\bar{x}\bar{y}z$	1	1	1	1	$\begin{pmatrix} -1 & 0 \\ 0 & -1 \end{pmatrix}$
$2\,C_4$	g_3	$\bar{y}xz$	1	1	-1	-1	$\begin{pmatrix} 0 & -1 \\ 1 & 0 \end{pmatrix}$
	g_4	$y\bar{x}z$	1	1	-1	-1	$\begin{pmatrix} 0 & 1 \\ -1 & 0 \end{pmatrix}$
$2\,J\,C^2_{4\perp}$	g_5	$\bar{x}yz$	1	-1	1	-1	$\begin{pmatrix} -1 & 0 \\ 0 & 1 \end{pmatrix}$
	g_6	$x\bar{y}z$	1	-1	1	-1	$\begin{pmatrix} 1 & 0 \\ 0 & -1 \end{pmatrix}$
$2\,J\,C_2$	g_7	$\bar{y}\bar{x}z$	1	-1	-1	1	$\begin{pmatrix} 0 & -1 \\ -1 & 0 \end{pmatrix}$
	g_8	yxz	1	-1	-1	1	$\begin{pmatrix} 0 & 1 \\ 1 & 0 \end{pmatrix}$

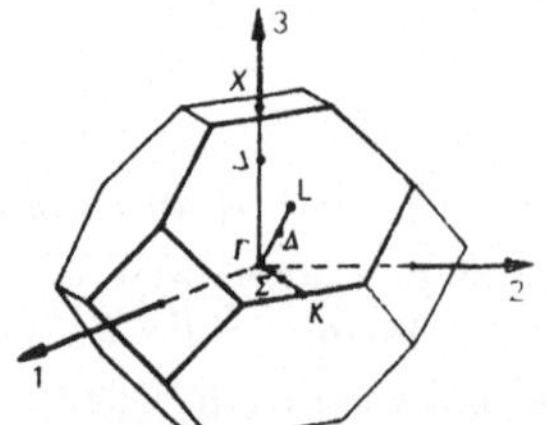

Fig. 1. Brillouin zone for the fcc- or rocksaltstructure

for the construction of the projectors (3.1). The conventional symbols are explained in standard texts such as [11] where also tables of group characters will be found. For the other groups therefore only the 2-dimensional representations are given here in the Tables 2 and 3.

Table 2. Two-dimensional irreducible representations of

$$G_\Lambda\left[\boldsymbol{k}=\frac{2\pi}{a}\,(\lambda,\lambda,\lambda),\quad 0<\lambda<\frac{1}{2}\right]$$

$$\text{and}\qquad G_L=G_\Lambda\otimes I\left[\boldsymbol{k}=\frac{2\pi}{a}\left(\frac{1}{2},\frac{1}{2},\frac{1}{2}\right);\quad I:\ \text{inversion}\right]$$

Class	Operation	L_3	$L_{3'}$	Λ_3
E	xyz	$\begin{pmatrix} 1 & 0 \\ 0 & 1 \end{pmatrix}$	$\begin{pmatrix} 1 & 0 \\ 0 & 1 \end{pmatrix}$	$\begin{pmatrix} 1 & 0 \\ 0 & 1 \end{pmatrix}$
$2\,C_3$	zxy	$-\frac{1}{2}\begin{pmatrix} 1 & \sqrt{3} \\ -\sqrt{3} & 1 \end{pmatrix}$	$-\frac{1}{2}\begin{pmatrix} 1 & \sqrt{3} \\ -\sqrt{3} & 1 \end{pmatrix}$	$-\frac{1}{2}\begin{pmatrix} 1 & \sqrt{3} \\ -\sqrt{3} & 1 \end{pmatrix}$
	yzx	$-\frac{1}{2}\begin{pmatrix} 1 & -\sqrt{3} \\ \sqrt{3} & 1 \end{pmatrix}$	$-\frac{1}{2}\begin{pmatrix} 1 & -\sqrt{3} \\ \sqrt{3} & 1 \end{pmatrix}$	$-\frac{1}{2}\begin{pmatrix} 1 & -\sqrt{3} \\ \sqrt{3} & 1 \end{pmatrix}$
$3\,C_2$	$\bar{y}\bar{x}\bar{z}$	$\frac{1}{2}\begin{pmatrix} -1 & \sqrt{3} \\ \sqrt{3} & 1 \end{pmatrix}$	$-\frac{1}{2}\begin{pmatrix} -1 & \sqrt{3} \\ \sqrt{3} & 1 \end{pmatrix}$	
	$\bar{z}\bar{y}\bar{x}$	$-\frac{1}{2}\begin{pmatrix} 1 & \sqrt{3} \\ \sqrt{3} & -1 \end{pmatrix}$	$\frac{1}{2}\begin{pmatrix} 1 & \sqrt{3} \\ \sqrt{3} & -1 \end{pmatrix}$	
	$\bar{x}\bar{z}\bar{y}$	$\begin{pmatrix} 1 & 0 \\ 0 & -1 \end{pmatrix}$	$\begin{pmatrix} -1 & 0 \\ 0 & 1 \end{pmatrix}$	
J	$\bar{x}\bar{y}\bar{z}$	$\begin{pmatrix} 1 & 0 \\ 0 & 1 \end{pmatrix}$	$\begin{pmatrix} -1 & 0 \\ 0 & -1 \end{pmatrix}$	
$2\,J\,C_3$	$\bar{z}\bar{x}\bar{y}$	$-\frac{1}{2}\begin{pmatrix} 1 & \sqrt{3} \\ -\sqrt{3} & 1 \end{pmatrix}$	$\frac{1}{2}\begin{pmatrix} 1 & \sqrt{3} \\ -\sqrt{3} & 1 \end{pmatrix}$	
	$\bar{y}\bar{z}\bar{x}$	$-\frac{1}{2}\begin{pmatrix} 1 & -\sqrt{3} \\ \sqrt{3} & 1 \end{pmatrix}$	$\frac{1}{2}\begin{pmatrix} 1 & -\sqrt{3} \\ \sqrt{3} & 1 \end{pmatrix}$	
$3\,J\,C_2$	yxz	$\frac{1}{2}\begin{pmatrix} -1 & \sqrt{3} \\ \sqrt{3} & 1 \end{pmatrix}$	$\frac{1}{2}\begin{pmatrix} -1 & \sqrt{3} \\ \sqrt{3} & 1 \end{pmatrix}$	$\frac{1}{2}\begin{pmatrix} -1 & \sqrt{3} \\ \sqrt{3} & 1 \end{pmatrix}$
	zyx	$-\frac{1}{2}\begin{pmatrix} 1 & \sqrt{3} \\ \sqrt{3} & -1 \end{pmatrix}$	$-\frac{1}{2}\begin{pmatrix} 1 & \sqrt{3} \\ \sqrt{3} & -1 \end{pmatrix}$	$-\frac{1}{2}\begin{pmatrix} 1 & \sqrt{3} \\ \sqrt{3} & -1 \end{pmatrix}$
	xzy	$\begin{pmatrix} 1 & 0 \\ 0 & -1 \end{pmatrix}$	$\begin{pmatrix} 1 & 0 \\ 0 & -1 \end{pmatrix}$	$\begin{pmatrix} 1 & 0 \\ 0 & -1 \end{pmatrix}$

3.3. Symmetrized Plane Waves

The basis functions in the spaces $\mathfrak{H}_i$ are constructed by symmetrizing the plane waves $\varphi_j(\boldsymbol{k},\boldsymbol{x})$ in such a way that they transform according to the rows of the irreducible representations of the group in question. The procedure will be explained for G_Λ only and is to be applied in the same way to the other cases.

Table 3. Two-dimensional irreducible representation of

$$G_X = G_\Delta \otimes I\left[\mathbf{k} = \frac{2\pi}{a}\,(0,0,1)\right]$$

Class	Operation	X_5	$X_{5'}$
E	xyz	$\begin{pmatrix} 1 & 0 \\ 0 & 1 \end{pmatrix}$	$\begin{pmatrix} 1 & 0 \\ 0 & 1 \end{pmatrix}$
$2\,C_{4\perp}^2$	$x\bar{y}\bar{z}$	$\begin{pmatrix} -1 & 0 \\ 0 & 1 \end{pmatrix}$	$\begin{pmatrix} 1 & 0 \\ 0 & -1 \end{pmatrix}$
	$\bar{x}y\bar{z}$	$\begin{pmatrix} 1 & 0 \\ 0 & -1 \end{pmatrix}$	$\begin{pmatrix} -1 & 0 \\ 0 & 1 \end{pmatrix}$
$C_{4\parallel}^2$	$\bar{x}\bar{y}z$	$\begin{pmatrix} -1 & 0 \\ 0 & -1 \end{pmatrix}$	$\begin{pmatrix} -1 & 0 \\ 0 & -1 \end{pmatrix}$
$2\,C_{4\parallel}$	$\bar{y}xz$	$\begin{pmatrix} 0 & -1 \\ 1 & 0 \end{pmatrix}$	$\begin{pmatrix} 0 & -1 \\ 1 & 0 \end{pmatrix}$
	$y\bar{x}z$	$\begin{pmatrix} 0 & 1 \\ -1 & 0 \end{pmatrix}$	$\begin{pmatrix} 0 & 1 \\ -1 & 0 \end{pmatrix}$
$2\,C_2$	$yx\bar{z}$	$\begin{pmatrix} 0 & -1 \\ -1 & 0 \end{pmatrix}$	$\begin{pmatrix} 0 & 1 \\ 1 & 0 \end{pmatrix}$
	$\bar{y}\bar{x}\bar{z}$	$\begin{pmatrix} 0 & 1 \\ 1 & 0 \end{pmatrix}$	$\begin{pmatrix} 0 & -1 \\ -1 & 0 \end{pmatrix}$
J	$\bar{x}\bar{y}\bar{z}$	$\begin{pmatrix} 1 & 0 \\ 0 & 1 \end{pmatrix}$	$\begin{pmatrix} -1 & 0 \\ 0 & -1 \end{pmatrix}$
$2\,J\,C_{4\perp}^2$	$\bar{x}yz$	$\begin{pmatrix} -1 & 0 \\ 0 & 1 \end{pmatrix}$	$\begin{pmatrix} -1 & 0 \\ 0 & 1 \end{pmatrix}$
	$x\bar{y}z$	$\begin{pmatrix} 1 & 0 \\ 0 & -1 \end{pmatrix}$	$\begin{pmatrix} 1 & 0 \\ 0 & -1 \end{pmatrix}$
$J\,C_{4\parallel}^2$	$xy\bar{z}$	$\begin{pmatrix} -1 & 0 \\ 0 & -1 \end{pmatrix}$	$\begin{pmatrix} 1 & 0 \\ 0 & 1 \end{pmatrix}$
$2\,J\,C_{4\parallel}$	$y\bar{x}\bar{z}$	$\begin{pmatrix} 0 & -1 \\ 1 & 0 \end{pmatrix}$	$\begin{pmatrix} 0 & 1 \\ -1 & 0 \end{pmatrix}$
	$\bar{y}x\bar{z}$	$\begin{pmatrix} 0 & 1 \\ -1 & 0 \end{pmatrix}$	$\begin{pmatrix} 0 & -1 \\ 1 & 0 \end{pmatrix}$
$2\,J\,C_2$	$\bar{y}\bar{x}z$	$\begin{pmatrix} 0 & -1 \\ -1 & 0 \end{pmatrix}$	$\begin{pmatrix} 0 & -1 \\ -1 & 0 \end{pmatrix}$
	yxz	$\begin{pmatrix} 0 & 1 \\ 1 & 0 \end{pmatrix}$	$\begin{pmatrix} 0 & 1 \\ 1 & 0 \end{pmatrix}$

From (3.2) it follows that the transformation of plane waves is

$$\left(\hat{U}(g_s)\,\varphi_j\right)(\mathbf{k}, \mathbf{x}) = \frac{1}{\Omega_0}\exp\{i\,(\mathbf{K}_j + \mathbf{k})\cdot(g_s^{-1}\mathbf{x})\}$$

$$= \frac{1}{\Omega_0}\exp\{i\,(g_s\mathbf{K}_j + \mathbf{k})\cdot\mathbf{x}\} = \varphi_{j'}(\mathbf{k}, \mathbf{x})\,.$$

The operation (3.1) applied to any φ_j admits in the sum only those waves $\varphi_{j'}$ whose wave vector is of the same modulus and the same (3)-component as $\mathbf{K}_j$. Therefore the essential feature is seen to be the permutation- and reflection character of the (1)- and (2)-components of the reciprocal lattice vectors as determined by the crystal lattice. Now fcc-symmetry restricts the reciprocal vectors to the following four types:

α) $(0, 0, w) : w$ even, 0 included,

β) $(u, 0, w) : u, w$ even, $w = 0$ included,

γ) $(u, u, w) : u, w$ even, $w = 0$ included
                 u, w odd,

δ) $(u, v, w) : u, v, w$ even, $w = 0$ included
                 u, v, w odd.

These four types in turn determine four different types of symmetrized functions. In the following we use the abbreviations $[u\,v] = \varphi_j(\boldsymbol{k}, \boldsymbol{x})$ with $\boldsymbol{K}_j = 2\pi/a\,(u, v, w)$ and $\boldsymbol{k} = 2\pi/a\,(0, 0, \lambda)$, as well as $\varphi_{\Delta_i}(u\,v\,w)$ for the symmetrized functions and $P_{\nu\nu}^{\Delta_i}$ for the corresponding projectors.

Wavefunctions of type β:

$$\varphi_{\Delta_1}(u\,0\,w) \sim P_{11}^{\Delta_1} \cdot [u\,0] = \tfrac{1}{8}\,([u\,0] + [\bar{u}\,0] + [0\,u] + [0\,\bar{u}] + [\bar{u}\,0] + [u\,0] + [0\,\bar{u}]$$
$$+ [0\,u])\,,$$
$$= \tfrac{1}{4}\,([u\,0] + [\bar{u}\,0] + [0\,u] + [0\,\bar{u}])\,,$$

$$(\varphi_{\Delta_{1'}}(u\,0\,w) \sim P_{11}^{\Delta_{1'}} \cdot [u\,0] = 0\,, \qquad \varphi_{\Delta_{2'}}(u\,0\,w) \sim P_{11}^{\Delta_{2'}} \cdot [u\,0] = 0)\,,$$

$$\varphi_{\Delta_2}(u\,0\,w) \sim P_{11}^{\Delta_2} \cdot [u\,0] = \tfrac{1}{4}\,([u\,0] + [\bar{u}\,0] - [0\,u] - [0\,\bar{u}])\,,$$

$$\varphi_{\Delta_5}^1(u\,0\,w) \sim P_{11}^{\Delta_5} \cdot [u\,0] = \tfrac{1}{4}\,([u\,0] - [\bar{u}0])\,,$$

$\varphi_{\Delta_5}^2$ must be generated from $[0\,u]$:

$$\varphi_{\Delta_5}^2(u\,0\,w) \sim P_{22}^{\Delta_5} \cdot [0\,u] = \tfrac{1}{4}\,([0\,u] - [0\,\bar{u}])\,.$$

Collecting terms the normalized functions read

$$\varphi_{\Delta_1}(u\,0\,w) = \left[\cos\left(\frac{2}{a}\,\pi\,u\,x\right) + \cos\left(\frac{2}{a}\,\pi\,u\,y\right)\right] f(z)\,,$$

$$\varphi_{\Delta_2}(u\,0\,w) = \left[\cos\left(\frac{2}{a}\,\pi\,u\,x\right) - \cos\left(\frac{2}{a}\,\pi\,u\,y\right)\right] f(z)\,,$$

$$\left.\begin{aligned}\varphi_{\Delta_5}^1(u\,0\,w) &= \sqrt{2}\,\sin\left(\frac{2}{a}\,\pi\,u\,x\right) f(z)\\[2mm]\varphi_{\Delta_5}^2(u\,0\,w) &= \sqrt{2}\,\sin\left(\frac{2}{a}\,\pi\,u\,y\right) f(z)\end{aligned}\right\} \quad \text{degenerate}\,,$$

$$(\varphi_{\Delta_{1'}}(u\,0\,w) = \varphi_{\Delta_{2'}}(u\,0\,w) = 0)\,,$$

$$f(z) = \exp\left\{i\,\frac{2}{a}\,\pi\,(w + \lambda)\,z\right\}\,.$$

(The common normalizing factor $\Omega_0^{-\frac{1}{2}}$ is omitted.)

For all degenerate functions the phase is omitted, because it is determined by the group itself and not by the different vector types.

Wavefunctions of type γ:

$$\varphi_{\Delta_1}(u\,u\,w) = \left[\cos\frac{2}{a}\pi(ux+uy) + \cos\frac{2}{a}\pi(ux-uy)\right]f(z)\,,$$

$$\varphi_{\Delta_2'}(u\,u\,w) = \left[\cos\frac{2}{a}\pi(ux+uy) - \cos\frac{2}{a}\pi(ux-uy)\right]f(z)\,,$$

$$\varphi^1_{\Delta_5}(u\,u\,w) = \left[\sin\frac{2}{a}\pi(ux+uy) + \sin\frac{2}{a}\pi(ux-uy)\right]f(z)\,,$$

$$\varphi^2_{\Delta_5}(u\,u\,w) = \left[\sin\frac{2}{a}\pi(ux+uy) - \sin\frac{2}{a}\pi(ux-uy)\right]f(z)\,.$$

Wavefunctions of type δ:

The corresponding 8-dimensional reducible representation contains 2 equivalent irreducible Δ_5-representations whose basis functions can be distinguished by writing $\varphi_{\Delta_5}(uvw)$ and $\varphi_{\Delta_5}(vuw)$, respectively.

$$\varphi_{\Delta_1}(u\,v\,w) = \frac{1}{\sqrt{2}}\left[\cos\frac{2}{a}\pi(ux+vy) + \cos\frac{2}{a}\pi(ux-vy) + \cos\frac{2}{a}\pi(vx+uy)\right.$$
$$\left. + \cos\frac{2}{a}\pi(vx-uy)\right]f(z)\,,$$

$$\varphi_{\Delta_1'}(u\,v\,w) = \frac{1}{\sqrt{2}}\left[\cos\frac{2}{a}\pi(ux+vy) - \cos\frac{2}{a}\pi(ux-vy) - \cos\frac{2}{a}\pi(vx+uy)\right.$$
$$\left. + \cos\frac{2}{a}\pi(vx-uy)\right]f(z)\,,$$

$$\varphi_{\Delta_2}(u\,v\,w) = \frac{1}{\sqrt{2}}\left[\cos\frac{2}{a}\pi(ux+vy) + \cos\frac{2}{a}\pi(ux-vy) - \cos\frac{2}{a}\pi(vx+uy)\right.$$
$$\left. - \cos\frac{2}{a}\pi(vx-uy)\right]f(z)\,,$$

$$\varphi_{\Delta_2'}(u\,v\,w) = \frac{1}{\sqrt{2}}\left[\cos\frac{2}{a}\pi(ux+vy) - \cos\frac{2}{a}\pi(ux-vy) + \cos\frac{2}{a}\pi(vx+uy)\right.$$
$$\left. - \cos\frac{2}{a}\pi(vx-uy)\right]f(z)\,,$$

$$\varphi^1_{\Delta_5}(u\,v\,w) = \left[\sin\frac{2}{a}\pi(ux+vy) + \sin\frac{2}{a}\pi(ux-vy)\right]f(z)\,,$$

$$\varphi^2_{\Delta_5}(u\,v\,w) = \left[\sin\frac{2}{a}\pi(ux+vy) - \sin\frac{2}{a}\pi(ux-vy)\right]f(z)\,,$$

$$\varphi^1_{\Delta_5}(v\,u\,w) = \left[\sin\frac{2}{a}\pi(vx+uy) + \sin\frac{2}{a}\pi(vx-uy)\right]f(z)\,,$$

$$\varphi^2_{\Delta_5}(v\,u\,w) = \left[\sin\frac{2}{a}\pi(vx+uy) - \sin\frac{2}{a}\pi(vx-uy)\right]f(z)\,.$$

As mentioned earlier group theory cannot predict anything about the orthogonality between $\varphi^i_{\Delta_5}(uvw)$ and $\varphi^i_{\Delta_5}(vuw)$, but in this particular case they are already orthogonal, because the projection operators admit for one function only those plane waves which are not contained in the other one. This e.g. is not the case in the next example. Consider G_Λ allowing for 6 different plane waves of type (uvw). The corresponding reducible representation decomposes into

$$D_\Lambda(uvw) = \Lambda_1 \oplus \Lambda_2 \oplus 2\,\Lambda_3\,.$$

Looking at functions transforming only according to the first row of Λ_3, the operations $P^{\Lambda_3}_{11}[uvw]$ and $P^{\Lambda_3}_{11}[vuw]$ lead to

$$\varphi^1_{\Lambda_3^{(1)}} = \frac{1}{\sqrt{12}}\,(2[uvw] - [wuv] - [vwu] - [vuw] - [wvu] + 2[uwv])\,,$$

$$\varphi^1_{\Lambda_3^{(2)}} = \frac{1}{\sqrt{12}}\,(-[uvw] - [wuv] + 2[vwu] + 2[vuw] - [wvu] - [uwv])\,.$$

These are not orthogonal, their scalar product being

$$s = (\varphi^1_{\Lambda_3^{(1)}},\ \varphi^1_{\Lambda_3^{(2)}}) = -\tfrac{1}{2}\,.$$

Therefore a superposition

$$\tilde{\varphi}^1_{\Lambda_3^{(2)}} = a\,\varphi^1_{\Lambda_3^{(1)}} + b\,\varphi^1_{\Lambda_3^{(2)}}$$

has to be found under the conditions

$$(\tilde{\varphi}^1_{\Lambda_3^{(2)}},\ \tilde{\varphi}^1_{\Lambda_3^{(2)}}) = 1 \quad \text{and} \quad (\tilde{\varphi}^1_{\Lambda_3^{(1)}},\ \tilde{\varphi}^1_{\Lambda_3^{(2)}}) = 0$$

leading to

$$a = \frac{-s}{\sqrt{1 - s^2}}\,, \qquad b = \frac{1}{\sqrt{1 - s^2}}$$

for the two weights. The new function finally is given as

$$\tilde{\varphi}^1_{\Lambda_3^{(2)}} = \tfrac{1}{2}\,(-[wuv] + [vwu] + [vuw] - [wvu])\,.$$

Similar cases are met with G_L. The complete tables for the symmetrized functions are contained in the appendix.

3.4. Compatibility Relations

Compatibility relations provide a useful tool for testing the correctness of a band calculation. It is helpful to classify the relations according to whether or not they give a one-to-one correspondence between the irreducible representations of different groups. One-to-one correspondence exists in the three cases

$$\Gamma \leftrightarrow \Delta\,, \quad \Gamma \leftrightarrow \Sigma \quad \text{and} \quad \Sigma \leftrightarrow K\,.$$

G_Γ	Γ_1	Γ_2	Γ_{12}	$\Gamma_{15'}$	$\Gamma_{25'}$
G_Δ	Δ_1	Δ_2	$\Delta_1 \oplus \Delta_2$	$\Delta_{1'} \oplus \Delta_5$	$\Delta_{2'} \oplus \Delta_5$
G_Σ	Σ_1	Σ_4	$\Sigma_1 \oplus \Sigma_4$	$\Sigma_2 \oplus \Sigma_3 \oplus \Sigma_4$	$\Sigma_1 \oplus \Sigma_2 \oplus \Sigma_3$

G_Γ	$\Gamma_{1'}$	$\Gamma_{2'}$	$\Gamma_{12'}$	Γ_{15}	Γ_{25}
G_Δ	$\Delta_{1'}$	$\Delta_{2'}$	$\Delta_{1'} \oplus \Delta_{2'}$	$\Delta_1 \oplus \Delta_5$	$\Delta_2 \oplus \Delta_5$
G_Σ	Σ_2	Σ_3	$\Sigma_2 \oplus \Sigma_3$	$\Sigma_1 \oplus \Sigma_3 \oplus \Sigma_4$	$\Sigma_1 \oplus \Sigma_2 \oplus \Sigma_4\,.$

All other cases provide relations which are not one-to-one:

G_Γ	Γ_1	Γ_2	Γ_{12}	$\Gamma_{15'}$	$\Gamma_{25'}$	$\Gamma_{1'}$	$\Gamma_{2'}$	$\Gamma_{12'}$	Γ_{15}	Γ_{25}
G_Λ	Λ_1	Λ_2	Λ_3	$\Lambda_2\oplus\Lambda_3$	$\Lambda_1\oplus\Lambda_3$	Λ_1	Λ_2	Λ_3	$\Lambda_1\oplus\Lambda_3$	$\Lambda_2\oplus\Lambda_3$

G_X	X_1	X_2	X_3	X_4	X_5	$X_{1'}$	$X_{2'}$	$X_{3'}$	$X_{4'}$	$X_{5'}$
G_Λ	Λ_1	Λ_2	$\Lambda_{2'}$	$\Lambda_{1'}$	Λ_5	$\Lambda_{1'}$	$\Lambda_{2'}$	Λ_2	Λ_1	Λ_5

G_L	L_1	L_2	L_3	$L_{1'}$	$L_{2'}$	$L_{3'}$
G_Λ	Λ_1	Λ_2	Λ_3	Λ_2	Λ_1	Λ_3

3.5. Analysis of Empty Lattice States

The free states for $v = v_\mathrm{C} + v_\mathrm{R} = 0$ are

$$T\,\varphi_j(\boldsymbol{k}, \boldsymbol{x}) = \varepsilon_j(\boldsymbol{k})\,\varphi_j(\boldsymbol{k}, \boldsymbol{x})\,, \qquad \varepsilon_j(\boldsymbol{k}) = (\boldsymbol{k} + \boldsymbol{K}_j)^2$$

giving rise to an already quite complex „free" band structure. In most directions the bands are degenerate, and the corresponding eigenfunctions form a basis of a reducible representation in general. By reduction one can give a complete classification in terms of irreducible representations, leading to an essential understanding of the case $v \neq 0$. There is no need for a calculation of the reducible representations and their characters, because all necessary information is contained already in the tables for the symmetrized functions (see Appendix). The classification of states in $\mathcal{H}_2^{(n)}$ for arbitrary n is complete, if all different types have been classified.

Summarizing the results of reduction we have the following relations between D^l-dimensional reducible representations $D^l(uvw)$ of type (uvw) and irreducible representations of the groups in consideration:

G_Γ: 1) $D_\Gamma^1(000) = \Gamma_1$,

2) $D_\Gamma^8(uuu) = \Gamma_1 \oplus \Gamma_{2'} \oplus \Gamma_{15} \oplus \Gamma_{25'}$,

3) $D_\Gamma^6(u00) = \Gamma_1 \oplus \Gamma_{12} \oplus \Gamma_{15}$,

4) $D_\Gamma^{12}(uu0) = \Gamma_1 \oplus \Gamma_{12} \oplus \Gamma_{15} \oplus \Gamma_{25'} \oplus \Gamma_{25}$,

5) $D_\Gamma^{24}(uuw) = \Gamma_1 \oplus \Gamma_{2'} \oplus \Gamma_{12} \oplus \Gamma_{12'} + 2\,\Gamma_{15} + \Gamma_{15'} \oplus 2\,\Gamma_{25'} \oplus \Gamma_{25}$.

The remaining two types $(uv0)$ and (uvw) do not occur among the first 65 plane waves.

G_Λ: 1) $D_\Lambda^1(000) = \Lambda_1$

2) $D_\Lambda^4(uuw) = \Lambda_1 \oplus \Lambda_{2'} \oplus \Lambda_5$,

3) $D_\Lambda^4(u0w) = \Lambda_1 \oplus \Lambda_2 \oplus \Lambda_5$,

4) $D_\Lambda^8(uvw) = \Lambda_1 \oplus \Lambda_{1'} \oplus \Lambda_2 \oplus \Lambda_{2'} \oplus 2\,\Lambda_5$.

G_Λ: 1) $D_\Lambda^1(uuu) = \Lambda_1$,

2) $D_\Lambda^3(uuv) = \Lambda_1 \oplus \Lambda_3$,

3) $D_\Lambda^6(uvw) = \Lambda_1 \oplus \Lambda_2 \oplus 2\,\Lambda_3$:

G_Σ: 1) $D_\Sigma^1(0\,0\,0) = \Sigma_1$,

2) $D_\Sigma^2(uuw) = \Sigma_1 \oplus \Sigma_3$,

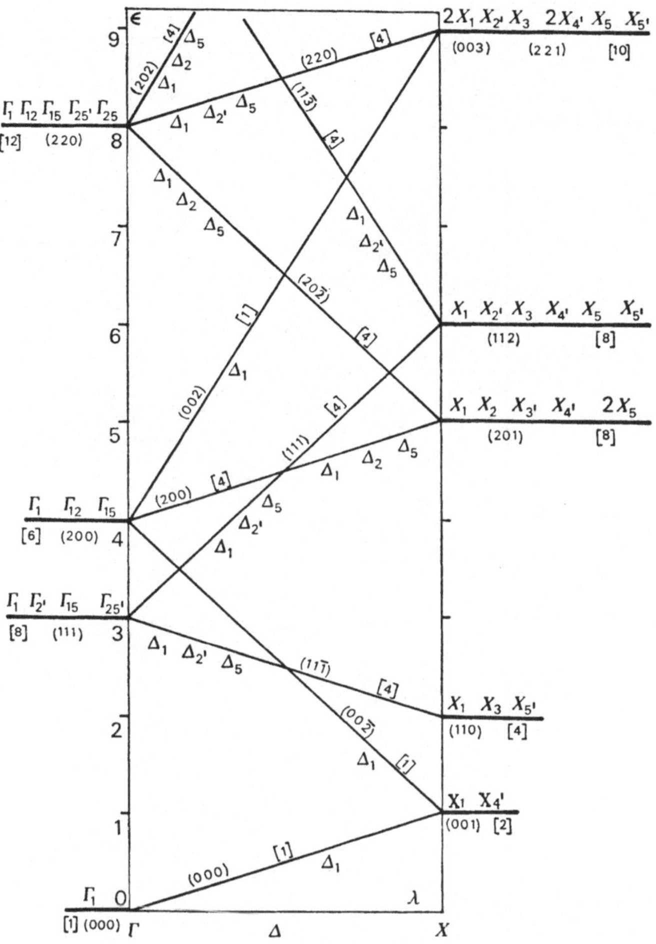

Fig. 2. Energy bands for the empty lattice in (001)-direction

$$3)\quad D^2_\Sigma(u\,v\,0) = \Sigma_1 \oplus \Sigma_4 ,$$

$$4)\quad D^4_\Sigma(u\,v\,w) = \Sigma_1 \oplus \Sigma_2 \oplus \Sigma_3 \oplus \Sigma_4 ;$$

$G_X:$

$$1)\quad D^2_X(0\,0\,w) = X_1 \oplus X_{4'} ,$$

$$2)\quad D^4_X(u\,u\,0) = X_1 \oplus X_3 \oplus X_{5'} ,$$

$$3)\quad D^8_X(u\,u\,w) = X_1 \oplus X_{2'} \oplus X_3 \oplus X_{4'} \oplus X_5 \oplus X_{5'} ,$$

$$4)\quad D^8_X(u\,v\,0) = X_1 \oplus X_2 \oplus X_3 \oplus X_4 \oplus 2X_{5'} ,$$

$$5)\quad D^8_X(u\,0\,w) = X_1 \oplus X_2 \oplus X_{3'} \oplus X_{4'} \oplus 2X_5 ,$$

$$6)\quad D^{16}_X(u\,v\,w) = X_1 \oplus X_2 \oplus X_3 \oplus X_4 \oplus 2X_5 \oplus X_{1'} \oplus X_{2'} \oplus X_{3'} \oplus X_{4'}$$
$$\oplus 2X_{5'} ,$$

$G_L:\quad (u'\,v'\,w') = \tfrac{1}{2}\,(u\,v\,w)$

$$1)\quad D^2_L(u'\,u'\,u') = L_1 \oplus L_{2'} ,$$

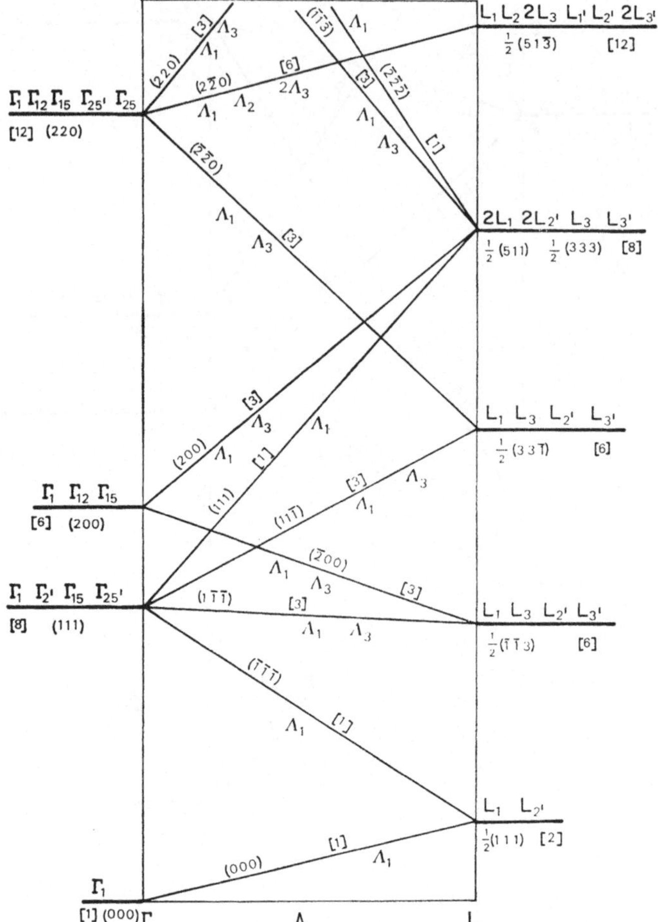

Fig. 3. Energy bands for the empty lattice in (111)-direction

2) $D_L^6(u'\,u'\,v') = L_1 \oplus L_{2'} \otimes L_3 \oplus L_{3'}$,

3) $D_L^{12}(u'\,v'\,w') = L_1 \oplus L_{1'} \oplus L_2 \oplus L_{2'} \oplus 2\,L_3 \oplus 2\,L_{3'}$.

In the point L (as well as in X) there are accidental degeneracies, e.g. for

$$\boldsymbol{K_j + k} = \begin{cases} \frac{1}{2}(1\,1\,5) \sim (u''\,u''\,v'') \\ \frac{1}{2}(3\,3\,3) \sim (u'\,u'\,u') \end{cases}.$$

In such a case the reduction is always trivial:

$$D_L^8(u'\,u'\,u' \mid u''\,u''\,v'') = D_L^2(u'\,u'\,u') \oplus D_L^6(u''\,u''\,v'') = 2\,L_1 \oplus 2\,L_{2'} \oplus L_3 \oplus L_{3'}.$$

G_K: The reduction is trivial: the reducible representations in every crossing point in which different Σ-bands come together decompose into l K_i-representations ($i = 1, 2, 3, 4$) if the incoming bands contain l Σ_i-representations.

The Figs. 2 and 3 schematically show the lowest band states in the empty lattice in a (001)- and a (111)-direction respectively. The representatives (uvw)

for different types are in parentheses, the degree of degeneracy in square brackets, and the results of reduction are given too.

3.6. Outlook for $v \neq 0$

Qualitatively speaking as soon as the potential is switched on the degenerate bands will split into their irreducible components, e.g. in the centre of the Brillouin zone the (111)-level is separated into Γ_1, $\Gamma_{2'}$, Γ_{15} and $\Gamma_{25'}$. A special consideration is demanded for the crossing points. Bands of the same symmetry never cross because the corresponding interaction matrix elements are $\neq 0$. These bands actually provide the only example in which the dispersion of the effective band differs drastically from the free one. However, this extraordinary behaviour can be deduced qualitatively from the empty lattice analysis too. In contrast bands of different symmetry can cross arbitrarily.

Now experience shows that a physical potential, as constructed approximately in the preceding chapter, is never able to change completely the qualitative character of a free band. This is also expected to be true from analytical reasons, since the only essential k-dependence of matrix elements stems from the kinetic diagonal term, whereas the off-diagonal elements vary very slowly with k. This can best be seen from the numerical tables for the orthogonality coefficients $R^{\sigma}_{nlj}(k)$ (Appendix of Part II) being slowly varying functions of the reduced wave vector; v_c even contributes k-independent terms.

Finally the question of whether selected symmetry levels obey a certain ordering quite independently of a particular potential is of central importance. This question is suggested by the prominent example of atomic levels ε_{nl} satisfying the inequality

$$\varepsilon_{nl} > \varepsilon_{nl'} \quad \text{for} \quad l > l'$$

as well as by the interesting situation in the one-dimensional lattice [22]. Saffren [23] in a study of two- and three-dimensional lattices has shown, that e.g. all possible Γ-levels are forming classes, and there are definite ordering relations for the levels within one and the same class, whereas the relative position of levels belonging to different classes is no consequence of symmetry, but depends on the particular potential. These results, however, are true for local potentials only.

One can gain insight into ordering tendencies by means of perturbation theory applied to the empty lattice states. Supposing the matrix elements to be of the order of average crystal field energies, i.e. in the scale used roughly of modulus 1, lowest order perturbation will cause the levels to split according to their symmetry: levels with wavefunctions with dominantly positive signs in the linear combination of plane waves or having the property of accumulating preferably the lowest Fourier coefficients with positive sign quite definitely are lowered strongly, which is particularly true for Γ_1 and Γ_{15}. In most ab-initio-calculations Fourier coefficients behaving like $\tilde{v}_c(q) \sim 1/q$ in the q-interval of interest are used, leading to the lowest order splittings shown in Fig. 4. Possible crossings are studied by means of the analytical expression for level differences, as for instance

$$E_{\Gamma_{15}(111)} - E_{\Gamma_1(111)} = -2\,\tilde{v}(200) - 4\,\tilde{v}(220) - \tilde{v}(222)\,.$$

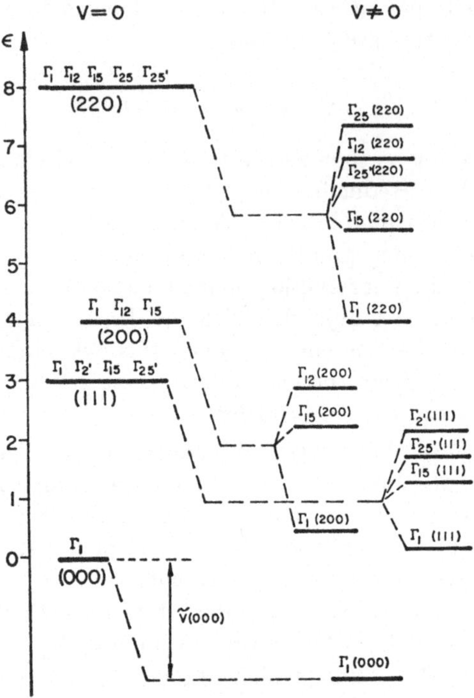

Fig. 4. Qualitative splittings of the lowest Γ-levels in lowest order perturbation theory

Any reasonable choice of coefficients will give

$$E_{\Gamma_1(111)} < E_{\Gamma_{15}(111)} \, .$$

The effect of second order perturbation is to split crossing bands of the same symmetry and to lower the lowest bands of each symmetry. Without going into details further, inequalities of the following type can be deduced:

$$\Gamma_1(000) < \Gamma_{15}(111) < \left\{ \begin{array}{c} \Gamma_1(111) \\ \Gamma_{25'}(111) \end{array} \right\} < \left\{ \begin{array}{c} \Gamma_{12}(200) \\ \Gamma_{2'}(111) \end{array} \right\} < \Gamma_1(200) \, .$$

In addition the contributions of v_{R} have to be analyzed separately. Shells of angular momentum l are able to push upwards only those crystal levels which are contained in the decomposition of the corresponding reducible representation of the rotation group.

On the basis of the above analysis finally we give a plot of the lowest s-, p-, d- and f-like bands in a (001)-direction in Fig. 5 (compare also with Figs. 2 and 3). Fig. 5b shows a typical case of hybridization of two Δ_1-bands. The position of the Γ_1-levels is extremely sensitive to details of the potential.

Like any plane wave method the OPW-procedure deals satisfactorily with s- and p-like valence states only. Any other situation will cause difficulties due to a complete interchange of the natural empty lattice levels by the repulsive part v_{R} as e.g. in the case of transition metals, where only a mixture of a tight-binding-

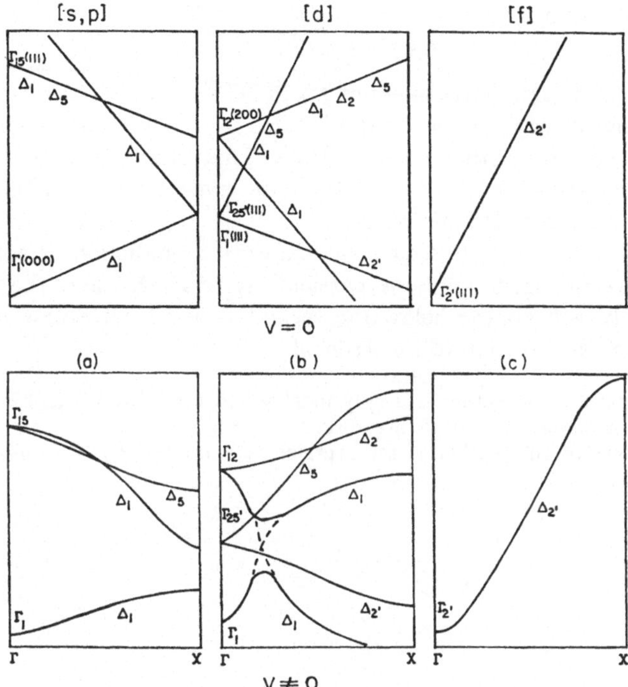

Fig. 5. Distortion of the lowest bands in (001)-direction

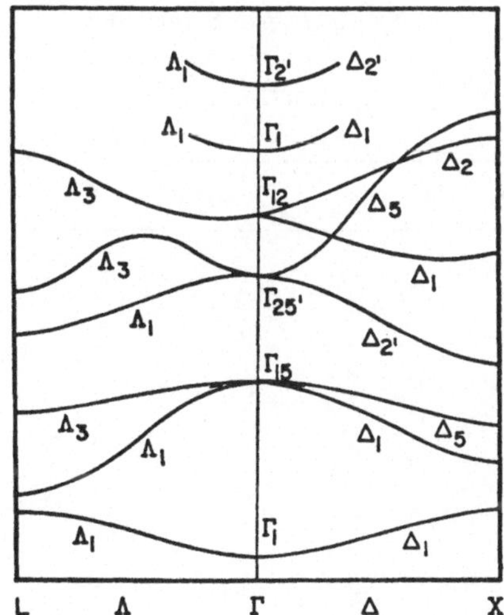

Fig. 6. Qualitative prediction of a band scheme for fcc- or rocksalt-structure

and plane wave method is able to produce reliable results [24]. It would be entirely hopeless trying to get the $4f$-level of Eu from the OPW-band calculation, as can be understood from the decomposition of the reducible representation for $l = 3$

K. Lendi

of the rotation group:

$$D_3 = \Gamma_{2'} \oplus \Gamma_{15} \oplus \Gamma_{25}.$$

In particular a correct treatment of Γ_{25} in $\mathscr{H}_2^{(n)}$ would imply $n \to \infty$. However, there is no reason why the localized $4f$-states, if included as core states, should not be in the region of other valence states. This situation is methodically consistent, because the energy scales of the core levels $\varepsilon_{nl}^{\sigma}$ and of the band levels $E_i^{\sigma}(\boldsymbol{k})$ are fixed relative to each other.

All the above arguments lead to a qualitative prediction of a band scheme like that shown in Fig. 6, applying particularly to EuO, where the empty d-like bands probably will appear below the second Γ_1-level. Of course no prediction can be made of the position of the $4f$-level.

Acknowledgement. The author is deeply indebted to Prof. Dr. W. Baltensperger for his continuous and generous support of this work.

Also an extensive information on the experimental situation by G. Güntherodt is gratefully acknowledged.

Appendix

Symmetrized Functions

$$G_{\Delta} \quad \boldsymbol{k} = \frac{2}{a}\pi(0,0,\lambda) \quad 0 < \lambda < 1$$

Type	Components	Δ_1	$\Delta_{1'}$	Δ_2	$\Delta_{2'}$	Δ_5	
(000)	000	+					
(uuw)	uuw	+			+	+	
	$u\bar{u}w$	+			−	+	
	$\bar{u}uw$	+			−	−	
	$\bar{u}\bar{u}w$	+			+	−	
(u0w)	$u0w$	+		+		+	
	$\bar{u}0w$	+		+		−	
	$0uw$	+		−		0	
	$0\bar{u}w$	+		−		0	
(uvw)	uvw	+	+	+	+	+	0
	$u\bar{v}w$	+	−	+	−	+	0
	$\bar{u}vw$	+	−	+	−	−	0
	$\bar{u}\bar{v}w$	+	+	+	+	−	0
	vuw	+	−	−	+	0	+
	$v\bar{u}w$	+	+	−	−	0	+
	$\bar{v}uw$	+	+	−	−	0	−
	$\bar{v}\bar{u}w$	+	−	−	+	0	−

$$G_\Lambda \quad k = \frac{2}{a}\pi(\lambda, \lambda, \lambda) \quad 0 < \lambda < \tfrac{1}{2}$$

Type	Components	Λ_1	Λ_2	Λ_3	
(uuu)	uuu	$+$			
(uuv)	uuv	$+$		$+$	
	uvu	$+$		$-$	
	vuu	$+$		0	
(uvw)	uvw	$+$	$+$	$-$	$+$
	wuv	$+$	$+$	$+2$	0
	vwu	$+$	$+$	$-$	$-$
	vuw	$+$	$-$	$+$	$+$
	wvu	$+$	$-$	-2	0
	uwv	$+$	$-$	$+$	$-$

The functions φ_{Λ_3} transform according to the second row of the irreducible representation.

$$G_\Sigma \quad k = \frac{2}{a}\pi(\lambda, \lambda, 0) \quad 0 < \lambda < \frac{3}{4}$$

Type	Components	Σ_1	Σ_2	Σ_3	Σ_4
(000)	000	$+$			
(uuw)	uuw	$+$		$+$	
	$uu\bar{w}$	$+$		$-$	
$(uv0)$	$uv0$	$+$			$+$
	$vu0$	$+$			$-$
(uvw)	uvw	$+$	$+$	$+$	$+$
	$vu\bar{w}$	$+$	$+$	$-$	$-$
	$uv\bar{w}$	$+$	$-$	$-$	$+$
	vuw	$+$	$-$	$+$	$-$

$$G_X \quad k = \frac{2}{a}\pi(0, 0, 1)$$

Type	Components	X_1	X_2	X_3	X_4	X_5	$X_{1'}$	$X_{2'}$	$X_{3'}$	$X_{4'}$	$X_{5'}$
$(00w)$	$00w$	$+$								$+$	
	$00\bar{w}$	$+$								$-$	
$(uu0)$	$uu0$	$+$		$+$							$+$
	$u\bar{u}0$	$+$		$-$							$-$
	$\bar{u}u0$	$+$		$-$							$+$
	$\bar{u}\bar{u}0$	$+$		$+$							$-$
	uuw	$+$		$+$		$+$	$+$			$+$	$+$
	$u\bar{u}\bar{w}$	$+$		$-$		$-$	$+$			$-$	$-$
	$\bar{u}u\bar{w}$	$+$		$-$		$+$	$+$			$-$	$+$

Type	Components	X_1	X_2	X_3	X_4	X_5	$X_{1'}$	$X_{2'}$	$X_{3'}$	$X_{4'}$	$X_{5'}$
(uuw)	$\bar u\bar u w$	+		+		−		+		+	−
	$\bar u u w$	+		−		−		−		+	+
	$u\bar u w$	+		−		+		−		+	−
	$u u \bar w$	+		+		−		−		−	+
	$\bar u\bar u\bar w$	+		+		+		−		−	−
$(uv0)$	$uv0$	+	+	+	+					+	0
	$u\bar v0$	+	+	−	−					−	0
	$\bar u v0$	+	+	−	−					+	0
	$\bar u\bar v0$	+	+	+	+					−	0
	$\bar v u0$	+	−	−	+					0	+
	$v\bar u0$	+	−	−	+					0	−
	$v u0$	+	−	+	−					0	+
	$\bar v\bar u0$	+	−	+	−					0	−
$(u0v)$	$u0v$	+	+			+	0		+	+	
	$u0\bar v$	+	+			−	0		−	−	
	$\bar u0\bar v$	+	+			+	0		−	−	
	$\bar u0v$	+	+			−	0		+	+	
	$0uv$	+	−			0	+		−	+	
	$0\bar u v$	+	−			0	−		−	+	
	$0u\bar v$	+	−			0	+		+	−	
	$0\bar u\bar v$	+	−			0	−		+	−	
(uvw)	uvw	+	+	+	+	+	0	+	+	+	0
	$u\bar v\bar w$	+	+	−	−	−	0	+	+	−	0
	$\bar u v\bar w$	+	+	−	−	+	0	+	+	+	0
	$\bar u\bar v w$	+	+	+	+	−	0	+	+	−	0
	$\bar u\bar v\bar w$	+	+	+	+	+	0	−	−	−	0
	$\bar u v w$	+	+	−	−	−	0	−	−	+	0
	$u\bar v w$	+	+	−	−	+	0	−	−	−	0
	$u v\bar w$	+	+	+	+	−	0	−	−	+	0
	$\bar v u w$	+	−	−	+	0	−	+	−	0	−
	$v\bar u w$	+	−	−	+	0	+	+	−	0	+
	$v u\bar w$	+	−	+	−	0	−	+	−	0	−
	$\bar v\bar u\bar w$	+	−	+	−	0	+	+	−	0	+
	$v\bar u\bar w$	+	−	−	+	0	−	−	+	0	+
	$\bar v u\bar w$	+	−	−	+	0	+	−	+	0	−
	$\bar v\bar u w$	+	−	+	−	0	−	−	+	0	+
	$v u w$	+	−	+	−	0	+	−	+	0	−

G_L $\boldsymbol{k} = \dfrac{2}{a}\,\pi\left(\dfrac{1}{2},\,\dfrac{1}{2},\,\dfrac{1}{2}\right)$ $(uvw) = \tfrac{1}{2}(u'\,v'\,w')$; $u',\,v',\,w'$ integers

Type	Components	L_1	L_2	L_3	$L_{1'}$	$L_{2'}$	$L_{3'}$
(uuu)	uuu	+				+	
	$\bar u\bar u\bar u$	+				−	
(uuv)	uuv	+		+		+	+
	vuu	+		0		+	0
	uvu	+		−		+	−
	$\bar u\bar u\bar v$	+		+		−	−
	$\bar v\bar u\bar u$	+		0		−	0
	$\bar u\bar v\bar u$	+		−		−	+

Type	Components	L_1	L_2	L_3		$L_{1'}$	$L_{2'}$	$L_{3'}$	
	uvw	+	+	+2	0	+	+	+2	0
	wuv	+	+	−	+	+	+	−	+
	vwu	+	+	−	−	+	+	−	−
	$\bar{v}\bar{u}\bar{w}$	+	−	+	+	+	−	−	−
	$\bar{w}\bar{v}\bar{u}$	+	−	+	−	+	−	−	+
(uvw)	$\bar{u}\bar{w}\bar{v}$	+	−	−2	0	+	−	+2	0
	$\bar{u}\bar{v}\bar{w}$	+	+	+2	0	−	−	−2	0
	$\bar{w}\bar{u}\bar{v}$	+	+	−	+	−	−	+	−
	$\bar{v}\bar{w}\bar{u}$	+	+	−	−	−	−	+	+
	vuw	+	−	+	+	−	+	+	+
	wvu	+	−	+	−	−	+	+	−
	uwv	+	−	−2	0	−	+	− 2	0

Remarks: $\pm$ means ± 1.

Recalling the abbreviation $[u\,v\,w] = \dfrac{1}{\sqrt{\Omega_0}}\exp\left\{i\,\dfrac{2\pi}{a}\,(u\,x + v\,y + w\,z)\right\}$ and the normalization $([u\,v\,w],[r\,s\,t]) = \delta_{ur}\,\delta_{vs}\,\delta_{wt}$, e.g. a normalized function $\varphi_{L_3}(uvw)$, transforming according to the second row of L_3, reads

$$\varphi_{L_3}(u\,v\,w) = \frac{1}{\sqrt{24}}\left(2\,[u\,v\,w] - [w\,u\,v] - [v\,w\,u] + [\bar{v}\,\bar{u}\,\bar{w}] + [\bar{w}\,\bar{v}\,\bar{u}] - 2\,[\bar{u}\,\bar{w}\,\bar{v}]\right.$$

$$\left. + 2\,[\bar{u}\,\bar{v}\,\bar{w}] - [\bar{w}\,\bar{u}\,\bar{v}] - [\bar{v}\,\bar{w}\,\bar{u}] + [v\,u\,w] + [w\,v\,u] - 2\,[u\,w\,v]\right)$$

$$= \frac{1}{\sqrt{6}}\left(2\cos\frac{2\pi}{a}\,(u\,x + v\,y + w\,z) - \cos\frac{2\pi}{a}\,(w\,x + u\,y + v\,z) - \cos\frac{2\pi}{a}\right.$$

$$\times\,(v\,x + w\,y + u\,z) + \cos\frac{2\pi}{a}\,(v\,x + u\,y + w\,z) + \cos\frac{2\pi}{a}\,(w\,x + v\,y + u\,z)$$

$$\left. - 2\cos\frac{2\pi}{a}\,(u\,x + w\,y + v\,z)\right).$$

References

1. Methfessel, S., Mattis, D. C.: Handbuch der Physik, Vol. XVIII, Berlin-Heidelberg-New-York: Springer 1968
2. Busch, G., Wachter, P.: Phys. kondens. Materie **5**, 232 (1968)
3. Feinleib, J., Scouler, W. J., Dimmock, J. O., Hanus, J., Reed, T.B., Pidgeon, C. R.: PRL **22**, 1385 (1969)
4. Günterodt, G., Wachter, P., Imboden, D. M.: Phys. kondens. Materie **12**, 292 (1971)
5. Kunz, A. B.: PR **175**, 1147 (1968).
6. Herman, F., Kortum, R. L., Kuglin, C. D., Shay, J. L.: II—VI Semiconducting Compounds, 1967 International Conference, p. 503. Benjamin 1967
7. Müller, F. M.: PR **153**, 659 (1967)
8. Heine, V.: PR **153**, 673 (1967)
9. Herman, F., Kortum, R. L., Kuglin, C. D., Van Dyke, J. P.: Methods in Computational Physics, Vol. 8, Academic Press 1968
10. Woodruff, T. O.: Solid State Physics, Vol. 4, Academic Press 1957
11. Callaway, J.: Energy Band Theory, Academic Press 1964
12. Herman, F., Skillman, S.: Atomic Structure Calculations. Prentice-Hall Inc. 1963
13. Lendi, K.: Thesis ETH Zürich, 1972

14. Robinson, J. E., Bassani, F., Knox, R. S., Schrieffer, J. R.: PRL **9**, 215 (1962)
15. Tosi, M.: Solid State Physics, Vol. 16. Academic Press 1964
16. Ballhausen, C. J.: Ligand Field Theory. McGraw-Hill 1962
17. Vetri, G., Bassani, F.: Nuovo Cimento **55**B, 504 (1968)
18. Calabrese, E.: Thesis, Lehigh University 1970
19. Loucks, T.: Augmented Plane Wave Method. Benjamin 1967
20. Wigner, E.: Group Theory. Academic Press 1959
21. Koster, G. F.: Solid State Physics, Vol. 5. Academic Press 1957
22. Joos, R.: Thesis ETH Zürich, 1971
23. Saffren, M. M.: PR **165**, 870 (1968)
24. Müller F. M.: PR **153** 659 (1967)

Dr. Karl Lendi
Eidgenössische Technische Hochschule
Institut für Theoretische Physik
Hönggerberg
CH-8049 Zürich, Switzerland

Phys. cond. Matter 17, 215—231 (1974)
© by Springer-Verlag 1974

Spinpolarized OPW-Band Structure of Magnetic Semiconductors with Application to EuO. Part II

Karl Lendi *

Institut für Theoretische Physik
Eidg. Technische Hochschule, Zürich, Switzerland

Received July 17, 1973

The OPW-method has been generalized for the calculation of spinpolarized energy bands in magnetic semiconductors and applied to the ferromagnetic semiconductor EuO.

Part II describes the numerical methods used in the application to EuO together with a careful consideration of numerical problems involved. Finally the results of the calculations are presented. With the aid of selection rules and Van Hove-singularities a list of allowed optical transitions has been deduced. These results compare favorably with recent experiments.

1. Numerical Methods and Results

With a few exceptions publications of band structure calculations do not give any insight into the numerical methods used and their reliability. It is therefore worthwhile to give a more detailed description and a connection to the corresponding mathematical literature.

1.1. Unrestricted Hartree-Fock-Slater Theory (HFS)

The restricted, unrelativistic HFS-theory, as applied to non-magnetic atoms, is described extensively in the book by Herman and Skillman [1] where also comments on numerical treatment and some Fortran programs are found. Beside some essential improvements concerning a more rapid convergence in the solution of the radial Schrödinger equation, which are seen to be particularly important for heavy atoms, these methods and programs have been generalized to multiplet configurations of pure spin states. The ground state configuration of Eu^{2+} is $^8S_{7/2}$, i.e. all seven $4f$-electrons have parallel spin ($\sigma = +1$), generating two different exchange potentials for spin up ($\sigma = +1$)- and spin down ($\sigma = -1$)-electrons. For the electrons of a filled shell there are two different radial solutions $P_{nl}^\sigma(r)$ of

$$\left[-\frac{d^2}{dr^2} + \frac{l(l+1)}{r^2} + v_a^\sigma(r) - \varepsilon_{nl}^\sigma \right] P_{nl}^\sigma(r) = 0, \qquad \sigma = \pm 1 \qquad (1.1)$$

and correspondingly two different charge distributions. Since in the ground state exchange has a negative sign, the density of spin up electrons is concentrated somewhat closer to the nucleus than the density of spin down electrons.

Fig. 1 shows (in arbitrary units) the difference ($[P_{50}^{+1}]^2 - [P_{50}^{-1}]^2$) for the $5s$-electrons in Eu^{2+}; the corresponding difference of level energies ($\varepsilon_{50}^{+1} - \varepsilon_{50}^{-1}$) amounts to -4.1 eV.

---

* *Present address:* Centro de Investigación del Instituto Politécnico Nacional, Departamento de Física, Apartado Postal 14—740; México 14, D.F. México.

K. Lendi

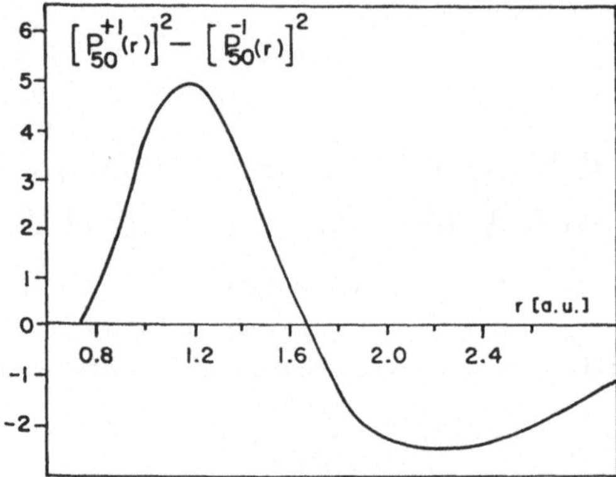

Fig. 1. Difference in $\tau(r)$ between the $(\sigma = +1)$ — and $(\sigma = -1)$ — contributions of the $5s$-electrons

If the potential of the neutral nonmagnetic atom is used as starting potential, the convergence to selfconsistency (within prescribed limits) is slower for v_a^{+1} than for v_a^{-1}. Beside that the convergence criteria described in [1] and a (441)-point mesh have been found to be appropriate also for the spinpolarized case.

No extra calculation has been performed for O^{2-} because of evident convergence difficulties which can be got rid of only by taking into account an external potential. This is not necessary because to a good approximation the data of the neutral O can be used, however, with the modified $(2p)^6$ occupation. The main difference between the neutral atom and the ion appears in the energy eigenvalue of the $2p$-state which does not enter the starting data for the OPW-calculation.

The level energies obtained for Eu^{2+} and O are listed in Table 1.

Table 1. HFS-eigenvalues $[- \varepsilon_{nl}^\sigma]$ for Eu^{2+} and O

Shell	$- \varepsilon_{nl}^\sigma$ (Ryd) $\sigma = +1$	for Eu^{2+} $\sigma = -1$	Shell	$- \varepsilon_{nl}$ (Ryd) for O
$1s$	3365.5	3365.5	$1s$	39.456
$2s$	537.65	537.63	$2s$	2.1439
$2p$	512.18	512.17	$2p$	1.0408
$3s$	119.54	119.15		
$3p$	108.05	107.70		
$3d$	86.469	86.210		
$4s$	24.719	23.991		
$4p$	20.213	19.490		
$4d$	12.134	11.422		
$5s$	4.3601	4.0440		
$5p$	3.1123	2.8596		
$4f$	2.2562	—		

A comparison of the present data with relativistic ones [2] suggests, that the unrelativistic treatment is a reasonable averaged approach.

1.2. Computation of Integrated Quantities

The evaluation of the orthogonality coefficients $R_{nlj}^{\sigma}(\mathbf{k})$ (1.15, I) and of the Fourier coefficients $\tilde{v}_e(q)$ (1.20, I) and $\tilde{v}_x^{\sigma}(q)$ (1.21, I) involves a numerical integration over the tabulated radial functions $P_{nl}^{\sigma}(r)$. The point mesh of the radial distance r is adapted to the oscillating behavior of the wave functions and depends therefore upon the nuclear charge. Several tests for Eu and O have shown that it is sufficient to use a 99-point mesh taking only every fourth point and 9 blocks of the original 441-point mesh [1]. By using Simpson-integration the results obtained differ by less than 1 ppm. If in an interval $[x_0, x_n]$ a total number of $(n+1)$ function values f_n in a constant distance h from each other is given, the continued Simpson rule reads

$$\int_{x_0}^{x_n} f(x)\, dx \cong \frac{h}{3}\left(f_0 + \sum_{l=1}^{n-1} w_l f_l + f_n\right) \qquad w_l = \begin{cases} 4, & l \text{ odd} \\ 2, & l \text{ even} \end{cases}.$$

Some of the numerical results for the Fourier coefficients $\tilde{v}^{\sigma}(q)$ are listed in Table 2.

Table 2. Fourier coefficients of the atomic potentials of Eu^{2+} and O^{2-} (Ryd)

$\left(\dfrac{a}{2\pi}\right)q$	$\tilde{v}^{\sigma}(q)$ for Eu^{2+} $\sigma = +1$	$\sigma = -1$	$\tilde{v}(q)$ for O^{2-}
(111)	-0.90173	-0.88293	-0.11378
(200)	-0.78756	-0.77199	-0.11946
(220)	-0.56549	-0.55689	-0.10686
(311)	-0.48069	-0.47503	-0.09423
(222)	-0.45908	-0.45410	-0.09045
(400)	-0.39161	-0.38875	-0.07755
(331)	-0.35443	-0.35254	-0.06975
(420)	-0.34383	-0.34231	-0.06754
(422)	-0.30776	-0.30707	-0.05975
(511)	-0.28573	-0.28551	-0.05489

The calculation of the $R_{nlj}^{\sigma}(\mathbf{k})$'s requires particular care because the integrand contains the spherical Bessel functions which provide an outstanding example for numerical instabilities. Any direct programming and evaluation of the analytical relations (1.2) for certain arguments and indices will lead to extremely wrong values, especially for functions with high l. Fortunately only l-values up to 3 are needed in the present application.

$$\left.\begin{aligned} j_0(x) &= \frac{\sin x}{x}, \\[2mm] j_1(x) &= \frac{\sin x}{x^2} - \frac{\cos x}{x}, \\[2mm] j_2(x) &= \left(\frac{3}{x^3} - \frac{1}{x}\right)\sin x - \frac{3}{x^2}\cos x, \\[2mm] j_3(x) &= \left(\frac{15}{x^4} - \frac{6}{x^2}\right)\sin x - \left(\frac{15}{x^3} - \frac{1}{x}\right)\cos x. \end{aligned}\right\} \qquad (1.2)$$

For increasing l the analytical expressions become more and more complicated suggesting to take advantage of the recursion relation

$$j_l(x) = \frac{2\,l-1}{x}\, j_{l-1}(x) - j_{l-2}(x)\,. \tag{1.3}$$

This procedure, however, is even more unstable for numerical evaluation and cannot be used in general. For instance for the argument $x_0 = 10^{-4}$ a machine with 12 digits and simple accumulator (for definitions see [3]) will give

$$j_3(x_0) = -\,1\,099.999\,948$$

whereas up to 6 decimals the exact value is zero. Despite these striking difficulties it has been found that in the present application a double precision calculation of the j_l's (on CDC 6400) yields all results up to the required accuracy for $x > 10^{-4}$ and $l \leq 3$. In order to judge this the Bessel functions have been calculated by means of a much more stable but more time consuming method proposed by Corbatò and Uretsky [4]. The results coincided after rounding to simple precision.

Some numerical values of the OPW-coefficients R_{nlj}^{σ} ($k = 0$) are listed in the appendix.

1.3. Solution of the General Eigenvalue Problem $(A - \lambda B)\,\boldsymbol{x} = 0$

It has been shown in Part I that all matrix elements are real, so the following discussion is restricted to $\boldsymbol{R}^n$.

The general eigenvalue problem

$$(A - \lambda B)\,\boldsymbol{x} = 0 \tag{1.4}$$

has to be solved for real-symmetric $(n \times n)$-matrices, where the eigenvectors $\boldsymbol{x}$ are to be identified with the column vectors of the matrix $S = \{s_{li}\}$ in (1.3) of Part I. We are looking for effective procedures yielding all eigenvalues and -vectors in order to have a test for the correctness of the programs through the compatibility relations. The ordinary numerical methods solve the special eigenvalue problem

$$(M - \lambda 1)\,\boldsymbol{y} = 0 \tag{1.5}$$

and therefore (1.4) has first to be reduced to (1.5). In principle this can be accomplished in three different ways:

(a) First B is diagonalized by means of an orthogonal transformation O:

$$D_B = O^{\mathrm{T}} B O\,. \tag{1.6}$$

If B is positive definite ($B > 0$) there exists the real decomposition

$$D_B = D_B^{\frac{1}{2}} D_B^{\frac{1}{2}} \tag{1.7}$$

and the reduction is achieved by the transformations

$$M = D_B^{-\frac{1}{2}} O^{\mathrm{T}} A\, O D_B^{-\frac{1}{2}}\,, \qquad \boldsymbol{x} = O D_B^{-\frac{1}{2}} \boldsymbol{y}\,. \tag{1.8}$$

(b) If $B > 0$ the reduction can be performed avoiding the diagonalization of B by means of a real Cholesky decomposition

$$B = R^{\mathrm{T}} R\,, \tag{1.9}$$

R being a regular right-triangular matrix. Using the abbreviation R^{-T} for $(R^{-1})^T = (R^T)^{-1}$, we have

$$M = R^{-T} A R \quad \text{and} \quad x = R^{-1} y . \tag{1.10}$$

(c) If $\det(B) \neq 0$, but not necessarily $B > 0$, the only possibility is to invert B, leading to

$$M = A B^{-1} \quad \text{and} \quad x = B^{-1} y \tag{1.11}$$

where x are the right-eigenvectors of M which in general is not symmetric any more. The procedures (a) and (b) solve (1.5) for an arbitrary symmetric A and a positive definite symmetric B. (a) has the advantage of computing the eigenvalues of B in addition and admits therefore of a detailed discussion of probable difficulties in finding stable solutions. If B possesses one or more eigenvalues close to zero, $D_B^{-\frac{1}{2}}$ in (1.8) correspondingly has some large elements. Assuming the matrix elements of A are roughly of the same order of magnitude, the transformation to M will cause some columns and rows to increase enormously in their norm with respect to the other ones. This mechanism in turn causes a worsening of the condition of some eigensolutions ([5], p. 344). Besides the multiplication by O and O^T, respectively, is of no influence [3]. In principle one has to distinguish two kinds of conditions, namely a condition (or a sensitivity) of the solutions with respect to slight changes in the input data, and a condition with respect to numerical errors inherent to the algorithm used for calculation. We always speak of the first one assuming that we work with numerically most reliable methods on a big machine. However in case (a) the amount of computing effort is much greater than for (b) and (c), and for practical purposes the accuracy obtained is very questionable if B has a highly degenerate spectrum [5] like in the OPW-method, because then the eigensystem O and therefore the matrix M might be affected by quite considerable errors. Finally procedure (b) is by far the most effective and numerically most stable one.

The condition of the eigenvalue problem (1.4) and in particular that of the lowest eigenvalues of main interest is intimately connected with the properties of the matrix B. Unfortunately the application of the OPW-method to heavy substances, presumably for nuclear charge $Z > 55$ will give rise to difficulties because the lowest eigenvalue of B is too close to zero or, what is even worse, the approximations used lead to a non-positive definite B in contradiction to the analytical estimates given in chapter 2 of Part I.

The „good" positive definiteness of B (actually all eigenvalues close to or equal to 1) gets lost with increasing number of occupied core states, as can be seen from

$$(\varphi_j, P_1 \varphi_j) = \sum_{n, l, m} |(\psi_{nlm}, \varphi_j)|^2$$

and

$$\hat{B} = 1 - P_1$$

(compare with (1.5a) and (2.1), Part I) and from the corresponding numerical values for the OPW-coefficients in the appendix. Whereas an estimate for Al [6] gives

$$\max_j (\varphi_j, P_1 \varphi_j) \approx 0.1$$

we have for EuO

$$\max_j (\varphi_j, \boldsymbol{P}_1 \varphi_j) \approx 0.7 \ .$$

The heavier the elements involved are, the more the matrix B differs from the unit matrix 1. In case of eigenvalues close to zero particular care is due to maintaining up to high accuracy the orthogonality relations

$$\int\limits_0^{r_0} P_{nl}^\sigma(r)\, P_{n'l}^\sigma(r)\, dr = \delta_{nn'} \ ,$$

otherwise even slightly negative eigenvalues can occur.

In order to overcome the above mentioned problems in obtaining the band structure of EuO it was best to omit the calculation of the lowest s-like band which usually is not of particular interest. This means that the first plane wave with reciprocal wave vector $\dfrac{2\pi}{a}\,(0, 0, 0)$ and therefore the largest contribution to the matrix elements of $\boldsymbol{P}_1$ has been eliminated. This is convenient for group theoretical reasons too, because this band does not fulfill any compatibility relations. Of course, the corresponding states (actually the 2s-levels of the oxygen ions) have to be included in the core states. After this ad hoc modification one is left with an eigenvalue problem of acceptable stability.

For illustration we consider a numerical example in 4 dimensions, computed on CDC 6400.

$$A = \begin{pmatrix} 2 & 1 & 3 & 4 \\ 1 & -3 & 1 & 5 \\ 3 & 1 & 6 & -2 \\ 4 & 5 & -2 & -1 \end{pmatrix} \qquad \left.\begin{aligned} \lambda_1 &= -8.0286 \\ \lambda_2 &= -1.5732 \\ \lambda_3 &= 5.6689 \\ \lambda_4 &= 7.9329 \end{aligned}\right\} \qquad (1.12)$$

The λ_i's are the eigenvalues rounded to 4 decimals.

Instead of B we give the diagonal elements d_{ii} of B_D and transform afterwards to general form:

$$B = O D_B O^\mathrm{T}, \qquad O = \tfrac{1}{2}\begin{pmatrix} 1 & -1 & -1 & 1 \\ 1 & -1 & 1 & -1 \\ 1 & 1 & -1 & -1 \\ 1 & 1 & 1 & 1 \end{pmatrix} .$$

Then B is of the form

$$B = \begin{pmatrix} u & v & v & -v \\ v & u & -v & v \\ v & -v & u & v \\ -v & v & v & u \end{pmatrix} .$$

In all following cases three eigenvalues are taken to be 1, i.e.

$$d_{22} = d_{33} = d_{44} = 1$$

and only d_{11} is going to be varied.

Consider the cases

$$\text{(I)} \quad d_{11} = 10^{-2}, \quad u = 0.7525, \quad\quad v = 0.2475,$$
$$\text{(II)} \quad d_{11} = 10^{-4}, \quad u = 0.750025, \quad\; v = 0.249975,$$
$$\text{(III)} \quad d_{11} = 10^{-6}, \quad u = 0.75000025, \quad v = 0.24999975.$$

u and v are exact in all cases, and it is important to note that all matrices have exact representations on the machine and are not affected already by input rounding errors [7]. Now the general eigenvalue problem $(A - \lambda B)\, x = 0$ has quite differing solutions listed in Table 3. The first and fourth eigenvalue are extremely ill-conditioned, whereas the remaining two are surprisingly stable.

Table 3. Eigenvalues of $(A - \lambda B)\, x = 0$

	I	II	III
$\lambda_1 =$	$-\ 51.7002$	$-\ 510.4221$	$-\ 5'099.5215$
$\lambda_2 =$	$-\ 2.7796$	$-\ 2.8004$	$-\ 2.8006$
$\lambda_3 =$	7.7990	7.8006	7.8006
$\lambda_4 =$	50.6807	509.4219	$5'098.5215$

It has to be emphasized strongly that in going from case (I) to (III) the elements of B are changing very slightly only, i.e. u changes by less than 3.5 ppm, v by 1 percent.

In order to show the influence of inaccurately known matrixelements of A, as an exampel, a_{12} is altered by 10 percent:

$$a_{12} = a_{21} = 1.1\,.$$

The results are given in Table 4 together with the modified spectrum of A alone.

Table 4. Eigenvalues of $(A - \lambda B)\, x = 0$ for $a_{12} = a_{21} = 1.1$

	$B = 1$	I	II	III
$\lambda_1 =$	$-\ 7.9942$	$-\ 54.4599$	$-\ 820.8081$	$-\ 50'520.6382$
$\lambda_2 =$	$-\ 1.6590$	$-\ 2.9004$	$-\ 2.9212$	$-\ 2.9214$
$\lambda_3 =$	5.6960	7.8341	7.8355	7.8356
$\lambda_4 =$	7.9571	48.5762	319.9437	519.7740

A comparison of the four columns of Table 4 makes evident the relative insensitivity of the special problem with $B = 1$, whereas the lowest eigenvalue for $B \neq 1$ shows extreme deviations. Finally let us consider in Table 5 a simulation

Table 5. Eigenvalues of $(A - \lambda B)\, x = 0$ for $a_{ii} \to a_{ii} + 0.5$ $(i = 1, 2, 3, 4,)$

	I	II	III
$\lambda_1 =$	$-\ 32.6270$	$-\ 52.3619$	$-\ 52.8911$
$\lambda_2 =$	$-\ 1.8976$	$-\ 1.9124$	$-\ 1.9126$
$\lambda_3 =$	8.3079	8.3092	8.3092
$\lambda_4 =$	81.7166	$5'051.4652$	$500'051.9951$

of the adjustment of the zeroth Fourier coefficient by shifting all diagonal elements by the amount $+0.5$.

Evidently the sensitivity of λ_1 has concentrated almost entirely on λ_4.

Of course one has to be sure that all reproduced results are not affected by additional numerical inaccuracies. For this reason the method of solution described under (b) has been employed in connection with the most reliable version of the Jacobi diagonalization procedure due to H. Rutishauser [8].

In conclusion the examples given above reflect in an extreme way the situation met in a straightforward application of the OPW-method to EuO.

1.4. Cholesky Decomposition and QR-Transformation

The procedure (b) will now be described in more detail.

A positive-definite, real-symmetric matrix B is decomposable into the product

$$B = R^{\mathrm{T}} R \tag{1.13}$$

called Cholesky decomposition, where R is a real and regular right-triangular matrix. Choosing the diagonal elements of R with positive sign e.g., the decomposition is unique. The elements r_{ik} of R successively are calculated from those of B by means of the following relations:

$$\left.\begin{aligned}
&r_{11} = +(b_{11})^{\frac{1}{2}}, \qquad r_{1k} = b_{1k}/r_{11} \qquad && k = 2, 3, \dots, n\,, \\
&r_{ii} = +(b_{ii} - \sum_{l=1}^{i-1} r_{li}^2)^{\frac{1}{2}}, && i = 2, 3, \dots, n\,, \\
&r_{ik} = (b_{ik} - \sum_{l=1}^{i-1} r_{li} r_{lk})/r_{ii}\,, && \begin{aligned} i &= 2, 3, \dots, n\,, \\ k &= i+1, i+2, \dots, n\,, \end{aligned} \\
&(r_{ik} = 0,\ k < i)\,.
\end{aligned}\right\} \tag{1.14}$$

In addition the inverse $P = R^{-1}$ is needed, yielding again a right-triangular matrix, whose elements are calculated descending from n without the necessity of searching for pivots [3]:

$$\left.\begin{aligned}
&p_{ii} = 1/r_{ii}\,, \qquad && i = n, = n, n-1, \dots, 1 \\
&p_{ik} = \sum_{l=i+1}^{k} (r_{il} r_{lk})\, p_{ii}\,, \qquad && k = n, n-1, \dots, i+1\,.
\end{aligned}\right\} \tag{1.15}$$

For having an effective diagonalization procedure applicable also in case (c), the QR-transformation has been chosen. The corresponding Fortran programs are based essentially on the careful work by Grad and Brebner [9]. More recent versions will be found in Wilkinson and Reinsch [10]. The idea is to transform a matrix M by iterated left-right-decomposition convergently into a triangular matrix containing all eigenvalues of M along its diagonal. Thus for real $M_1 (= M)$

$$M_1 = Q_1 R_1 \tag{1.16}$$

in a first step. Including also non-symmetric matrices, M_1 leads to unitary Q_1 and right-triangular R_1 in general. The decomposition is made unique by choosing positive signs, for instance, for the diagonal elements of R_1. The algorithm is defined through

$$M_s = Q_s R_s\,, \qquad M_{s+1} = R_s Q_s \tag{1.17}$$

guaranteeing that M_{s+1} is unitary equivalent to M_s and therefore to M_1. In general M_s converges to a right-triangular matrix R with increasing s. Convergence proofs and a stability- and error analysis can be found in [5].

Once the spectrum of M is known, the eigenvectors are calculated by inverse vector iteration

$$Z_i^{(k+1)} = M^{-1} Z_i^{(k)} \qquad k = 0, 1, \ldots \tag{1.18}$$

using an arbitrary starting vector of the form

$$Z_i^{(0)} = \sum_{l \geqq i} c_{il} y_l \tag{1.19}$$

where $c_{ii} \neq 0$. Assuming the lowest eigenvalue of M to be non-degenerate, the procedure is started for the corresponding eigenvector and then continued for the other ones. It can be shown that the sequence $\{Z_i^{(k)}\}$ converges with increasing k to a multiple of y_i. Details of the method, modifications in case of nearly coinciding or even degenerate eigenvalues are found together with convergence proofs in [5] and [11].

In order to execute the above algorithms in an effective way with least numerical errors additional finesses have to be taken into account. First the matrix M is balanced by means of diagonal similarity transformations [12] and normalized such that

$$\| M \| = \beta \| \tilde{M} \| = \beta$$

where $\| \ldots \|$ denotes the Frobenius norm. The columns and rows of $\tilde{M}$ fulfill then approximately

$$\sum_{\substack{i=1 \\ i \neq k}}^{n} | \tilde{m}_{ik} | \approx \sum_{\substack{i=1 \\ i \neq k}}^{n} | \tilde{m}_{ki} | \qquad k = 1, 2, \ldots, n$$

which is most desirable for reasons of numerical stability. After that only the scaled matrix $\tilde{M}$ will be used. It is most effective then to bring $\tilde{M}$ first into upper Hessenberg form [13] H which is similar to $\tilde{M}$ but $\frac{1}{2}(n^2 - 3n + 2)$ elements vanish, i.e.

$$h_{ik} = 0 \quad \text{for} \quad i > k + 1.$$

Finally the QR-transformation is applied to the matrix H until all subdiagonal elements satisfy

$$|h_{i+1,i}| \leqq 2^{-t} \| H \|, \quad i = 1, 2, \ldots, n - 1$$

where t is the number of digits in the mantissa of a binary floating point number.

A satisfactory test of this diagonalization procedure has been based on a comparison of results for a pathological matrix $M = 8A - 5A^2 + A^3$ (A symmetric tridiagonal with $a_{i,i} = 2$; $a_{i,i+1} = 1$) with those obtained by Rutishauser [8].

2. Band Structure and Optical Transitions for EuO

2.1. The Band Structure

The numerical results for the band structure obtained by the methods described in the preceding chapter are plotted in the Figs. 2, 3 and 4. The bands have all qualitatively predicted properties although one could expect stronger

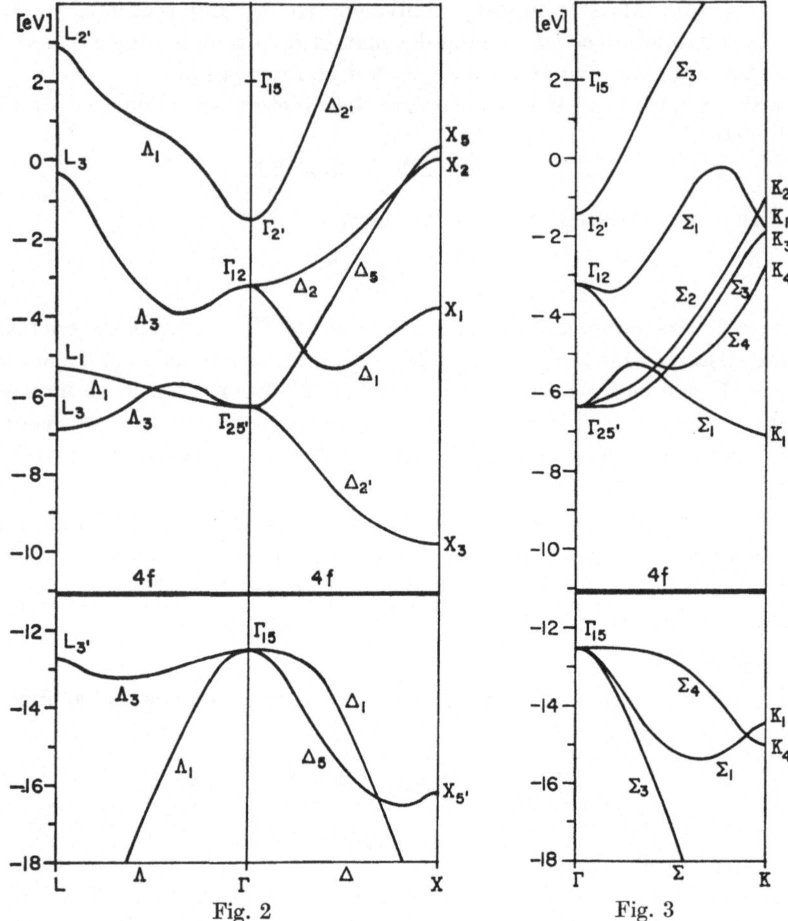

Fig. 2. Energy band structure of EuO ($\sigma = +1$) in (001)- and (111)-directions

Fig. 3. Energy band structure of EuO ($\sigma = +1$) in (110)-direction

deviations because of the non-local repulsive part of the potential. The most striking feature is the high lying magnetic $4f$-level actually separating valence- and conduction bands. It must be emphasized that this extraordinary position is no consequence of any artificial adjustment of parameters. The only parameter for adjustment is the zeroth Fourier-coefficient finally varied to reproduce the gap. Then automatically the $4f$-level falls into the energy gap.

It is somewhat surprising that the energy of Γ_1 is so high (actually outside the picture), whereas in other calculations it is at least comparable to that of the d-like states. However, Γ_1 is most sensitive to details in the potential and is not forced to obey strict ordering relations, therefore its position will be somewhat uncertain. Furthermore, the dispersion of the lowest valence bands belonging to the fully symmetric group representations should not be given too much confidence because of the earlier discussed instabilities. In the final calculation roughly 65 plane waves were used. Convergence tests up to using 113 plane waves manifested satisfactory stability of the lowest conduction bands (≈ 0.1 eV).

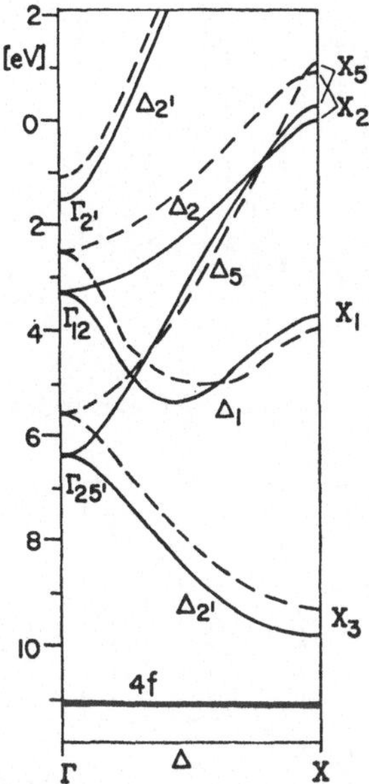

Fig. 4. Spin-split energy bands of EuO in (001)-direction: —— : $\sigma = +1$; -----: $\sigma = -1$

Since for the interpretation of optical transitions spin is taken to be conserved the Figs. 2 and 3 only show bands for electrons with spin parallel ($\sigma = +1$) to that of the $4f$-electrons. Fig. 4 then gives the spinpolarized splittings of conduction bands in a (001)-direction. For $\sigma = -1$ the zeroth Fourier coefficient has been adjusted to reproduce the experimentally observed splitting at X_3 [14]. It is interesting to note the possible crossings of spinpolarized bands suggesting that there are magnetic materials showing red shift [15], blue shift or even no shift of the absorption edge with decreasing temperature.

2.2. Selection Rules and Optical Transitions

First of all an analysis of interband transitions has to be made on the basis of selection rules for dipole transitions [16]. The dipole operator transforms according to the vector representation D_1 of the rotation group, and D_1 must be decomposed in terms of the irreducible representations of the group belonging to the point of the Brillouin zone in which the transition is considered. The following figures schematically indicate by arrows the allowed transitions for all groups. The reduction of the $4f$-level in terms of crystal levels is also given schematically although the splittings will be almost zero.

In addition to selection rules the theoretical predictions of transition probabilities and -intensities are of importance. Dominant contributions are found at those points in the Brillouin zone where the band tangents are parallel. Most directly connected with the band structure is the imaginary part ε_2 of the dielectric function [16] given by

$$\varepsilon_2(\omega) \sim \frac{1}{\omega^2} \int_{\mathrm{BZ}} d^3k \, |M_{\mathrm{vc}}(\mathbf{k})|^2 \, \delta(E_{\mathrm{c}}(\mathbf{k}) - E_{\mathrm{v}}(\mathbf{k}) - \omega) \, .$$

The integration is over the Brillouin zone.

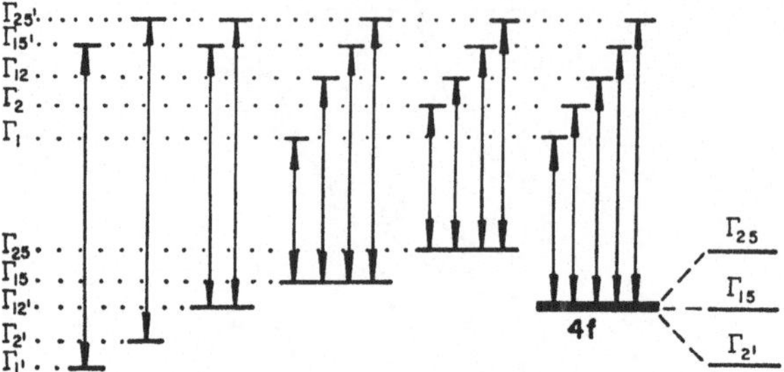

Fig. 5. Allowed direct transitions for G_Γ

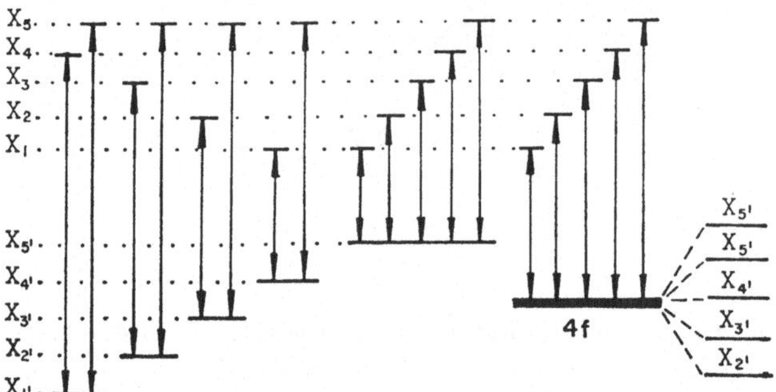

Fig. 6. Allowed direct transitions for G_X

At critical (extremal) points on the energy surface S_ω defined by

$$E_c(\mathbf{k}) - E_v(\mathbf{k}) = \omega$$

the matrix element $M_{vc}(\mathbf{k})$ for valence-conduction band transitions usually is a slowly varying function of $\mathbf{k}$ and is replaced by its average $\tilde{M}_{vc}$. After some transformations we can write

$$\varepsilon_2(\omega) \sim \frac{1}{\omega^2} \, | \tilde{M}_{vc} |^2 \, I_{vc}(\omega)$$

where $I_{vc}(\omega)$ is the so-called interband density of states

$$I_{vc}(\omega) = \int_{S_\omega} \frac{dS_\omega}{\dfrac{\partial}{\partial \mathbf{k}} \left[E_c(\mathbf{k}) - E_v(\mathbf{k}) \right]} .$$

explaining why parallel tangents are important. The structure of I_{vc} for 3-dimensional lattices is given in [16, 17] showing the different types of Van Hove singularities.

We now describe the most important transitions for EuO up to 12 eV, taking for convenience the energies at the edges.

In the centre of the Brillouin zone (Fig. 5) there is a first transition at 4.8 eV from $4f \to \Gamma_{25'}$. Due to the initially flat structure of Σ_3 and Λ_1 a lot of transitions near $\mathbf{k} = 0$ at the same energy will be involved. $\Gamma_{15} \to \Gamma_{25'}$ takes place at 6.2 eV, whilst the differences of 7.9 eV and 9.3 eV belong to $4f \to \Gamma_{12}$ and $\Gamma_{15} \to \Gamma_{12}$. The splitting of d-like states denoted conventionally by $10\,Dq$ amounts to 3.1 eV.

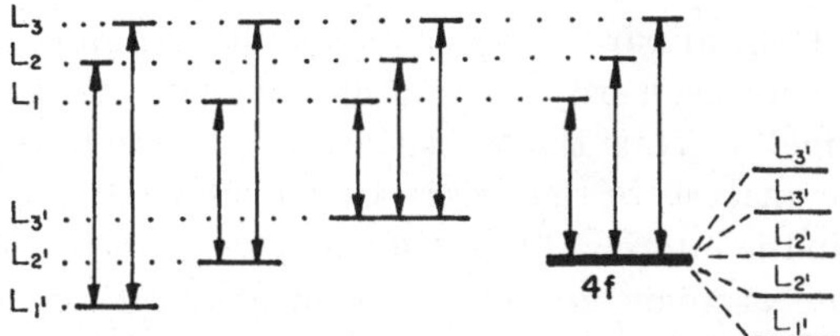

Fig. 7. Allowed direct transitions for G_L

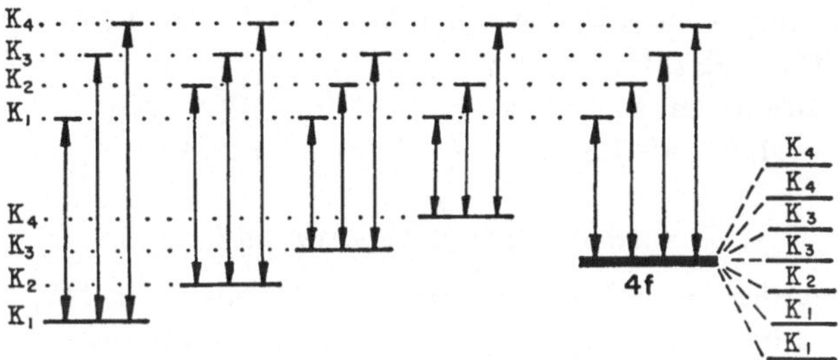

Fig. 8. Allowed direct transitions for G_K (or G_Σ)

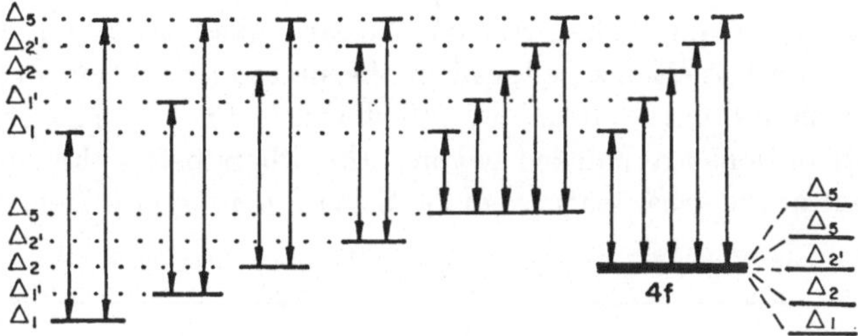

Fig. 9. Allowed direct transitions for G_Δ

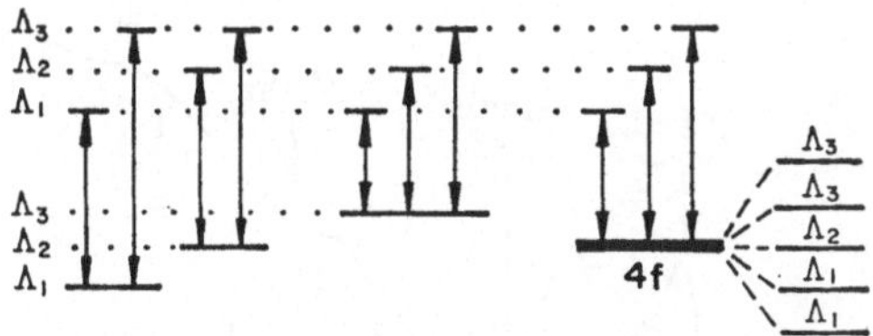

Fig. 10. Allowed direct transitions for G_Λ

The gap $4f \rightarrow X_3$ is adjusted to 1.2 eV, the experimental value being 1.12 eV [18]. Other transitions (Fig. 6) are $X_{5'} \rightarrow X_3$ at 6.2 eV, $4f \rightarrow X_1$ at 7.3 eV, $4f \rightarrow X_2$ at 11.1 eV, $4f \rightarrow X_5$ at 11.5 eV and $X_{5'} \rightarrow X_1$ at 12.4 eV.

In the point L (Fig. 7) $4f \rightarrow L_3$ at 4.2 eV, $4f \rightarrow L_1$ at 5.8 eV, $4f \rightarrow L_3$ at 10.8 eV are direct transitions from the atomic level. In addition there are three valence-band-conductionband transitions, namely $L_{3'} \rightarrow L_3(1)$ at 5.9 eV, $L_{3'} \rightarrow L_3(2)$ at 12.4 eV and $L_{3'} \rightarrow L_1$ at 7.4 eV. Parity is a good quantum number in the points Γ, X and L and there occur transitions between levels of different parity only.

At the point K (Fig. 8) parity is no good quantum number and therefore the band tangents are not necessarily horizontal and there are no Van Hove singularities due to symmetry. Thus the transitions will be weak in general.

As shown in the analysis of the empty lattice, crossing bands can give rise to band extrema also in the interior of the Brillouin zone. Because of the flat structure of the $4f$-level these extrema are always connected with singular points. In a (110)-direction one finds as a consequence the transitions $4f \rightarrow \Sigma_4(2)$ at 5.7 eV, $4f \rightarrow \Sigma_1(2)$ at 5.8 eV and $4f \rightarrow \Sigma_1(3)$ at 7.7 eV.

The main contributions in a (001)-direction (Fig. 9) will arise from $4f \rightarrow \Delta_1(2)$ at 5.7 eV and due to parallel but non-horizontal tangents $\Delta_1(1) \rightarrow \Delta_1(2)$ at 7.7 eV and $\Delta_5(1) \rightarrow \Delta_1(2)$ at 9.0 eV.

Direct transitions in a (111)-direction (Fig. 10) finally are $4f \rightarrow \Lambda_3(2)$ at 5.4 eV, $\Lambda_3(1) \rightarrow \Lambda_3(2)$ at 7.3 eV, $4f \rightarrow \Lambda_3(3)$ at 7.2 eV and $\Lambda_3(1) \rightarrow \Lambda_3(3)$ at 9.2 eV.

2.3. Interpretation of Experimental Results

The problematics of such interpretations is discussed in [17, 19]. The most reliable test for a band structure investigation is the comparison of the calculated imaginary part $\varepsilon_2(\omega)$ of the dielectric function with the corresponding experimental quantity. This has not yet been done for two reasons: first the computing effort would be enormous and second, the experiments must be carried out with extremely pure substances, as is known from the early analysis of Si and Ge. Instead of this we plot the transitions deduced in the preceding section in the recently obtained experimental curve for $\varepsilon_2(\omega)$ [20] shown in Fig. 11. The probably most dominant contributions are marked beyond, the others below the curve. There is a fair agreement of the gross features of ε_2. Several points, however, deserve some more remarks.

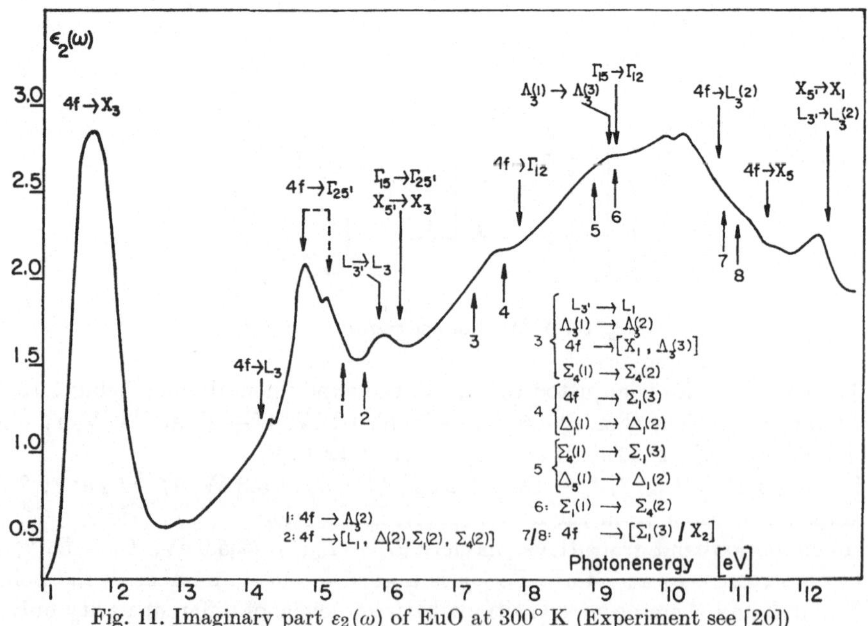

Fig. 11. Imaginary part $\varepsilon_2(\omega)$ of EuO at 300° K (Experiment see [20])

First of all any interpretation based on a one-electron band theory does not take into account the effects of the hole left behind in a flat band after an electron has been excited. That means that neither excitons nor multiplet structure can be analyzed directly in such an approach. In the case of EuO an electron excited from the $4f$-shell leaves behind a ^7FI $(I = 0, 1, \ldots, 6)$ configuration of Eu^{3+}, and one would expect to see somehow the structure of this multiplet whose energy range is roughly 0.6 eV for EuF_2 [21]. Whereas this structure seems to be smeared out in EuO, most recent experiments have been able to resolve it in EuS [20]. It must be realized for the above reasons that the present tentative assignments cannot give quite correct transition energies. Disregarding excitons the symmetries of the assignments, however, should be correct at least for energywise well separated transitions. The remaining puzzle is the first strong peak in ε_2 whose form and height can be explained only if the transition matrix element shows an extraordinary dependence on energy. It is more probable that an excitonic-like transition is involved with properties not known up to now for ordinary excitons. In particular there seems to be no temperature variation of the height of this peak [21]. It has to be emphasized that the adjustment of the gap does not depend on the interpretation of the nature of this first peak because the consistent assignments for the rest of ε_2 necessarily place the $(4f \rightarrow X_3)$-transition at about 1 eV, independent of whether or not the main contribution arises from an exciton.

The shoulder beyond the $(4f \rightarrow \Gamma_{25'})$-transition is probably due to spin-orbit splitting. Because of overlapping of a lot of transitions these splittings will not be resolvable in most cases. The first valence-conduction band transition is found at about 6 eV, which at low temperatures should show a splitting due to spin-polarization. If one assumes that the Γ_{15}-level arising from the oxygen $(2p)$-level is almost degenerate in spin because of the short-range dominant part of the exchange potential centered at Eu, the splitting, if observed, should be nearly that of $\Gamma_{25'}$ alone. An even better candidate for the observation of this splitting is the $(\Gamma_{15} \rightarrow \Gamma_{12})$-transition.

The spinpolarization of Γ_{15} has not been given due to the uncertainties mentioned in chapter 1. Besides, any spinpolarized band structure calculation must be taken to be tentative, because within the usual approximations it is quite impossible to treat long-range exchange effects satisfactorily. This model at least is supposed to simulate the situation at zero temperature.

In conclusion one must say that only a theoretical determination of $\varepsilon_2(\omega)$ based on a completely relativistic band structure calculation will allow for more reliable interpretations.

Concluding Remarks

The application of the OPW-method to the heavy ferromagnetic semiconductor EuO revealed the difficulties in applying the method to heavy substances in general. Finally the band structure could be obtained only after modifications and a careful selection of reliable numerical procedures. Despite some discrepancies between the present results and those of Cho [22] there is agreement in the assignments of the most important optical transitions. Due to the extraordinary position of the $4f$-level there is a big manifold of transitions, a lot of them occurring at the same or nearly the same energy. In addition relativistic and exchange splittings complicate the situation considerably such that any detailed resolutions in experiments and corresponding exact theoretical assignments for this substance appear to be rather illusory.

Acknowledgement. The author is deeply indebted to Prof. Dr. W. Baltensperger for his continuous and generous support of this work.

Also an extensive information on the experimental situation by G. Güntherodt is gratefully acknowledged.

Appendix

OPW-coefficients for Eu $^{2+}$ (orthogonality coefficients)

$\dfrac{a}{2\pi} K_j$ $(k=0)$			(000)	(111)	(200)	(220)	(311)
$n=$	$l=$	$\sigma=$					
1	0	$+1$	0.00191	0.00191	0.00191	0.00191	0.00191
		-1	0.00191	0.00191	0.00191	0.00191	0.00191
2	0	$+1$	-0.01217	-0.01210	-0.01207	-0.01197	-0.01190
		-1	-0.01217	-0.01209	-0.01207	-0.01197	-0.01189
2	1	$+1$	0.	0.00100	0.00115	0.00162	0.00189
		-1	0.	0.00100	0.00115	0.00162	0.00188
3	0	$+1$	0.04352	0.04203	0.04154	0.03966	0.03831
		-1	0.04340	0.04192	0.04144	0.03956	0.03822
3	1	$+1$	0.	-0.01026	-0.01174	-0.01601	-0.01827
		-1	0.	-0.01021	-0.01169	-0.01593	-0.01818
3	2	$+1$	0.	0.00116	0.00154	0.00296	0.00396
		-1	0.	0.00115	0.00152	0.00293	0.00392
4	0	$+1$	-0.14053	-0.11842	-0.11183	-0.08882	-0.07453
		-1	$-0\,14118$	$-0\,11881$	$-0\,11216$	-0.08897	-0.07459
4	1	$+1$	0.	0.07502	0.08259	0.09695	0.09931
		-1	0.	0.07567	0.08325	0.09750	0.09973
4	2	$+1$	0.	-0.03260	-0.04081	-0.06431	-0.07491
		-1	0.	-0.03374	-0.04213	-0.06585	0.07633
4	3	$+1$	0.	0.06470	0.77740	0.10209	0.10694
5	0	$+1$	0.54066	0.19658	0.13759	0.01576	-0.01864
		-1	0.57275	0.19474	0.13322	0.01094	-0.02157
5	1	$+1$	0.	-0.38431	-0.33450	-0.16995	-0.09534
		-1	0.	-0.39630	-0.33803	-0.15142	-0.08684

OPW-coefficients for O^{2-}

$\dfrac{a}{2\pi} K_j$ $(k=0)$		(000)	(111)	(200)	(220)	(311)
$n=$	$l=$					
1	0	0.04602	0.04391	0.04325	0.04072	0.03898
2	0	-0.50782	-0.21463	-0.17005	-0.07618	-0.04448

References

1. Herman, F., Skillman, S.: Atomic Structure Calculations. Prentice-Hall Inc. 1963
2. Liberman, D.: Comp. Phys. Commun. **2**, 107 (71)
3. Wilkinson, J. H.: Rounding Errors in Algebraic Processes. London 1963
4. Corbató, F. J., Uretsky, J. L.: J. Assoc. Comp. Mach. **6**, 366 (1959)
5. Wilkinson, J. H.: The Algebraic Eigenvalue Problem. Oxford University Press 1965
6. Harrison, W. A.: Pseudopotentials in the Theory of Metals. Benjamin 1966
7. Forsythe, G.: Computer Solution of Linear Algebraic Systems. Prentice-Hall Inc. 1967
8. Rutishauser, H.: Numer. Math. **9**, 1 (1966)
9. Grad, J., Brebner, M. A.: Commun Assoc. Comp. Mach. (CACM) **11**, 820 (1968)
10. Wilkinson, J. H., Reinsch, C.: Linear Algebra. Berlin-Heidelberg- New York:Springer 1971
11. Schwarz, H. R., Rutishauser, H., Stiefel, E.: Matrizen-Numerik. Stuttgart: Teubner 1968
12. Parlett, P. N., Reinsch, C.: Numer. Math. **13**, 293 (1969)
13. Martin, R. S., Wilkinson, J. H.: Numer. Math. **12**, 349 (1968)
14. Methfessel, S., Mattis, D. C.: Handbuch der Physik, Vol. XVIII Berlin-Heidelberg-New York : Springer 1968
15. Busch, G., Wachter, P.: Phys. kondens. Materie **5**, 232 (1968)
16. Bassani, F.: The Optical Properties of Solids. Academic Press 1966
17. Phillips, J. C.: Solid State Physics, Vol. 18. Academic Press 1966
18. Güntherodt, G., Wachter, P., Imboden, D. M.: Phys. kondens. Materie **12**, 292 (1971)
19. Greenaway, D. C., Harbeke, G.: Optical Properties and Band Structure of Semiconductors. Pergamon Press 1968
20. Güntherodt, G. Wachter, P.: to be published in the AIP Conference Proceedings **6**, 1973
21. Feinleib, J., Scouler, W. J., Dimmock, J O., Hanus, J., Reed, T. B., Pidgeon, C. R.: PRL **22**, 1385 (1969)
22. Cho, S. J.: PR B**1**, 4589 (1970)

Dr. Karl Lendi
Institut für Theoretische Physik
Eidgen. Technische Hochschule Hönggerberg
CH-8049 Zurich, Switzerland

Hinweise für die Autoren

1. *Manuskripte* werden auf einseitig beschriebenen Blättern erbeten. Sie müssen formal wie inhaltlich so durchgearbeitet sein, daß Änderungen in den Korrekturabzügen unnötig sind. Nachträgliche (vom Manuskript abweichende) Korrekturen müssen den Autoren in Rechnung gestellt werden.

Für den Text und die Formeln ist Maschinenschrift (mit doppeltem Zeilenabstand) zu verwenden. Vorkommende gotische (Fraktur) oder griechische Buchstaben sowie einander ähnliche Zeichen, z.B. 0 (Null) und O (Buchstabe) oder 1 (Eins) und l (Buchstabe), sind stets besonders (mit verschiedenfarbigen Stiften) zu kennzeichnen; außerdem ist zwischen Groß- und Kleinbuchstaben (durch doppeltes bzw. einfaches Unterstreichen) zu unterscheiden.

2. Die *Abbildungen* sind auf das sachlich unbedingt Notwendige zu beschränken. Bereits veröffentlichte Figuren sowie farbige Bilder können in der Regel keine Aufnahme finden.

Für Kurven oder Schwarzweißzeichnungen, die als Strichätzung reproduziert werden können, werden als Vorlagen in einheitlicher Linienstärke angelegte schwarze Tuschezeichnungen erbeten. Unter besonderen Umständen werden klar leserliche Skizzen auch vom Verlag umgezeichnet. Bei Halbtonbildern (Photos, Mikrophotos, Röntgenbildern) sind saubere, scharfe und tonwertreiche Originalvorlagen (Hochglanzabzüge von Originalphotographien) notwendig, für Halbtonzeichnungen die Originale. Besondere Wünsche des Autors hinsichtlich des linearen Verkleinerungs- und Vergrößerungsmaßstabes der Vorlagen sollen vermerkt werden.

Die *Beschriftung* der Abbildungen mit Buchstaben oder Ziffern erfolgt durch den Verlag. Die Hinweise hierfür sollen keinesfalls in der Abbildung selbst, sondern auf einem darüberliegenden transparenten Deckblatt angegeben werden. Dort sind auch Abstriche oder gewünschte Bildausschnitte anzugeben und besonders wichtige Abbildungspartien zu kennzeichnen, damit die Kunstanstalt bei der Ätzung ihr besonderes Augenmerk darauf richten kann.

Zu jeder Figur wird eine knappe, klare Unterschrift erbeten. Die Legenden sollen getrennt von den Figuren auf einem besonderen Blatt dem Textmanuskript beigefügt werden.

Die gleichzeitige Wiedergabe in einer graphischen Darstellung *und* einer Tabelle ist zu vermeiden.

3. Bei Originalarbeiten wird eine *Zusammenfassung* (nicht über 100 Worte) in Englisch erbeten.

4. Die *Literatur*zitate erscheinen nicht als Fußnoten, sondern als geschlossenes Verzeichnis am Ende der Arbeit. Sie sollen Initialen und Namen aller Autoren, den nach den internationalen Regeln gekürzten Zeitschriftentitel, Band-, Seiten- und Jahreszahl enthalten. Bücher werden mit Autor, vollem Titel, Auflage, Ort, Verlag und Jahr zitiert.

5. Von jeder Arbeit werden 75 *Sonderdrucke* kostenlos zur Verfügung gestellt. Weitere Separata können von den Autoren zum Selbstkostenpreis bezogen werden.

Indications aux auteurs

1. Les *manuscrits* doivent être dactylographiés seulement au recto de la feuille. Ils doivent être présentés sous leur forme définitive, afin d'éviter toute modification ultérieure des étreuves. Des corrections supplémentaires (c'est-à-dire divergent du manuscrit) seront portées au débit des auteurs.

Le texte et les notations seront dactylographiés à double interligne. Devront en outre être distingués les uns des autres (au moyen de crayons de couleurs différentes) les caractères gothiques, les caractères grecs ainsi que les caractères semblables de sens différent, comme, par example, 0 (zéro) et O (lettre), ou 1 (chiffre)et l (lettre); de plus, afin de marquer la différence entre les caractères minuscules et majuscules, les uns seront soulignés une fois, les autres deux.

2. Les *illustrations* seront limitées au strict nécessaire. En règle générale ni les illustrations déjà publiées, ni les photos en couleur ne seront acceptées.

Les dessins au trait et les graphiques devront être d'un tracé régulier et uniforme, à l'encre de Chine, sur du papier blanc lisse ou du papier »Bristol«. Dans certains cas, les éditeurs pourront faire redessiner des croquis préparés avec netteté. Pour les reproductions en demi-ton (photos, microphotos, radios), les auteurs sont priés de faire parvenir les épreuves net-tes et précises des originaux, faites sur papier glacé, pour les dessins en demi-ton les originaux eux-mêmes, et d'indiquer tout désir particulier concernant leur agrandissement ou leur réduction linéaire.

La maison d'édition se charge des *légendes* (lettres et chiffres). Les indications nécessaires devront être portées sur une feuille couvrante transparente et non sur l'illustration même. Devront être également mentionnées sur cette feuille des sections marginales inutiles, les parties importantes des illustrations ou celles que les auteurs désirent mettre particulièrement en relief.

On est prié de joindre à chaque illustration, une légende claire et concise, présentée sur une feuille du manuscrit, separée des illustrations.

On est égelment prié d'éviter de représenter es mêmes faits à la fois sous forme de tableau *et* sous forme graphique.

3. A tout travail original sera adjoint un *résumé* (ne dépassant pas 100 mots) en anglais

4. Les *références*, ne seront pas citées en nota bene mais réunies à la fin de l'article. Elles se présenteront comme suit: initiales et noms de taus les auteurs, le titre de la revue, abrégé selon les règles internationales, tome, numéro de la page et année. Les livres seront cités précédés du nom de l'auteur, titre complet, édition, lieu du publication, éditeur et date de parution.

5. 75 *tirages à part* de chaque article seront remis gratuitement aux auteurs. Les auteurs pourront acquérir au prix de revient tout exemplaire supplémentaire.

Notes for Contributors

1. Please submit *manuscripts*, typed on one side only. The paper must be in its final form in order to avoid corrections of the proofs. Changes in content and style are to be avoided; expenses accruing from such additional corrections will be charged to the authors.

The text and the formulae should be typed (double-spaced). Gothic letters, Greek letters and signs similar to one another e.g. 0 (zero) and O (letter) or 1 (one) and l (letter) should be marked specially with coloured pencils; furthermore the difference between small and capital letters should be made evident, by underlining once or twice.

2. *Illustrations* should be restricted to those that are essential. Illustrations, previously published, or coloured illustrations are usually not admitted. Line drawings and graphs should be drawn with Indian ink in clean, uniform lines on smooth white paper or Bristol board. Under special circumstances, clearly prepared sketches may be redrawn by the publish-ers. Sharp, neatly traced high-gloss prints of original photos should be submitted for half-tane illustrations (photos, micro-photos, X-ray photos), for half-tone drawings the originals. Any special desires concerning the enlargement or reduction of the originals should be mentioned.

The publisher will attend to *lettering* (letters, ciphers, arrows); the necessary information should be given not on the original photo or drawing itself but on a transparent cover sheet. Desired particular sections of an illustration or dispensable marginal treas should likewise be marked on a cover sheet.

Please provide illustrations with brief and clear legends, which should be given on separate sheets of the manuscript, detached from the illustrations.

The same item should not appear simultaneously in a table *and* a diagram.

3. For original investigations one *summary* (not more than 100 words) in English should be submitted.

4. *References* should be listed at the end of the paper. Every article cited should include the names and initials of all authors, the title of the journal according to the official internationa. abbreviation, volume, page number, and year. For books, the name(s) of the author(s) the title, edition, the publisher's name and address, and the year of publication should be included.

5. 75 *offprints* of each paper are provided free of charge. Additional copies may be purchased at cost price.

Phys. cond. Matter 17, 233—248 (1974)
© by Springer-Verlag 1974

Electronic Wave Functions of F- and F′-Centres in Ionic Crystals

W. Renn

Institut für Theoretische Physik der Universität Tübingen *

Received July 16, 1973

Using a theory developed by Stumpf [1], the wave functions and energy values of the F- and $F′$-centre are calculated. The main feature of the theory is the atomistic description of the electronic and ionic polarization of the host crystal and their influence on the impurity centre electrons. The wave functions are evaluated by minimizing the energy expectation value of the total electronic energy of the crystal. Numerical results are given for all alkali halides with NaCl-structure and compared with the experimental values of the F and $F′$ absorption energies. Fair agreement is obtained.

The F- and $F′$-centres have been subjected to many theoretical investigations and it would exceed the scope of an introduction to discuss all the different approaches. Therefore, we only refer to the appropriate literature. Calculations of F-centre wave functions were listed and discussed by Gilbert and Markham [2] and recently by Schmid [3][1]. $F′$-centre calculations can be found in a paper by Bennett [4].

In all those calculations a common difficulty arises in the correct treatment of the polarization of the host crystal caused by the impurity centre. Returning to the conception of the early work of Pekar [5], Stumpf [1][2] derived a model which fully takes into account the discrete structure of the crystal and describes both the electronic polarization and the distortion of the lattice in a microscopic, atomistic manner.

By the aid of this theory, the wave functions of the F-centre electron in alkali halides were calculated in paper II. Numerical results for the spatial extent of the wave function, the spin density at the nuclei surrounding the vacancy and the static displacements of the ions were given and compared with experiment and literature.

In this paper we are interested in the wave functions and energy levels of both the F- and $F′$-centre. We follow the treatment of I and II and improve the method used.

After summarizing all abbreviations used throughout the paper in Sect. 1, we derive the functional of the total electronic energy of the crystal in Sect. 2. This energy functional depends on the wave function of the impurity centre electrons, and on the dipoles representing the electronic polarization and the displacements of the ions. Sect. 3 contains the elimination procedure of the polarization dipoles and the displacements. Sects. 4 and 5 present the calculation

---

* Present address see end of the paper.

1 Hereafter, we shall refer to this as paper II.

2 Hereafter, we shall refer to this as paper I.

of the wave functions and the numerical results. In the final conclusion we discuss connections with other theories and give an outlook on further applications of our model.

1. Definitions

For ease of reference, all abbreviations are collected in this first section. We shall use, whenever possible, the notation of paper I. The slight deviations in the definition of the polarizabilities should not give rise to any confusion.

a) Notation of the Crystal Structure

e := charge of the electron;

r_i := position vector of the i-th impurity electron;

l := index of the l-th elementary cell of the ideal lattice;

μ := index of the ions in the elementary cell;

$e_{l\mu}$:= charge of the $l\mu$-th nucleus;

$R^0_{l\mu}$:= ideal lattice position of the $l\mu$-th nucleus;

$R_{l\mu}$:= lattice position of the $l\mu$-th nucleus in the distorted lattice;

$\alpha_{l\mu}$:= index of the electrons at the $l\mu$-th ion;

$n_{l\mu}$:= $\sum\limits_{\alpha_{l\mu}}$ number of core electrons of the $l\mu$-th ion;

$$a_{l\mu} := (e_{l\mu} + e\,n_{l\mu}) \text{ charge of the } l\mu\text{-th ion;} \tag{1.1}$$

$$\mathop{\neq\!\!\sum}\limits_{i,k} := \sum\limits_{i,k} \text{ with } i \neq k; \tag{1.2}$$

$$\sum\limits_{l\mu}{}' := \text{ sum over all lattice points excluding } l = 0,\ \mu = 0; \tag{1.3}$$

$C(x, v) := |x - v|^{-1}, \quad C(r, l\mu) := |r - R^0_{l\mu}|^{-1}$ and

$$C(l\mu, k\varrho) := |R^0_{l\mu} - R^0_{k\varrho}|^{-1} \text{ Coulomb potentials;} \tag{1.4}$$

$$C(x, v)^n := b\,|x - v|^{-n} \text{ phenomenological repulsive potential;} \tag{1.5}$$

$$d := \text{next nearest neighbor distance;} \tag{1.6}$$

$$F(x, v) := d^3/4\pi \sum\limits_{l\mu}{}' \nabla_{l\mu} C(x, l\mu)\, \nabla_{l\mu} C(v, l\mu); \tag{1.7}$$

$$F(x, v)^n := b\,d^3/4\pi \sum\limits_{l\mu}{}' \nabla_{l\mu} C(x, l\mu)\, \nabla_{l\mu} C(v, l\mu)^n \text{ polarization}$$
$$\text{potentials;} \tag{1.8}$$

$$\alpha_M := \text{Madelung constant.} \tag{1.9}$$

b) Quantum Mechanical Expectation Values

$$H_1(r) := -\tfrac{1}{2}\Delta + \sum\limits_{l\mu}{}' e\,e_{l\mu} C(r, R_{l\mu}) \text{ kinetic energy and one-electron}$$
$$\text{interaction with the nuclei} \tag{1.10}$$

$$V'_0 := \tfrac{1}{2} \mathop{\neq\!\!\sum}\limits_{l\mu, k\varrho} e_{l\mu} e_{k\varrho} C(R_{l\mu}, R_{k\varrho}) \text{ nucleus-nucleus Coulomb interaction;} \tag{1.11}$$

$M \equiv$ number of the impurity centre electrons;   (1.12)

$\Phi_n(r_1, \ldots, r_M) :=$ impurity centre wave function;   (1.13)

$\varrho_n(r) := M \int |\Phi_n(r, r_2, \ldots, r_M)|^2 \, dr_2 \ldots dr_M$ probability density of
$\qquad$ the impurity centre electrons;   (1.14)

$\Psi^0_{\alpha_{l\mu}}(r) :=$ wave function of the $\alpha_{l\mu}$-th electron in the ideal crystal;   (1.15)

$\Psi_{\alpha_{l\mu}}(r) :=$ wave function of the $\alpha_{l\mu}$-th electron in the disturbed crystal;   (1.16)

$\varrho_{\alpha_{l\mu}}(r) := |\Psi_{\alpha_{l\mu}}(r)|^2$ probability density of the $\alpha_{l\mu}$-th electron;   (1.17)

$\xi(r) \quad := e\,\varrho_n(r) - a_{00}\,\delta(r - R_{00})$ total impurity centre charge
$\qquad$ density;   (1.18)

$m_{l\mu} \quad := e \sum\limits_{\alpha_{l\mu}} \int \varrho_{\alpha_{l\mu}}(r)\,(r - R_{l\mu})\,dr$ electronic dipole moment of the
$\qquad l\mu$-th ion;   (1.19)

$$H_{l\mu} \equiv H_{l\mu}(\Psi_{\alpha_{l\mu}}) := \sum\limits_{\alpha_{l\mu}} \int \Psi^*_{\alpha_{l\mu}}(r)\,[-\tfrac{1}{2}\Delta + e\,e_{l\mu}\,C(r, R_{l\mu})]\,\Psi_{\alpha_{l\mu}}(r)\,dr$$
$$+\tfrac{1}{2} \neq \sum\limits_{\alpha_{l\mu},\,\beta_{l\mu}} e^2 \int \Psi^*_{\alpha_{l\mu}}(r)\,\Psi^*_{\beta_{l\mu}}(r')$$
$$\times\, C(r, r')[\Psi_{\alpha_{l\mu}}(r)\,\Psi_{\beta_{l\mu}}(r') - \Psi_{\alpha_{l\mu}}(r')\,\Psi_{\beta_{l\mu}}(r)]\,dr\,dr'$$
internal energy of the $l\mu$-th ion in the disturbed crystal;   (1.20)

$H^0_{l\mu} \quad := H_{l\mu}(\Psi^0_{\alpha_{l\mu}})$ internal energy of the ions in the ideal crystal;   (1.21)

$$H^{\mathrm{ex}}_{l\mu k\varrho} := -\sum\limits_{\alpha_{l\mu},\,\beta_{k\varrho}} e^2 \int \Psi^*_{\alpha_{l\mu}}(r)\,\Psi^*_{\beta_{k\varrho}}(r')\,C(r, r')\,\Psi_{\alpha_{l\mu}}(r')\,\Psi_{\beta_{k\varrho}}(r)\,dr\,dr'$$
exchange interaction between the $l\mu$-th and $k\varrho$-th ion;   (1.22)

$$H_n(M) := M \int \Phi^*_n(r_1, \ldots, r_M)\,[-\tfrac{1}{2}\Delta(r_1)]\,\Phi_n(r_1, \ldots, r_M)\,dr_1 \ldots dr_M$$
$$+\frac{e^2}{2}\,M(M-1) \int \Phi^*_n(r_1, \ldots, r_M)\,C(r_1, r_2)$$
$$\times\, \Phi_n(r_1, \ldots, r_M)\,dr_1 \ldots dr_M$$ kinetic energy of the impurity
centre electrons and their mutual interaction;   (1.23)

$$H^{\mathrm{ex}}_{\mathrm{I}} \equiv H^{\mathrm{ex}}_{\mathrm{I}}(\Phi, \Psi_{\alpha_{l\mu}}) := -e^2 M \sum\limits_{l\mu\alpha_{l\mu}}{}' \int [\int \Phi^*_n(r, \ldots, r_M)$$
$$\times\, \Phi_n(r', \ldots, r_M)\,dr_2 \ldots dr_M]\,C(r, r')\,\Psi^*_{\alpha_{l\mu}}(r')\,\Psi^*_{\alpha_{l\mu}}(r)\,dr\,dr'$$
exchange interaction between the impurity electrons and
the core electrons;   (1.24)

$U_{\mathrm{ideal}} \quad := \sum\limits_{l\mu} H^0_{l\mu} + \tfrac{1}{2} \sum\limits_{l\mu,\,k\varrho} [a_{l\mu}\,a_{k\varrho}\,C(l\mu, k\varrho) + C(l\mu, k\varrho)^\eta]$ energy of
$\qquad$ the ideal crystal;   (1.25)

c) Calculation Aids

$M_{l\mu} \quad := (R_{l\mu} - R^0_{l\mu})$ displacement of the $l\mu$-th ion;   (1.26)

$P_{l\mu} \quad := 1 + (M_{l\mu}\nabla_{l\mu});$ (1.27)

$$D_{l\mu} := [a_{l\mu} + (\boldsymbol{m}_{l\mu}\,\nabla_{l\mu}) + \alpha_{l\mu}\,(\boldsymbol{M}_{l\mu}\,\nabla_{l\mu})]; \tag{1.28}$$

$$\gamma := \text{average polarizability of the ions}; \tag{1.29}$$

$$\alpha^e := 4\pi\,d^{-3}\,\gamma \quad \text{dimensionless polarizability of the ions}; \tag{1.30}$$

$$\tau_{\mathrm{r}} := [\tfrac{1}{18}\,\alpha_M\,e^2\,d^{\eta-1}\,(\eta-1)\sum_{l\mu} R_{l\mu}^{-(\eta+2)}]^{-1} \quad \text{polarizability of the lattice}; \tag{1.31}$$

$$\tau_{\mathrm{e}} := \left(\frac{\alpha^e}{1+\alpha^e}\,e^2\,\frac{d^3}{2\pi}\sum_{l\mu} R_{l\mu}^{-6}\right)^{-1} \quad \text{contribution of the electronic polarization to the lattice polarizability}; \tag{1.32}$$

$$\tau := (\tau_{\mathrm{r}}^{-1} - \tau_{\mathrm{e}}^{-1})^{-1} \quad \text{total lattice polarizability}; \tag{1.33}$$

$$\alpha^d := 4\pi\,e^2\,d^{-3}\,\tau \quad \text{dimensionless lattice polarizability}; \tag{1.34}$$

$$\Gamma_{l\mu k\varrho} := [\delta_{l\mu k\varrho} + \gamma\,\nabla_{l\mu}\otimes\nabla_{k\varrho}\,C(l\mu,k\varrho)(1-\delta_{l\mu k\varrho})]^{-1} \quad \text{Green's function of the electronic dipole moments}; \tag{1.35}$$

$$V_{l\mu k\varrho} := \tau\,\nabla_{l\mu}\otimes\nabla_{k\varrho}\,C(l\mu,k\varrho)^\eta\,(1-\delta_{l\mu k\varrho}); \tag{1.36}$$

$$G_{l\mu k\varrho} := \left[\delta_{l\mu k\varrho} + \tau\,\frac{a_{l\mu}\,a_{k\varrho}}{(1+\alpha^e)}\,\nabla_{l\mu}\otimes\nabla_{k\varrho}\,C(l\mu,k\varrho)\,(1-\delta_{l\mu k\varrho}) + V_{l\mu k\varrho}\right]^{-1}$$
$$\text{Green's function of the ion displacements}; \tag{1.37}$$

$$c := \left[\frac{\alpha^e}{1+\alpha^e} + \frac{\alpha^d}{(1+\alpha^e)(1+\alpha^e+\alpha^d)}\right] \equiv \frac{\alpha^e+\alpha^d}{1+\alpha^e+\alpha^d} \tag{1.38}$$

$$c' := \frac{\alpha^d}{1+\alpha^e+\alpha^d} \quad \text{screening factors}. \tag{1.39}$$

2. The Energy Expectation Value

We begin with the Hamiltonian for the electrons of a crystal with an anion vacancy. Most of the electrons are attached to the ions of the crystal, which we call core electrons, and very few are localized at the impurity centre, called impurity electrons. Let M be the number of the impurity electrons then $M = 1$ refers to the F-centre and $M = 2$ to the F'-centre. To simplify the calculations we use some familiar approximations: We neglect interactions of the impurity centre with other defects and with the crystal surface. The problem is treated in the adiabatic approximation and Coulomb interactions are only taken into account.

With the definitions (1.4), (1.10) and (1.11) the Hamiltonian has the form

$$H = \sum_i H_1(\boldsymbol{r}_i) + \tfrac{1}{2} \mathrel{\reflectbox{$+$}} \sum_{i,k} e^2\,C(\boldsymbol{r}_i,\boldsymbol{r}_k) + {\sum_{i,l\mu\alpha_{l\mu}}}' e^2\,C(\boldsymbol{r}_i,\boldsymbol{r}_{\alpha_{l\mu}})$$
$$+ {\sum_{l\mu\alpha_{l\mu}}}' H_1(\boldsymbol{r}_{\alpha_{l\mu}}) + \tfrac{1}{2} \mathrel{\reflectbox{$+$}} {\sum_{\substack{l\mu\alpha_{l\mu}\\ k\varrho\beta_{k\varrho}}}}' e^2\,C(\boldsymbol{r}_{\alpha_{l\mu}},\boldsymbol{r}_{\beta_{k\varrho}}) + V_0'. \tag{2.1}$$

To indicate the missing anion at $l = 0$, $\mu = 0$ we mark every summation and abbreviation which contains energies of the ions with a prime. As a trial function for the variational calculation of the impurity electron wave function we use a linear combination of Slater determinants

$$\Phi_n(\xi_1, \ldots, \xi_M) := (M!)^{-1/2} \sum_s c^s \det \left| \varphi_1^s(\xi_1) \cdots \varphi_M^s(\xi_M) \right|, \qquad (2.2)$$

with suitable coefficients c^s to guarantee the normalization of the wave function.

The core electrons of the crystal can be represented by a single Slater determinant because their wave functions are eliminated later. If the total number of the electrons in the crystal is N we have

$$\Psi(\xi_{M+1}, \ldots, \xi_N) = [(N - M)!]^{-1/2} \det \left| \Psi_{M+1}(\xi_{M+1}) \cdots \Psi_N(\xi_N) \right|. \qquad (2.3)$$

The variables ξ in (2.2) and (2.3) contain the position vectors r as well as the spin variables.

The one-electron wave functions in (2.2) and (2.3) are assumed to be orthogonal to each other as is usually done in Hartree-Fock calculations. With these wave functions we can construct the total wave function of all crystal electrons as an antisymmetrized product in the form

$$X_n := A \left\{ \Phi_n(\xi_1, \ldots, \xi_M) \Psi(\xi_{M+1}, \ldots, \xi_N) \right\}. \qquad (2.4)$$

The antisymmetrizer A is given by

$$A := \left[\frac{M!(N - M)!}{N!} \right]^{1/2} \sum_P (-1)^P P, \qquad (2.5)$$

where the sum runs over all permutations which consist of an interchange of coordinates between the two functions in the curled brackets of (2.4).

If we take the expectation value of the Hamiltonian (2.1) with respect to the total electronic wave function (2.4) we get the average electronic energy of the crystal. The actual calculation, though tedious, is straightforward and much the same as the standard procedure of Hartree-Fock calculations. With the definitions (1.20)–(1.24) and the probability densities (1.14), (1.17) we get the result

$$\begin{aligned}
U_n &:= \langle X_n | H | X_n \rangle \\
&= H_n(M) + \sideset{}{'}\sum_{l\mu} e \int \varrho_n(r) \left[e_{l\mu} C(r, R_{l\mu}) + e \sum_{\alpha_{l\mu}} \int \varrho_{\alpha_{l\mu}}(r') C(r, r') \, dr' \right] dr \\
&\quad + H_I^{\mathrm{ex}} + \sideset{}{'}\sum_{l\mu} H_{l\mu} + \mp \sideset{}{'}\sum_{l\mu, k\varrho} \sum_{\alpha_{l\mu}} e \int \varrho_{\alpha_{l\mu}}(r) \\
&\quad \times \left[e_{k\varrho} C(r, R_{k\varrho}) + \frac{e}{2} \sum_{\beta_{k\varrho}} \int \varrho_{\beta_{k\varrho}}(r') C(r, r') \, dr' \right] dr + \mp \sum_{l\mu, k\varrho} H_{l\mu k\varrho}^{\mathrm{ex}} + V_0'.
\end{aligned} \qquad (2.6)$$

For the exchange energies in (2.6) we use suitable approximations because it is impossible to calculate them exactly. The exchange interactions among the core electrons are described by a phenomenological repulsive potential between the ions. The actual form of this potential is not important for our method and we choose for the sake of simplicity the potential defined in (1.5)

$$H_{l\mu, k\varrho}^{\mathrm{ex}} = C(l\mu, k\varrho)^n. \qquad (2.7)$$

For the following treatment it is assumed that the exchange interactions of the impurity electrons with the core electrons can be calculated representing the core electrons by the wave functions of the ideal crystal, i.e. with (1.24) and (1.15)

$$H_I^{ex} \equiv H_I^{ex}(\Phi, \Psi_{\alpha_{l\mu}}) \approx H_I^{ex}(\Phi, \Psi_{\alpha_{l\mu}}^0) \,. \tag{2.8}$$

Now we observe that every deviation of the ideal crystal in the ground state, i.e. every excitation or impurity centre, causes a polarization of the ions and a distortion of the lattice in the host crystal. The amount of these effects can be represented by the electronic dipoles $m_{l\mu}$ (1.19) and the displacements of the ions $M_{l\mu}$ (1.26), if the deviation of the ideal crystal is small. We assume throughout this paper that this condition is satisfied. Therefore, one can expand the interactions between different ions of (2.6) in a multipole series and neglect the quadrupole and higher moments in the expansion. The details of this procedure are given in paper I and we only show here the result as the calculation is straightforward. With the definitions (1.26)−(1.29) and (1.31) we have

$$\begin{aligned} U_n = H_n(M) + H_I^{ex} &+ \sum_{l\mu}{}' e \int \varrho_n(r) \, D_{l\mu} \, C(r, l\mu) \, dr \\ &+ \sum_{l\mu}{}' (H_{l\mu}^0 + \tfrac{1}{2}\gamma^{-1} m_{l\mu}^2 + \tfrac{1}{2}\tau_r^{-1} M_{l\mu}^2) \\ &+ \tfrac{1}{2} \mp \sum_{l\mu, k\varrho}{}' [D_{l\mu} D_{k\varrho} \, C(l\mu, k\varrho) + P_{l\mu} P_{k\varrho} \, C(l\mu, k\varrho)^\eta] \,. \end{aligned} \tag{2.9}$$

Of peculiar interest are the self-energies of the dipoles in the second row of the equation above: The difference between the internal energies of the ions in the ideal and disturbed crystal is described by a phenomenological polarization law in the form

$$H_{l\mu} = H_{l\mu}^0 + \tfrac{1}{2}\gamma^{-1} m_{l\mu}^2 \,, \tag{2.10}$$

where $H_{l\mu}^0$ (1.21) is the internal energy of the ions in the ideal crystal and γ is the average polarizability of the ions. Assuming that the internal energies of the ions are independent of the displacements $M_{l\mu}$, the lattice polarizability τ_r is given by (1.31), where we have used the result derived in the appendix of paper II. By its derivation τ_r is the polarizability of a lattice consisting of rigid ions.

We can simplify the expression (2.9) if we use some symmetry relations: By means of group theory it is easy to prove that in cubic crystals the following two relations hold

$$\sum_{k\varrho} a_{k\varrho} \nabla_{l\mu} C(l\mu, k\varrho) = 0 \quad \text{and} \quad \sum_{k\varrho} \nabla_{l\mu} C(l\mu, k\varrho)^\eta = 0 \,. \tag{2.11a, b}$$

In addition, we use the dipole m_{00} and the displacement M_{00} of the missing ion as auxiliary quantities. These do not appear in the energy of the crystal but one can introduce them in the calculation and put them zero in the final result. Thus, if we always remember

$$m_{00} = 0 \quad \text{and} \quad M_{00} = 0 \tag{2.12a, b}$$

and take into account (2.11), we may extend the summations in the last two rows of (2.9) over the perfect crystal lattice and get, after compensating the Coulomb and exchange interaction of the missing ion,

$$\begin{aligned} U_n = H_n(M) + H_I^{ex} &+ \sum_{l\mu}{}' [\int \xi(r) \, D_{l\mu} \, C(r, l\mu) \, dr - P_{l\mu} \, C(0\,0, l\mu)^\eta] \\ &+ \tfrac{1}{2} \sum_{l\mu} [\gamma^{-1} m_{l\mu}^2 + \tau_r^{-1} M_{l\mu}^2] + \tfrac{1}{2} \mp \sum_{l\mu, k\varrho} \{[(m_{l\mu} \nabla_{l\mu})(\nabla_{k\varrho} m_{k\varrho}) \end{aligned}$$

$$+ 2a_{k\varrho}(\boldsymbol{m}_{l\mu}\,\nabla_{l\mu})(\nabla_{k\varrho}\boldsymbol{M}_{k\varrho}) + a_{l\mu}\,a_{k\varrho}(\boldsymbol{M}_{l\mu}\,\nabla_{l\mu})(\nabla_{k\varrho}\boldsymbol{M}_{k\varrho})]\,C(l\,\mu,\,k\,\varrho) \qquad (2.13)$$

$$+ (\boldsymbol{M}_{l\mu}\,\nabla_{l\mu})(\nabla_{k\varrho}\boldsymbol{M}_{k\varrho})\,C(l\,\mu,\,k\,\varrho)^n\} + U_{\text{ideal}} - H_{00}^0 \,.$$

The charge density $\xi(\boldsymbol{r})$ (1.18) is the total charge density of the impurity centre, i.e. impurity electrons and the missing ion together. U_{ideal} (1.25) represents the energy of the ideal crystal and H_{00}^0 the internal energy of the missing ion.

3. Elimination of the Electronic Dipoles and the Ion Displacements

In the preceding section we have derived the total electronic energy of the crystal depending on the wave function of the impurity electrons, the electronic dipoles and the displacements of the ions. In the static equilibrium, i.e. when phonons are absent, the crystal energy is fixed by the three minimum conditions

$$\frac{\partial U_n}{\partial \boldsymbol{m}_{l\mu}} = 0, \qquad \frac{\partial U_n}{\partial \boldsymbol{M}_{l\mu}} = 0, \qquad \frac{\partial U_n}{\partial \varPhi_n} = 0. \qquad (3.1\,\text{a--c})$$

We shall solve these equations successively.

Application of the condition for the electronic dipoles to (2.13), using (1.35), yields

$$\sum_{k\varrho} \Gamma_{l\mu k\varrho}^{-1}\,\boldsymbol{m}_{k\varrho} = -\,\gamma\,\nabla_{l\mu}\,[\int \xi(\boldsymbol{r})\,C(\boldsymbol{r},\,l\,\mu)\,d\boldsymbol{r} + \sum_{j\nu} a_{j\nu}(\boldsymbol{M}_{j\nu}\,\nabla_{j\nu})\,C(l\,\mu,\,j\,\nu)]\,. \qquad (3.2)$$

This equation can be solved approximately, as shown in paper I, with the aid of the relation

$$F(\boldsymbol{x},\,\boldsymbol{v}) := \frac{d^3}{4\pi}\sum_{l\mu} \nabla_{l\mu}\,C(\boldsymbol{x},\,l\,\mu)\,\nabla_{l\mu}\,C(\boldsymbol{v},\,l\,\mu) \approx C(\boldsymbol{x},\,\boldsymbol{v})\,. \qquad (3.3)$$

The approximation hereby consists of replacing the summation by an integration, so that the equality holds exactly. But we have to emphasize that we are not supposed to apply (3.3) if $\boldsymbol{x}$ and $\boldsymbol{v}$ are equal or refer to the same lattice point because in these cases one would get infinite self-energies, as seen later.

Following the method of I we get the solution of the above system of equations as

$$\boldsymbol{m}_{l\mu} = -\,\frac{\gamma}{1+\alpha^e}\,\nabla_{l\mu}\,[\int \xi(\boldsymbol{r})\,C(\boldsymbol{r},\,l\,\mu)\,d\boldsymbol{r} + \sum_{j\nu} a_{j\nu}(\boldsymbol{M}_{j\nu}\,\nabla_{j\nu})\,C(l\,\mu,\,j\,\nu)]\,, \qquad (3.4)$$

where α^e (1.30) is the dimensionless polarizability of the ions.

If one inserts (3.4) in the original equation (3.2) and applies the approximation (3.3), one can immediately prove that the equation is satisfied. The physical meaning of this solution is obvious. As would be expected, the electronic dipoles are caused by the local perturbation field which is produced by the total impurity centre charge density and the displacements of the ions. We now insert the expression for the electronic dipoles in the energy of the crystal and use (3.3) for every interaction between different ions. The electronic dipoles are then eliminated and we have

$$U_n = H_n(M) + H_{\text{I}}^{\text{ex}} + \sum_{l\mu}{}' a_{l\mu}\,e \int \varrho_n(\boldsymbol{r})\,C(\boldsymbol{r},\,l\,\mu)\,d\boldsymbol{r}$$

$$-\,\frac{1}{2}\,\frac{\alpha^e}{1+\alpha^e}\int \xi(\boldsymbol{r})\,F(\boldsymbol{r},\,\boldsymbol{r}')\,\xi(\boldsymbol{r}')\,d\boldsymbol{r}\,d\boldsymbol{r}'$$

$$+ \sum_{l\mu}{}' (\boldsymbol{M}_{l\mu} \nabla_{l\mu}) \left[\frac{\alpha_{l\mu}}{1 + \alpha^e} \int \xi(\boldsymbol{r}) \, C(\boldsymbol{r}, l\,\mu) \, d\boldsymbol{r} - C(0\,0, l\,\mu)^\eta \right]$$

$$+ \tfrac{1}{2} (\tau_{\mathrm{r}}^{-1} - \tau_{\mathrm{e}}^{-1}) \sum_{l\mu} \boldsymbol{M}_{l\mu}^2 \tag{3.5}$$

$$+ \tfrac{1}{2} \sum_{l\mu, k\varrho}^{\neq} (\boldsymbol{M}_{l\mu} \nabla_{l\mu}) (\nabla_{k\varrho} \boldsymbol{M}_{k\varrho}) \left[\frac{a_{l\mu} \, a_{k\varrho}}{1 + \alpha^e} C(l\,\mu, k\,\varrho) + C(l\,\mu, k\,\varrho)^\eta \right]$$

$$+ U_{\mathrm{ideal}} - H_{00}^0 - \sum_{l\mu}{}' \left[a_{00} \, a_{l\mu} \, C(0\,0, l\,\mu) + C(0\,0, l\,\mu)^\eta \right].$$

As seen in the above expression, the electronic polarization of the host crystal causes three effects: The first is the appearance of a self-energy of the impurity charge density. Second, the interaction of the displacement dipoles are screened by the factor $(1 + \alpha^e)^{-1}$ which equals ε_∞^{-1}, where ε_∞ is the high frequency dielectric constant. The connection between α^e and ε_∞ will be discussed in Sect. 5. The last effect is a self-energy of the displacement dipoles which contributes to the polarizability of the lattice. This is given by

$$\tau_{\mathrm{e}}^{-1} \mathbf{1} = \frac{\alpha^e \, e^2}{1 + \alpha^e} \, \nabla_{k\varrho} \otimes \nabla_{j\nu} \, F(k\,\varrho, j\,\nu)\big|_{k\varrho = j\nu} . \tag{3.6}$$

We cannot use (3.3) here because this self-energy would become infinite but we can simplify the expression with the aid of group theory. The method is similar to that used for the calculation of τ_{r}^{-1} in II and we only give the result

$$\tau_{\mathrm{e}}^{-1} \mathbf{1} = \left(\frac{a^e \, e^2}{1 + \alpha^e} \, \frac{d^3}{2\pi} \sum_{l\mu} R_{l\mu}^{-6} \right) \mathbf{1} . \tag{3.7}$$

The physical meaning of this contribution to the lattice polarizability is that the lattice consists of polarizable ions rather than rigid ones. In the remaining calculation we will use the total lattice polarizability τ, defined in (1.33).

The next step is the calculation of the displacements of the ions. By means of the minimum condition (3.1 b) applied to (3.5) we get

$$\sum_{k\varrho} \boldsymbol{G}_{l\mu}^{-1} \boldsymbol{M}_{k\varrho} = - \tau \, \nabla_{l\mu} \left[\frac{a_{l\mu}}{1 + \alpha^e} \int \xi(\boldsymbol{r}) \, C(\boldsymbol{r}, l\,\mu) \, d\boldsymbol{r} - C(0\,0, l\,\mu)^\eta \right], \tag{3.8}$$

where $\boldsymbol{G}$, defined in (1.37), is the Green's function of the ion displacements. The techniques of solving these equations are nearly the same as in the case of the electronic dipoles. In paper II, the details are discussed. Together with the integral approximation (3.3), one has to use a second one

$$\sum_{k\varrho} V_{l\mu k\varrho} \nabla_{k\varrho} \frac{a_{k\varrho}}{1 + \alpha^e} \int \xi(\boldsymbol{r}) \, C(\boldsymbol{r}, k\,\varrho) \, d\boldsymbol{r} \approx 0 , \tag{3.9}$$

where $\boldsymbol{V}$ (1.36) is the part of the Green's function $\boldsymbol{G}$ which originates from the repulsive interaction between the ions. This relation is again exact, if one replaces the summation by an integration, as discussed in II.

With this approximation and (3.3) one can prove by means of an expansion of $\boldsymbol{G}$ that the following relation holds

$$\sum_{k\varrho} G_{l\mu k\varrho} \, a_{k\varrho} \, \nabla_{k\varrho} \int \xi(\boldsymbol{r}) \, C(\boldsymbol{r}, k\,\varrho) \, d\boldsymbol{r} = \frac{(1 + \alpha^e) \, a_{l\mu}}{1 + \alpha^e + \alpha^d} \, \nabla_{l\mu} \int \xi(\boldsymbol{r}) \, C(\boldsymbol{r}, l\,\mu) \, d\boldsymbol{r} . \tag{3.10}$$

The dimensionless polarizability appearing here is given by (1.34). Using (3.10), we get the solution of (3.8) as

$$M_{l\mu} = -\tau \frac{a_{l\mu}}{1 + \alpha^e + \alpha^d} \nabla_{l\mu} \int \xi(r)\, C(r, l\,\mu)\, dr + \tau \sum_{k\varrho} G_{l\mu k\varrho} \nabla_{k\varrho} C(0\,0, k\,\varrho)^n,$$

$$(3.11)$$

which shows that the displacements of the ions are caused by the field of the impurity charge density and the repulsive force of the missing ion. In the last term of (3.11) we have not eliminated the Green's function G as this is not necessary for the calculation of the wave function. If one is interested in the calculation of the ion displacements one has to expand G and compute the lattice sums numerically as was done in paper II.

We insert the displacements (3.11) in the energy expression (3.5) and get with the aid of (3.10)

$$U_n = H_n(M) + H_I^{\mathrm{ex}} + \sum_{l\mu} a_{l\mu}\, e \int \varrho_n(r)\, C(r, l\,\mu)\, dr$$

$$-\frac{1}{2}\left[\frac{\alpha^e}{1 + \alpha^e} + \frac{\alpha^d}{(1 + \alpha^e)(1 + \alpha^e + \alpha^d)}\right] \int \xi(r)\, F(r, r')\, \xi(r')\, dr\, dr'$$

$$+ \frac{\alpha^d}{1 + \alpha^e + \alpha^d} \int \xi(r)\, F(r, 0\,0)^n\, dr - \frac{1}{2}\alpha^d \qquad (3.12)$$

$$+ \sum_{l\mu, k\varrho} \nabla_{l\mu} C(l\,\mu, 0\,0)^n\, G_{l\mu k\varrho}\, \nabla_{k\varrho} C(k\,\varrho, 0\,0)^n$$

$$+ U_{\mathrm{ideal}} - H_{00}^0 - \sum_{l\mu} [a_{00}\, a_{l\mu}\, C(l\,\mu, 0\,0) + C(l\,\mu, 0\,0)^n],$$

where the polarization potential $F(x, v)^n$ is defined by (1.8). The physical meaning of our final expression for the total electronic energy becomes clear if we write the impurity charge density explicitly and introduce the actual values for e and a_{00}, which are both equal to -1 in atomic units. We can now divide U_n into three parts

$$U_n = U_n(\Phi) + U_{\mathrm{vac}} + U_{\mathrm{ideal}}. \qquad (3.13)$$

U_{ideal} (1.25) is the energy of the ideal crystal and is unimportant in the present context.

$U_n(\Phi)$ is that part of the total energy which depends on the wave function of the impurity electrons and is given by

$$U_n(\Phi) := H_n(M) + H_I^{\mathrm{ex}} - \sum_{l\mu}{}' a_{l\mu} \int \varrho_n(r)\, C(r, l\,\mu)\, dr$$

$$- \tfrac{1}{2}\mathrm{c} \int \varrho_n(r)\, F(r, r')\, \varrho_n(r')\, dr\, dr' + \mathrm{c} \int \varrho_n(r)\, F(r, 0\,0)\, dr \qquad (3.14)$$

$$- \mathrm{c}' \int \varrho_n(r)\, F(r, 0\,0)^n\, dr.$$

The factors c and c' are defined in (1.38).

The first row of (3.14) contains the kinetic energy and the mutual interaction of the impurity electrons, collected in $H_n(M)$, the exchange interaction with the core electrons H_I^{ex} and the Coulomb interaction with the ions. One would get all these energies if one would neglect the polarization of the host crystal. In the rest we have the energies due to these effects. The first is a self-energy and represents the interaction of the impurity electrons with the polarization of the crystal produced by their own charge density. This energy is negative and leads to a self-trapping of the electrons. The last two terms are the interactions of the electrons with the polarization which is caused by the Coulomb and repulsive force of the missing ion. One can already see at this stage of the calculation that this Coulomb term is repulsive and screens the interaction of the electrons with the ions. In the last section we will discuss the polarization potentials $F(x, v)$ and $F(x, v)^\eta$.

Finally we have the energy U_{vac}, which is the energy of the vacancy in the absence of the impurity electrons. This energy has the form

$$
\begin{aligned}
U_{\text{vac}} = &- H_{00}^0 + \sum_{l\mu}{}' \left[a_{l\mu} C(l\,\mu, 0\,0) - C(l\,\mu, 0\,0)^\eta \right] \\
&- \frac{1}{2} \frac{d^3}{4\pi} \sum_{l\mu}{}' \left[c\, R_{l\mu}^{0-4} - 2\, c'\, b\, a_{l\mu}\, R_{l\mu}^{0-(\eta+3)} \right. \\
&\left. + \tau\, b^2 \sum_{k\varrho}{}' \nabla_{l\mu} C(l\,\mu, 0\,0)^\eta\, G_{l\mu k\varrho}\, \nabla_{k\varrho}\, C(k\,\varrho, 0\,0)^\eta \right].
\end{aligned}
\tag{3.15}
$$

The first row of (3.15) contains the energies of the vacancy without polarization effects and the remainder the self-energies of the vacancy due to the polarization of the host crystal. These self-energies have a clear physical meaning. If one takes a ion out of a crystal the lattice around the vacancy relaxes and the energy of this relaxation is given by those self-energies. U_{vac} is not important for the remaining calculation as it is independent of the electronic states, but if one is interested in the formation energy of a vacancy in a crystal one can use (3.15) for numerical calculations.

4. The Wave Functions of the F- and F'-Centres

After the elimination of the electronic dipoles and the displacements of the ions, the calculation of the impurity electron wave functions can be carried out. To do this, we use a simple variational method representing the electrons by trial functions which contain variational parameters. The minimum condition, (3.1c), is now replaced by a minimum condition with respect to the variational parameters. As the energies U_{ideal} and U_{vac} are independent of the wave functions of the impurity electrons the minimum of the total electronic energy is given by the minimum value of $U_n(\Phi)$.

For the sake of simplicity and to keep the computational effort in reasonable limits we use some approximations: The exchange interaction between the impurity and core electrons H_I^{ex} as well as the orthogonality of their wave functions is neglected in the numerical calculation. In addition, we describe each impurity electron by a single variational parameter only. These approximations are not vital to our method and can be avoided by a higher computational effort.

The test functions applied in the minimizing procedure have the following form and are well known in F-centre calculations.

a) F-Centre

$$
\Phi_g^F \equiv \varphi_{1s}(r, \alpha) := \alpha^{3/2}(7\pi)^{-1/2}(1 + \alpha\, r) \exp(-\alpha\, r),
\tag{4.1a}
$$

$$\Phi^{\mathrm{F}}_{\mathrm{ex}} \equiv \varphi_{2p}(\boldsymbol{r},\beta) := \beta^{5/2}\,\pi^{-1/2}\,r\cos\vartheta\exp(-\beta\,r)\,, \tag{4.1b}$$

where α and β are variational parameters.

The ground state of the F-centre is Pekar's [5] function with s-symmetry and the excited state, to which the optical transition takes place, is a hydrogen-like $2p$-function.

If we take the expectation value (3.14) with respect to the above test functions, the energy functional, as a function of the variational parameter, may be represented by

$$U^{\mathrm{F}}_{1s}(\Phi^{\mathrm{F}}_{\mathrm{g}}) \equiv U_{1s}(\alpha)\,, \tag{4.2a}$$

$$U^{\mathrm{F}}_{2p}(\Phi^{\mathrm{F}}_{\mathrm{ex}}) \equiv U_{2p}(\beta)\,, \tag{4.2b}$$

where the number of the impurity electrons M equals 1.

b) F'-Centre

At the F'-centre we have $M = 2$ and triplet and singlet states are possible. The ground state, however, is a singlet state as the F'-centre is analogous to the hydrogen negative ion H^-. In view of this we write the trial wave function, omitting the spin functions, in the form

$$\Phi^{\mathrm{F}'}_{\mathrm{g}} \equiv \Phi_{\mathrm{g}}(\boldsymbol{r},\boldsymbol{r}',\gamma) := \varphi_{1s}(\boldsymbol{r},\gamma)\cdot\varphi_{1s}(\boldsymbol{r}',\gamma)\,. \tag{4.3a}$$

Correlations between the electrons are neglected to simplify the calculations. In addition, we have used only one parameter for the two electrons because the problem is symmetric with respect to both electrons. Calculations with two parameters were carried out and these were found to be equal.

The excited state of the F'-centre to which the optical transition can be made is now considered. Its existence is assumed in this section and discussed after obtaining the numerical results. From the atomic calculations one expects that the excited state of interest is a singlet state in which one electron is in a $1s$-state and the other one in a $2p$-state. Thus the trial function for the excited state is given by

$$\Phi^{\mathrm{F}'}_{\mathrm{ex}} \equiv \Phi_{\mathrm{ex}}(\boldsymbol{r},\boldsymbol{r}',\delta_1,\delta_2) := 2^{-1/2}\,[\varphi_{1s}(\boldsymbol{r},\delta_1)\,\varphi_{2p}(\boldsymbol{r}',\delta_2)$$
$$+\,\varphi_{1s}(\boldsymbol{r}',\delta_1)\,\varphi_{2p}(\boldsymbol{r},\delta_2)]\,. \tag{4.3b}$$

The wave functions φ_{1s} and φ_{2p} are assumed to have the same functional form as in the F-centre case.

The energy expectation value of the F'-centre in the ground and excited state can be represented in terms of one electron energies and the interaction between the electrons. With (4.3) applied to (3.14) one derives

$$U^{\mathrm{F}'}_{\mathrm{g}}(\Phi^{\mathrm{F}'}_{\mathrm{g}}) = 2\cdot U_{1s}(\gamma) + \int|\varphi_{1s}(\boldsymbol{r},\gamma)|^2\,C(\boldsymbol{r},\boldsymbol{r}')\,|\varphi_{1s}(\boldsymbol{r}',\gamma)|^2\,d\boldsymbol{r}\,d\boldsymbol{r}'$$
$$-\,c\int|\varphi_{1s}(\boldsymbol{r},\gamma)|^2\,F(\boldsymbol{r},\boldsymbol{r}')\,|\varphi_{1s}(\boldsymbol{r}',\gamma)|^2\,d\boldsymbol{r}\,d\boldsymbol{r}'\,, \tag{4.4a}$$

$$U^{\mathrm{F}'}_{\mathrm{ex}}(\Phi^{\mathrm{F}'}_{\mathrm{ex}}) = U_{1s}(\delta_1) + U_{2p}(\delta_2) + \int|\varphi_{1s}(\boldsymbol{r},\delta_1)|^2\,C(\boldsymbol{r},\boldsymbol{r}')\,|\varphi_{2p}(\boldsymbol{r}',\delta_2)|^2\,d\boldsymbol{r}\,d\boldsymbol{r}'$$
$$-\,c\int|\varphi_{1s}(\boldsymbol{r},\delta_1)|^2\,F(\boldsymbol{r},\boldsymbol{r}')\,|\varphi_{2p}(\boldsymbol{r}',\delta_2)|^2\,d\boldsymbol{r}\,d\boldsymbol{r}' \tag{4.4b}$$
$$+\,\int\varphi^*_{1s}(\boldsymbol{r},\delta_1)\,\varphi_{2p}(\boldsymbol{r},\delta_2)\,C(\boldsymbol{r},\boldsymbol{r}')\,\varphi^*_{2p}(\boldsymbol{r}',\delta_2)\,\varphi_{1s}(\boldsymbol{r}',\delta_1)\,d\boldsymbol{r}\,d\boldsymbol{r}'\,.$$

The functional form of the one electron energies is the same as that of the F-centre energies (4.2) due to our choice of the wave functions. Of interest is that the self-energy of the impurity electrons contributes to the interaction between the electrons. These energies are determined by the polarization potential $F(x, v)$ and screen the Coulomb interaction among the impurity electrons.

5. Numerical Calculation

The input parameters of our model are the next nearest neighbor distances d, the exponent η of the repulsive potential and the polarizability of the ions γ or α^e respectively. By the aid of the equilibrium condition for the ideal crystal the constant b in the repulsive potential is obtained to be

$$b = \alpha_M \, e^2 \, d^{\eta-1}/6\eta, \tag{5.1}$$

where α_M is the Madelung constant. The lattice polarizability α^d can be calculated according to (1.31) and (1.32).

Of peculiar interest are the polarizabilities and their connection to the dielectric constants. In a macroscopic polarization theory the local field, due to the polarization dipoles surrounding the considered ion, is replaced by the field of a spherical cavity in a uniformly polarized medium. With this approximation the Lorentz-Lorenz formula is derived. In our model we explicitly take into account the atomistic distribution of the dipoles and their mutual interaction as shown in (2.9). Therefore, we need no additional Lorentz correction, and the electronic polarizability is given by the so-called Drude formula [6]

$$\varepsilon_\infty = 1 + 4\pi \, d^{-3}\, \gamma \equiv 1 + \alpha^e, \tag{5.2}$$

where ε_∞ is the high frequency dielectric constant.

The lattice polarizability is related to the static dielectric constant as is well known from the theory of lattice vibrations [7]. With our notation the difference between the static and the high frequency dielectric constant is given by the dimensionless lattice polarizability α^d (1.34). We have together with (5.2)

$$\varepsilon = 1 + \alpha^e + \alpha^d, \tag{5.3}$$

where ε is the static dielectric constant.

In Table 1, we have listed all the input parameters. There we have also compared the calculated value for α^d with the experimental one. The agreement — except for the Lithium salts — is fairly good and can possibly be improved by taking a more sophisticated repulsive potential instead of (2.7). For the remaining calculation we use the experimental values of α^d to eliminate the uncertainty in the repulsive forces.

The lattice summations appearing in our theory can be carried out by direct summation over the lattice as these sums converge fairly well. The only exception is the sum which describes the interaction of the electrons with the ions — the third term in (3.14). This sum is similar to the Madelung sum and its convergence is rather poor. To get reliable results, we apply a method proposed by Evjen [9] which provides excellent convergence.

The integrals contained in the energy expectation values (4.2), (4.4) can be calculated analytically. A computer program is used to carry out the lattice sums and to minimize the energy expectation values. The ions surrounding the vacancy

Table 1. Parameters for the numerical calculation

	d [a] [Å]	η [a]	α^{e} [b]	α^{d} [b]	$\alpha^{dc}_{cal.}$
LiF	2.014	6	0.92	7.35	4.60
LiCl	2.565	7	1.75	8.30	4.17
LiBr	2.751	7.5	2.16	8.94	3.93
LiI	3.000	8.5	2.80	7.23	3.45
NaF	2.310	7	0.74	4.26	3.82
NaCl	2.820	8	1.25	3.37	3.47
NaBr	2.989	8.5	1.62	3.37	3.31
NaI	3.236	9.5	1.91	3.69	2.94
KF	2.678	8	0.85	4.20	3.35
KCl	3.146	9	1.13	2.55	3.01
KBr	3.298	9.5	1.33	2.45	2.86
KI	3.533	10.5	1.69	2.25	2.59
RbF	2.820	8.5	0.93	3.98	3.16
RbCl	3.291	9.5	1.19	2.81	2.84
RbBr	3.427	10	1.33	2.67	2.70
RbI	3.671	11	1.63	2.37	2.45

[a] Sherman, J. [8]; [b] α^{e}, α^{d} are given by (5.2) and (5.3) and the dielectric constants by Mott, N. F., Gurney, R. W. [6]; [c] our calculation.

Table 2. Results for the F-centre

	α_{1s}	U^{F}_{1s}	β_{2p}	U^{F}_{2p}	ΔU_{cal} [a]	ΔU_{exp} [b]	ΔU_{Ivey} [c]
LiF	1.32	− 4.44	0.63	− 1.05	3.39	5.08	4.86
LiCl	1.11	− 3.92	0.80	− 1.26	2.66	—	3.10
LiBr	1.05	− 3.75	0.80	− 1.31	2.44	—	2.73
LiI	0.99	− 3.56	0.78	− 1.39	2.17	—	2.33
NaF	1.20	− 4.37	0.82	− 1.34	3.03	3.72	3.66
NaCl	1.04	− 3.92	0.80	− 1.55	2.37	2.74	2.62
NaBr	0.99	− 3.74	0.78	− 1.56	2.18	2.35	2.36
NaI	0.93	− 3.53	0.75	− 1.57	1.96	—	2.03
KF	1.08	− 4.05	0.82	− 1.50	2.55	2.86	2.88
KCl	0.96	− 3.76	0.77	− 1.70	2.06	2.30	2.14
KBr	0.92	− 3.62	0.75	− 1.70	1.92	2.05	1.96
KI	0.87	− 3.42	0.72	− 1.69	1.73	1.85	1.72
RbF	1.02	− 3.93	0.80	− 1.55	2.38	2.42	2.61
RbCl	0.93	− 3.63	0.76	− 1.70	1.93	2.04	1.98
RbBr	0.89	− 3.52	0.74	− 1.70	1.82	1.85	1.82
RbI	0.85	− 3.34	0.71	− 1.70	1.64	1.68	1.62

Energies are in eV; parameters of the wave functions in Å^{-1}.
[a] Calculated energy difference between the $1s$- and $2p$-state.
[b] Maximum of the absorption band after Compton and Rabin [11].
[c] Absorption energy given by Ivey's law [10].

have been taken into account up to a cubus with a volume of 10,648 d^3 and the variational parameters have been varied in steps of 0.01 Å^{-1}.

In Table 2, we give the results for the F-centre and compare the energy difference between the ground and excited state with the observed absorption energy. In addition, we list the energy values given by Ivey's law [10].

The good agreement between the calculated and experimental energies suggests that the excited state of the F-centre, to which the optical transition from the ground state takes place, is well described by a p-symmetrical wave function. The remaining deviations are due to relaxation effects. With the minimum conditions (3.1) we have calculated the relaxed excited state which is lower in energy than the nonrelaxed state.

The results for the F'-centre are shown in Table 3. In contrast to the F-centre, there have been relatively few experimental investigations of the F'-centre. We compare our results with the average value of the F'-band measured by Lynch and Robinson [12] and get a reasonable agreement. It is anticipated that our calculations may prove useful in the identification of F'-centres in other alkali halides; but we have to emphasize that in the framework of the static calculation

Table 3. Results for the F'-centre

	γ	$U_{\mathrm{g}}^{F'}$	δ_1	δ_2	$U_{\mathrm{ex}}^{F'}$	ΔU_{cal} [a]	ΔU_{exp} [b]
LiF	1.19	− 7.10	1.33	0.26	− 4.56	2.54	—
LiCl	1.01	− 6.35	1.12	0.26	− 4.08	2.27	—
LiBr	0.95	− 6.11	1.07	0.26	− 3.93	2.18	—
LiI	0.89	− 5.75	1.00	0.25	− 3.74	2.01	—
NaF	1.08	− 6.74	1.21	0.24	− 4.48	2.26	—
NaCl	0.92	− 6.01	1.05	0.23	− 4.05	1.96	2.70
NaBr	0.87	− 5.77	1.00	0.23	− 3.88	1.89	—
NaI	0.83	− 5.48	0.95	0.23	− 3.68	1.80	—
KF	0.96	− 6.25	1.09	0.24	− 4.18	2.07	—
KCl	0.84	− 5.63	0.97	0.22	− 3.88	1.75	1.80
KBr	0.80	− 5.44	0.93	0.21	− 3.75	1.69	1.70
KI	0.75	− 5.16	0.89	0.21	− 3.56	1.60	1.55
RbF	0.92	− 6.07	1.05	0.23	− 4.07	2.00	—
RbCl	0.81	− 5.48	0.94	0.22	− 3.76	1.72	1.60
RbBr	0.77	− 5.32	0.91	0.22	− 3.66	1.66	—
RbI	0.73	− 5.04	0.86	0.21	− 3.48	1.56	1.30

Energies are in eV; parameters of the wave functions in Å^{-1}.

[a] Calculated energy difference between the ground and excited state.

[b] Average value of the F'-band measured by Lynch and Robinson [12].

we have carried out the comparison between theory and experiment is only qualitative. For a quantitative check of the theory one has to take into account the electron-phonon interaction and to calculate the shape of the absorption band. This problem is treated by Stumpf [13] theoretically and numerical calculations are in preparation.

Of interest is that we have got an excited state of the F'-centre which is given by the wave function (4.3b) together with the parameters δ_1, δ_2 listed in Table 3. These parameters show that one electron is strongly localized and its wave function is that of an electron in the ground state of the F-centre. The wave function of the other electron is widely spread and its optical binding energy — the difference between the energies of the F-centre ground state and the F'-centre excited state — is $0.11-0.18$ eV for all alkali halides. This result agrees with the thermal binding energy $0.04-0.08$ eV of the excited F-centre state in KBr suggested by Crandall [14]. Until now, we cannot decide wether our excited state is a real bound state or a resonant one consisting of an F-centre electron in the ground state and a polaron in the conduction band. This last possibility seems to be more likely. To give a definite statement one has to calculate the energy of a polaron resting at the bottom of the conduction band.

6. Conclusion

Finally, we discuss some connections with other theories. The screening factor c (1.38) is essential for the polarization energies. It consists of two parts: The first originates from the electronic and the second one from the ionic polarization of the host crystal. By the aid of the Eqs. (5.2) and (5.3) one can express this factor in terms of the static and high frequency dielectric constants ε and ε_∞. This yields

$$c = [(1 - \varepsilon_\infty^{-1}) + (\varepsilon_\infty^{-1} - \varepsilon^{-1})] = (1 - \varepsilon^{-1}) . \tag{6.1}$$

In this form both the constituents of c are well known from the theory of "electronic" and "lattice" polarons developed by Toyozawa [15] and Fröhlich [16] respectively. We want to emphasize that we have got these quantities as a result of our calculation and have not inserted them in the interaction potentials from the beginning by means of phenomenological arguments as it is done in other theories.

The polarization potential $F(x, v)$ contains a lattice sum which has to be calculated with the aid of a computer. To explain its dependence of the variables and its physical meaning, we discuss two limiting cases: If x and v are small compared with the lattice constant d, $c \cdot F(x, v)$ becomes constant and equal to the self-energy of the vacancy — the fourth term in (3.15). Thus, if the F-centre electron is strongly localized, all energies containing $F(x, v)$ in the total energy U_n cancel each other and there is no polarization of the crystal. In the other case, if the distance $|x - v|$ is large, the approximation (3.3) becomes valid and one can replace the polarization potential $c \cdot F(x, v)$ by the Coulomb potential $c \cdot C(x, v)$. This yields a screening of the Coulomb potentials by the static dielectric constant well known in the macroscopic polarization theory. For intermediate values of the variables our polarization potential provides a smooth transition between the discussed limiting cases.

There are two other polarization potentials with the same property. One was derived by Haken [18] for the theory of excitons and the other one by Wang et al. [19] for the theory of "electronic" polarons. These potentials, however, do not exhibit the atomistic structure of the host crystal.

The potential $F(x, v)^\eta$ caused by the repulsive potential of the missing ion has no analogous one in other theories. As numerical calculations show, it has a very short range and contributes less then 10% to the binding energy of the F-centre electron. Therefore, this potential can be neglected in F-centre calculations but might be essential for electrons trapped at isoelectronic impurities like AgBr:I [20].

As we have treated the polarization effects in a general manner, our theory is not confined to color centres but can be used for exciton and polaron problems as well. An extension to impurity centres in covalent semiconductors is in preparation [21].

Acknowledgements. The author wishes to thank Prof. H. Stumpf for suggesting the topic of this work and for valuable advices. He is grateful to Dr. J. Schmid and W. Heinzel for many stimulating discussions. Finally, he is indebted to the British Science Research Council for financial support. The computational part of this work was carried out in the Computer Centre of the University of Oxford.

References

1. Stumpf, H.: Phys. kondens. Materie 13, 9 (1971)
2. Gilbert, R. L., Markham, J. J.: J. Phys. Chem. Solids 30, 2699 (1969)
3. Schmid, J.: Phys. kondens. Materie 15, 119 (1972)
4. Bennett, H. S.: Phys. Rev. B 1, 1702 (1970)
5. Pekar, S. I.: Untersuchungen über die Elektronentheorie der Kristalle. Berlin: Akademie-Verlag 1954
6. Mott, N. F., Gurney, R. W.: Electronic processes in ionic crystals. Oxford: Clarendon Press 1953
7. Born, M., Huang, K.: Dynamical theory of crystal lattices. Oxford: Clarendon Press 1954
8. Sherman, J.: Chem. Rev. 11, 93 (1932)
9. Evjen, H. M.: Phys. Rev. 39, 675 (1932)
10. Ivey, H. F.: Phys. Rev. 72, 341 (1947)
11. Compton, W. D., Rabin, H.: Solid State Physics 16, 121 (1964)
12. Lynch, D. W., Robinson, D. A.: Phys. Rev. 144, 670 (1966)
13. Stumpf, H.: Phys. kondens. Materie 13, 101 (1971)
14. Crandall, R. S.: Phys. Rev. 138, A 1242 (1965)
15. Toyozawa, Y.: Progr. theor. Phys. 12, 421 (1954)
16. Fröhlich, H. in ref. [17]
17. Polarons and Excitons, C. G. Kuper and G. D. Whitfield (eds.), London: Oliver and Boyd 1962
18. Haken, H. in ref. [17]
19. Wang, S., Mahutte, C. K., Matsuura, M.: Phys. status solidi 51, 11 (1972)
20. Czaja, W.: Festkörperprobleme, Vol. XI, 65 (1971)
21. Renn, W.: Report for the British Science Research Council (1973) (unpublished)

Dr. Walter Renn
Department of Engineering Science
University of Oxford
Parks Road
Oxford
Great Britain

Phys. cond. Matter 17, 249—265 (1974)

Thermodynamische Untersuchung der Systeme
Kalzium-Strontium, Kalzium-Barium und Strontium-Barium

Bruno Predel und Ferdinand Sommer

Institut für Metallkunde der Universität Stuttgart
und Max-Planck-Institut für Metallforschung, Stuttgart

Eingegangen am 8. Oktober 1973

Thermodynamic Investigations of the Systems Ca-Sr, Ca-Ba and Sr-Ba

Regarding the thermodynamic properties of the liquid alloys of the earth alkali metals (Calcium, Strontium, Barium), due to considerable experimental difficulties up to now scarcely any investigations have been undertaken. These systems, however, similar to the alloys of the alkali metals, are peculiarly suitable for theoretical considerations about the energetics of alloy formation, as their electronic properties are easily to comprehend. Therefore it was necessary to perform experiments on liquid alloys of the earth alkali metals in order to get reliable data for the enthalpies of mixing as a basis for some theoretical calculations.

Zur Messung der Mischungsenthalpien in den Systemen Ca-Sr, Ca-Ba und Sr-Ba wurde ein für diese Mischphasen geeignetes Kalorimeter konstruiert, das bis 1000 °C arbeitsfähig ist. Es erfolgte eine thermodynamische Auswertung der Meßergebnisse. Eine thermodynamische Modellrechnung ergab für die untersuchten Systeme und die flüssigen Alkalimetall-Legierungen zusätzlich die Überschußfunktionen ΔG^E und ΔS^E. Mit Hilfe der Pseudopotentialtheorie gelang eine quantitative Berechnung der Mischungsenthalpien. Die Änderung der Bindungsverhältnisse und die Verzerrungsenergie konnten als die beiden die Legierungsbildung der flüssigen Phase bestimmenden Größen mit ihren Abhängigkeiten bestimmt werden.

Einführung

Wegen der erheblichen experimentellen Schwierigkeiten sind flüssige Legierungen der Erdalkalimetalle Kalzium, Strontium und Barium bisher in thermodynamischer Hinsicht kaum untersucht worden. Für theoretische Betrachtungen der Energetik der Legierungsbildung sind diese Systeme, ähnlich wie die Legierungen der Alkalimetalle, wegen der leicht überschaubaren elektronischen Gegebenheiten indessen besonders gut geeignet. Es lag daher nahe, die für einige Modellrechnungen erforderliche Basis durch experimentelle Ermittlung der Mischungsenthalpien flüssiger Erdalkalimetall-Legierungen zu schaffen.

Experimentelle Hinweise

Zur Bestimmung der integralen Mischungsenthalpien wurde nach dem Prinzip der Differential-Thermoanalyse ein Kalorimeter konstruiert, das Messungen bis zu Temperaturen von 1000 °C gestattet. Die Versuchsanordnung erlaubt eine direkte Messung der Mischungsenthalpie beim Vermischen der flüssigen Komponenten. Die große Reaktionsfähigkeit und der hohe Dampfdruck der flüssigen Erdalkalimetalle machten eine besondere Konstruktion erforderlich (vgl. Abb. 1).

Als Kalorimetergefäß dienten zwei ineinandergeschraubte Eisentiegel. Boden und Deckel des oberen Tiegels bestehen aus einer Tantalfolie. Vor dem Versuch

                                        B. Predel und F. Sommer

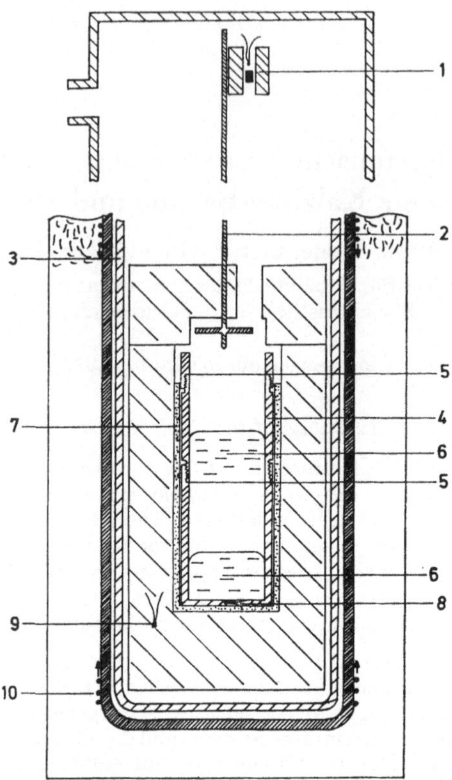

Fig. 1. Schematische Darstellung des Hochtemperatur-Kalorimeters. Erläuterung: (1) Probenofen; (2) Rührer; (3) Quarzglasrohr; (4) Eisentiegel; (5) Tantalfolie; (6) Erdalkalimetalle;
(7) Korundtiegel; (8) Thermoelement; (9) Referenzthermoelement; (10) Heizwicklung

wird in jedem Tiegel eine Legierungskomponente untergebracht, nach Einstellen
der Versuchstemperatur mit einem Rührer die Tantalfolie (Deckel und Boden)
durchstoßen und der Vermischungsvorgang der flüssigen Metalle in Gang gesetzt.
Die durch die Mischungsenthalpie bedingte Temperaturdifferenz zwischen dem
Eisentiegel und einem Referenzblock aus Kupfer wird als Funktion der Zeit
registriert. Die Fläche unter der ΔT-Zeit-Kurve ist ein Maß für die Wärmetönung der Reaktion. Ein Quadratzentimeter der Fläche eines registrierten Effekts
entsprach — je nach Versuchsführung — 2 bis 3 Kalorien. Die Eichung erfolgt
nach dem Meßvorgang durch Einwerfen eines Tantal-Zylinders von vorgegebener
Ausgangstemperatur, die in dem Probenofen (vgl. Fig. 1) eingestellt werden kann.
Die für die Eichung erforderlichen Wärmeinhaltswerte für Tantal sind dem
Tabellenwerk von R. Hultgren u. Mitarb. [1] entnommen worden.

Die Einwaage der Legierung und die Füllung der Eisentiegel erfolgte unter
Argon als Schutzgas. Das beschickte Kalorimeter wurde evakuiert und aus-
geheizt, bis ein Druck von 10^{-4} Torr erreicht war. Dann wurde mit Argon ge-
füllt (ca. 1 Atm. bei der Meßtemperatur). Zur Beseitigung letzter Sauerstoff-
spuren diente Tantalfolie als Gettermaterial.

Fehlerquellen

Die Zuverlässigkeit des Kalorimeters wurde durch Bestimmung der Mischungs-enthalpien hinreichend genau untersuchter Legierungen sichergestellt (Bi-Pb [2]; Bi-Ag [3]). Die erzielten Meßergebnisse an den Erdalkalimetall-Legierungen können mit Fehlern behaftet sein, die auf folgende Ursachen zurückzuführen sind:

a) Fehler des Schreibers, der planimetrischen Bestimmung der Fläche unter der ΔT-Zeit-Kurve und bei der Einhaltung einer konstanten Temperatur. Beim Planimetrieren und besonders beim Schreiber lagen die Fehlermöglichkeiten unter 1%. Abweichungen von der konstanten Temperatur für die über den ge-samten Konzentrationsbereich eines Systems durchgeführten Messungen lagen bei $\pm$ 3 °C. Diese Unterschiede spielen keine Rolle, da die Mischungsenthalpien näherungsweise als temperaturunabhängig angesehen werden können. Die Tem-peratur konnte absolut auf $\pm$ 1 °C bestimmt werden.

b) Verunreinigungen der Metalle. Die Reinheit von Kalzium, Strontium und Barium betrug 99,5%. Eine Abschätzung, die für die Verunreinigung eine mittlere maximale Mischungsenthalpie von 500 cal/g-Atom annimmt — dieser Wert ist auf jeden Fall nicht zu klein, da die Verunreinigungen größtenteils aus Erdalkalimetallen bestehen — ergibt bei den untersuchten Systemen selbst für eine Legierungskonzentration von 20 At.-% Größen, die weniger als 5% des Meß-wertes betragen.

c) Verfälschung der Wärmetönung durch Umoxydation. Die Oxydations-wärme der Erdalkalimetalle beträgt für Kalzium 152 kcal/g-Atom, für Strontium 141,2 kcal/g-Atom und für Barium 133,4 kcal/g-Atom. Die Umoxydation spielt besonders in den Systemen Ca-Sr und Ca-Ba eine Rolle. Es wurde stets darauf

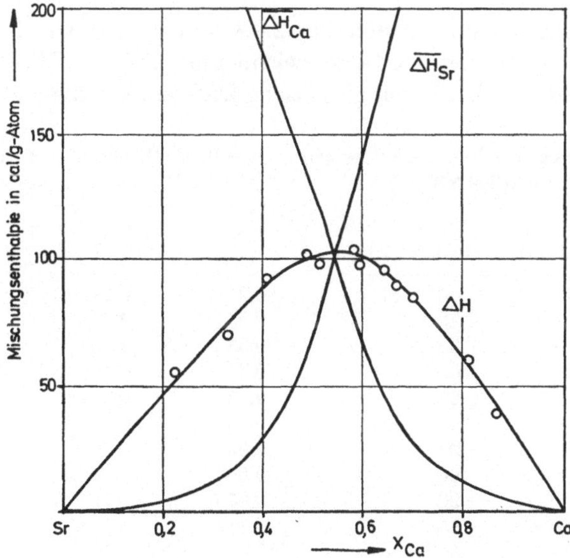

Fig. 2. Integrale und partielle Mischungsenthalpie flüssiger Ca-Sr-Legierungen bei 870 $\pm$ 3 °C

17 *

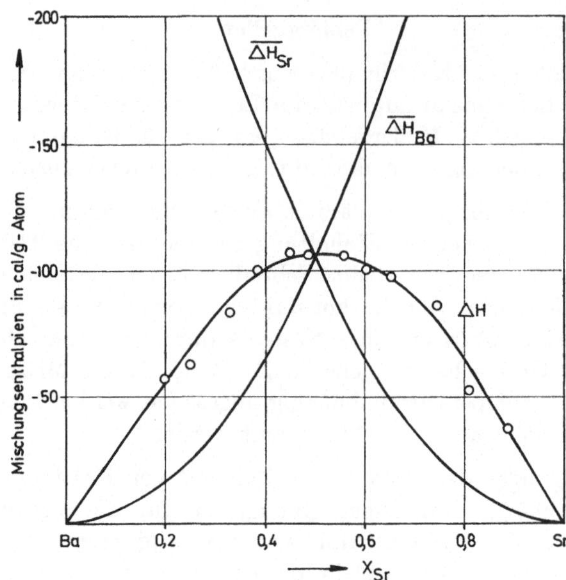

Fig. 3. Integrale und partielle Mischungsenthalpie flüssiger Sr-Ba-Legierungen bei 790 ± 3 °C

geachtet, daß der Oxidgehalt der eingesetzten Materialien so gering war, daß eine Verfälschung der Mischungsenthalpie durch die Wärmetönung bei der Umoxydation merklich unter einer Fehlergrenze von 5% blieb.

Die von uns erhaltenen Mischungsenthalpien der flüssigen Erdalkalimetall-Legierungen können maximal mit einem Fehler von 5% behaftet sein.

Experimentelle Ergebnisse

Die Meßergebnisse sind in den Fig. 2—4 und in den Tabellen 1—3 wiedergegeben. Die in den Bildern mit-eingezeichneten partiellen Mischungsenthalpien der Komponenten wurden durch graphische Differentiation der ΔH-x-Kurven er-

Tabelle 1. Mischungsenthalpien der flüssigen Ca-Sr-Legierungen bei 870 ± 3 °C

x_{Ca}	n (g-Atom)	ΔH (cal/g-Atom)
0.225	0.25	55
0.33	0.175	70
0.41	0.23	92
0.49	0.165	102
0.515	0.21	98
0.58	0.22	104
0.595	0.24	98
0.645	0.14	96
0.67	0.2	90
0.7	0.145	85
0.81	0.15	60
0.865	0.13	39

Tabelle 2. Mischungsenthalpien der flüssigen Sr-Ba-Legierungen bei 790 ± 3 °C

x_{Ca}	n (g-Atom)	ΔH (cal/g-Atom)
0.2	0.21	− 57
0.25	0.22	− 62
0.33	0.25	− 83
0.39	0.16	− 100
0.45	0.2	− 106
0.49	0.19	− 105
0.56	0.16	− 105
0.605	0.14	− 100
0.655	0.12	− 97
0.745	0.1	− 86
0.81	0.1	− 51
0.89	0.09	− 37

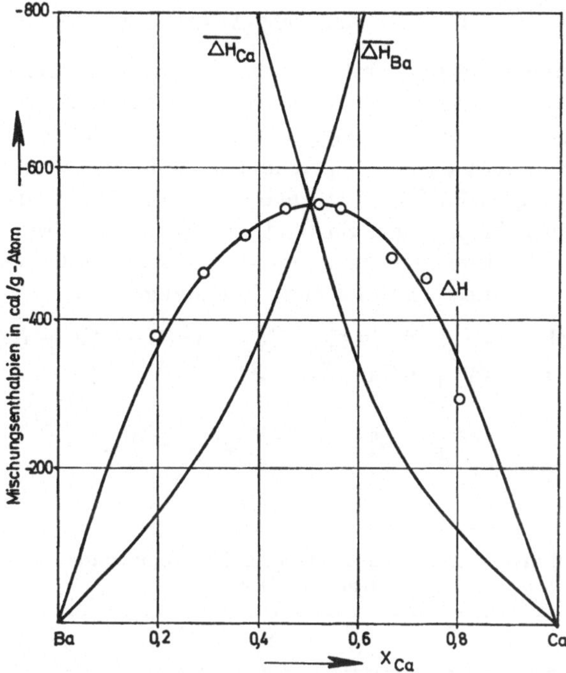

Fig. 4. Integrale und partielle Mischungsenthalpie flüssiger Ca-Ba-Legierungen bei $870 \pm 3\,°C$

Tabelle 3. Mischungsenthalpien der flüssigen Ca-Ba-Legierungen bei $870 \pm 3\,°C$

x_{Ca}	n (g-Atom)	ΔH (cal/g-Atom)
0.195	0.22	− 385
0.29	0.32	− 460
0.37	0.25	− 510
0.45	0.28	− 545
0.52	0.18	− 550
0.56	0.3	− 546
0.66	0.18	− 480
0.73	0.13	− 455
0.8	0.14	− 295

halten. Der maximale Betrag der integralen Mischungsenthalpie wird im System Kalzium-Strontium mit $\Delta H_{max} = 103$ cal/g-Atom bei $x_{Ca} = 0,55$ erreicht. Die Mischungsenthalpien der flüssigen Sr-Ba-Legierungen sind — anders als die der Ca-Sr-Legierungen — negativ. Das Maximum des fast streng symmetrischen ΔH-Verlaufs liegt bei $x = 0,5$ mit $\Delta H_{max} = − 105$ cal/g-Atom (vgl. Fig. 3 und Tabelle 2). Ebenfalls negativ sind die Mischungsenthalpien im System Ca-Ba. $\Delta H_{max} = − 550$ cal/g-Atom liegt bei $x = 0,5$.

Thermodynamische Auswertung

Wie weiter unten dargelegt wird, sind die Überschußentropien der hier interessierenden flüssigen Legierungen außerordentlich klein. Sie können in erster Näherung bei unseren Betrachtungen vernachlässigt werden. Gleiches kann auch für die zugehörigen Mischkristalle angenommen werden. Unter der dann gerechtfertigten Anwendung des Modells der regulären Lösung können aus den ermittelten Mischungsenthalpien im flüssigen Zustand (ΔH^L) und den bekannten Schmelzgleichgewichten die Mischungsenthalpien der Mischkristalle (ΔH^S) berechnet werden. Es gelten für diesen Fall folgende Beziehungen:

$$G_A^{0S} + A^S(1 - x_A^S)^2 + RT \ln x_A^S = G_A^{0L} + A^L(1 - x_A^L)^2 + RT \ln x_A^L,$$
$$G_B^{0S} + A^S(1 - x_B^S)^2 + RT \ln x_B^S = G_B^{0L} + A^L(1 - x_B^L)^2 + RT \ln x_B^L, \tag{1}$$

$$G_{A,B}^{0L} - G_{A,B}^{0S} = \frac{L_{A,B}}{T_{A,B}} (T_{A,B} - T). \tag{2}$$

Tabelle 4. Schmelzenthalpie und Schmelzpunkt der Erdalkalimetalle und der Parameter des regulären Modells

	Ca	Sr	Ba
L (cal/g-Atom)	2040 ± 100	(1978)	1852 ± 250
T (°K)	1112 ± 2	1041 ± 2	1002 ± 2

	Ca-Sr	Sr-Ba	Ca-Ba
A^L	412	-420	-2200
A^S	1000 ± 100	1200 ± 200	-1000 ± 120

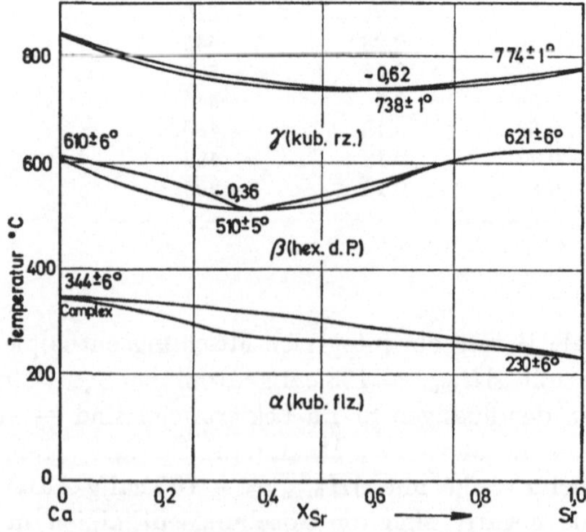

Fig. 5. Zustandsdiagramm des Systems Ca-Sr [4]

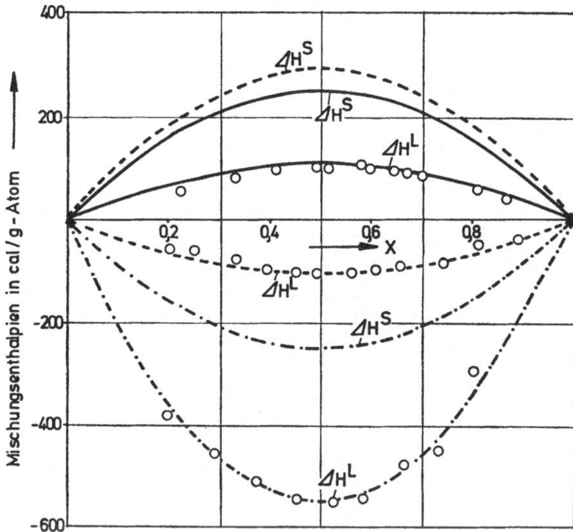

Fig. 6. ΔH^S und ΔH^L nach dem regulären Modell für die Systeme Ca-Sr ———, Ba-Sr ———,
Ba-Ca —·—·—, ooo experimentelle Werte für ΔH^L

Der Auswertung sind die von M. Hansen und K. Anderko [4] angegebenen
Schmelzgleichgewichte zugrunde gelegt worden. Für die Schmelzenthalpien L [1]
und die Schmelztemperaturen T [1] der reinen Komponenten sowie für die Wech-
selwirkungsparameter A^L des regulären Modells wurden die in Tabelle 4 angegebenen Werte benutzt, wobei A^L aus den experimentellen ΔH^L-Werten bei
$x = 0,5$ berechnet wurde. Zur Bestimmung von A^S wurden für jeweils 10 Temperaturen, bei denen Schmelzen mit Mischkristallen im Gleichgewicht stehen, die
Liquidus- und Soliduskonzentrationen aus dem Zustandsdiagramm entnommen.
Die sich daraus ergebenden mittleren A^S-Werte sind in Tabelle 4 mit aufgeführt.

Das Schmelzpunktminimum bei lückenloser Mischkristallbildung, die das
Zustandsdiagramm des Systems Ca-Sr (vgl. Fig. 5) zeigt, deutet darauf hin, daß
die Verzerrungsenergie bei der Legierungsbildung eine bedeutsame Rolle spielt.
Dies wird durch das positive Vorzeichen der von uns ermittelten Mischungsenthalpien ΔH^L und durch die mit dem regulären Modell erhaltenen ΔH^S-Werte
erhärtet (Fig. 6). Die ΔH^S-Werte sind etwa um den Faktor 2 größer als die ΔH^L-
Werte. Diese Befunde werden durch Rechnungen mit Hilfe der Pseudopotentialtheorie (vgl. Fig. 17—19), die eine geringe Änderung der Bindungsverhältnisse
und einen Verzerrungsenergieanteil von 130 cal/g-Atom bei $x = 0,5$ für ΔH^L ergeben, weiter gestützt [5]. Die geringe Asymmetrie der ΔH^L-x-Kurve dürfte
damit zusammenhängen, daß das Kalzium einen kleineren Kompressionsmodul
aufweist als das Strontium.

Eine Überstrukturbildung im System Ca-Ba (vgl. Fig. 7) weist auf einen im
Vergleich zu den Systemen Ca-Sr und Sr-Ba stärkeren Einfluß durch die Änderung
der Bindungsverhältnisse hin. Diese Annahme wird durch die gemessenen nega-
tiven ΔH^L-Werte gestützt (vgl. Fig. 6). Eine weitere Bestätigung liefert die be-

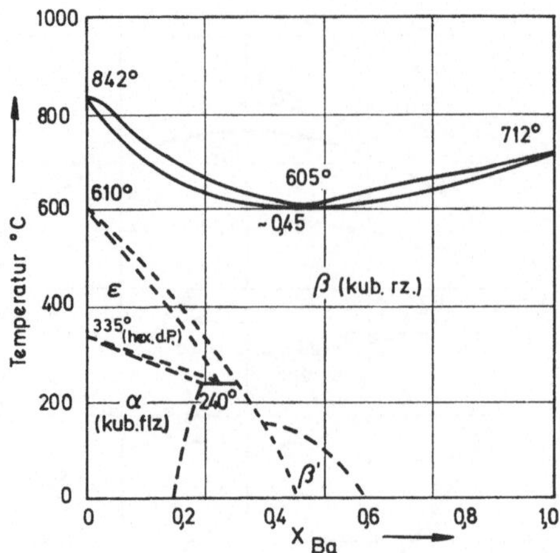

Fig. 7. Zustandsdiagramm des Systems Ca-Ba [4]

rechnete, durch Ladungstransfer hervorgerufene Änderung der Bindungsverhältnisse (vgl. Fig. 19). Dieser negative Anteil an der Mischungsenthalpie ist im System
Ca-Ba wesentlich größer als in den Systemen Ca-Sr und Sr-Ba.

Anwendung der „Significant-Structure"-Theorie zur Erschließung weiterer thermodynamischer Größen

Die von H. Eyring u. Mitarb. [6] entwickelte „Significant-Structure"-Theorie
hat sich bei einer Reihe verschiedenartiger thermodynamischer Überlegungen bewährt (z.B. Berechnung von Kompressibilitäten, thermischen Ausdehnungskoeffizienten und Viskositäten [7]). Eine Anwendung auf flüssige metallische
Mischphasen war indessen bisher noch nicht erfolgt. Für die Ermittlung der
Überschußentropien in den hier interessierenden Systemen der Erdalkalimetalle
scheint sie nützlich zu sein.

Die flüssigen Metalle besitzen ein Überschußvolumen $V - V_s$, wobei V und
V_s das molare Volumen des flüssigen und festen Zustandes ist. Da der Abstand
nächster Nachbarn, wie Röntgenstrahlbeugungsmessungen zeigen, durch den
Schmelzvorgang kaum geändert wird, ist das Überschußvolumen durch Löcher
ausgefüllt. Diese Löcher in einer Flüssigkeit sollen sich wie Atome in einem Gas
verhalten. $(V - V_s)/V$ ist damit der Anteil der Freiheitsgrade mit gasähnlichem
Verhalten. Die Zustandsfunktion Z_N für ein Mol Flüssigkeit kann damit als

$$Z_N = (Z_s)^{N(V_s/V)} (Z_g)^{N(V-V_s/V)} \tag{3}$$

angesetzt werden. Z_s und Z_g sind die Zustandsfunktionen der festkörperähnlichen
und gasähnlichen Freiheitsgrade. Als Zustandsfunktion für flüssige Metalle erhalten Eyring u. Mitarb. [6]

$$Z_N = \left[\frac{\exp\{E_s/RT\}}{(1 - \exp\{-\Theta/T\})^3} \left[1 + n\,\frac{V - V_s}{V_s}\, \exp - \frac{a\,E_s\,V_s}{RT(V - V_s)} \right] \right]^{N(V_s/V)}$$
$$\times \left[\frac{e\,V}{\Lambda^3} \right]^{N(V-V_s/V)} \tag{4}$$

E_s = Sublimationsenergie,

V_m = Volumen am Schmelzpunkt,

Λ = $(2\pi\, m\, k\, T)^{1/2}/h$,

Θ = Debyetemperatur des Metalls,

Z = Koordinationszahl,

n, a = Parameter.

Die Parameter n und a sind durch

$$n = Z\frac{V_s}{V_m}$$

und

$$a = \frac{n-1}{Z}\,\frac{1}{2}\,\frac{(V_m - V_s)^2}{V_m\, V_s}$$

festgelegt. Sie können auch als anpassungsfähige Größen behandelt werden. Für binäre Lösungen erhalten wir als Zustandsfunktion

$$
\begin{aligned}
Z_M = \Bigg\{ &\frac{(x_1 + x_2)!}{x_1!\, x_2!}\Bigg[\frac{\exp\{E_s/RT\}}{(1 - \exp\{-\Theta_1/T\})^3} \\
&\times \Bigg[1 + n\,\frac{V - V_s}{V_s}\exp\Big(-\frac{a E_s V_s}{RT(V - V_s)}\Big)\Bigg]\Bigg]^{x_1(V_s/V)}\Bigg[\frac{e\,V}{\Lambda_1^3}\Bigg]^{x_1(V - V_s/V)} \\
&\times \Bigg[\frac{\exp\{E_s/RT\}}{(1 - \exp\{-\Theta_2/T\})^3}\Bigg[1 + n\,\frac{V - V_s}{V}\exp\Big(-\frac{a E_s V_s}{RT(V - V_s)}\Big)\Bigg]\Bigg]^{x_2(V_s/V)} \\
&\times \Bigg[\frac{e\,V}{\Lambda_2^3}\Bigg]^{x_2(V - V_s/V)}\Bigg\}
\end{aligned}
\tag{5}
$$

x_1 und x_2 sind die Molenbrüche der beiden Legierungspartner. Durch den ersten Faktor berücksichtigen wir die zufällige Mischung der beiden Atomsorten. Sind x_1 oder x_2 null, so wird die Zustandsfunktion der Mischung auf die der reinen Metalle reduziert. Die Größen E_s, V_s, V, a und n sind für die Mischung charakteristisch. Die Zustandsfunktion für die ideale Mischung hat die folgende Form

$$Z_{id} = \Bigg[\frac{(x_1 + x_2)!}{x_1!\, x_2!}\, Z_1^{x_1} Z_2^{x_2}\Bigg]^N .\tag{6}$$

Der Zusammenhang zwischen der Zustandsfunktion und den thermodynamischen Größen ist durch

$$G = -RT\ln Z, \qquad S = RT\,\frac{\delta\ln Z}{\delta T} + R\ln Z + R\tag{7}$$

gegeben. Die Überschußfunktionen sind definiert als

$$\Delta G^E = G_M - G_{id}, \qquad \Delta S^E = S_M - S_{id}.\tag{8}$$

Numerische Behandlung und Ergebnisse

Die Zustandsfunktionen der reinen Metalle Z_1 und Z_2 wurden über die Gl. (4) berechnet. Die hierfür nötigen Größen sind aus verschiedenen Quellen [8] gewonnen und sind in Tabelle 5 aufgeführt. Für die Legierungen sind diese Größen

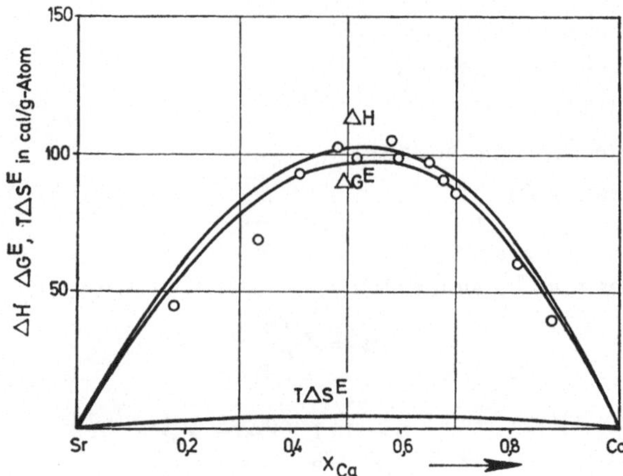

Fig. 8. Überschußfunktionen des Systems Ca-Sr bei $870 \pm 3\,°C$ mit $\delta_E = 0{,}0002$

zur Berechnung von Z_M nicht vorhanden. Sie müssen auf verschiedene Art festgelegt werden. Für die Alkali- und Erdalkalimetallsysteme wurde die Vegardsche
Regel angewandt, die, wie die Rechnungen [5] zeigen, hier gut erfüllt ist.

$$V(x) = x_1\,V_1 + x_2\,V_2, \qquad V_s(x) = x_1\,V_{1s} + x_2\,V_{2s}. \tag{9}$$

Tabelle 5. Zur Berechnung der Zustandsfunktionen in Erdalkalimetall- und in Alkalimetall-
systemen notwendige Größen [8]

	Na	K	Rb	Cs		Ca	Sr	Ba
Z	9,5	9,5	9,5	9		10	10	9
n	9,23	9,1	9,21	8,55		8,8	8,4 8,8	8,2 8,24
a	0,00035	0,00077	0,00039	0,0011	bei 1063 °K bei 1143 °K	0,006	0,01 0,006	0,0034 0,003
E_s cal/g-At	25490	21330	19330	18180		42600	42800	43500
Θ °K	156	91,1	55,5	39,5		230	170	116
V_s cm³/g-At	24,17	45,9	56,5	69,88		26,5	32	38,3
V cm³/g-At bei 383 °K	24,89	47,89	58,24	73,53	bei 1063 °K bei 1143 °K 30		38 36,4	42 41,8

Für die Parameter a und n wird Additivität vorausgesetzt:

$$n = x_1\,n_1 + x_2\,n_2, \qquad a = x_1\,a_1 + x_2\,a_2. \tag{10}$$

Für die Sublimationsenergie der Mischung wurde der häufig benutzte quadratische
Ansatz herangezogen und mit einem variablen Parameter δ_E versehen.

$$E_s = x_1^2\,E_{s1} + x_2^2\,E_{s2} + 2x_1\,x_2\,\sqrt{E_{s1}\,E_{s2}}\,(1 \pm \delta_E). \tag{11}$$

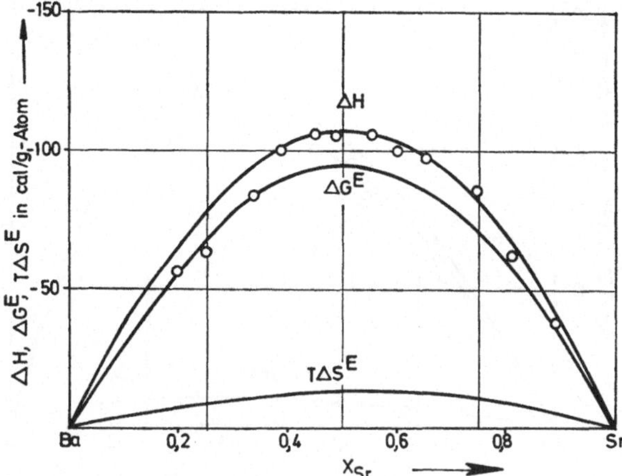

Fig. 9. Überschußfunktionen des Systems Sr-Ba bei $790 \pm 3\,°C$ mit $\delta_E = 0,0035$

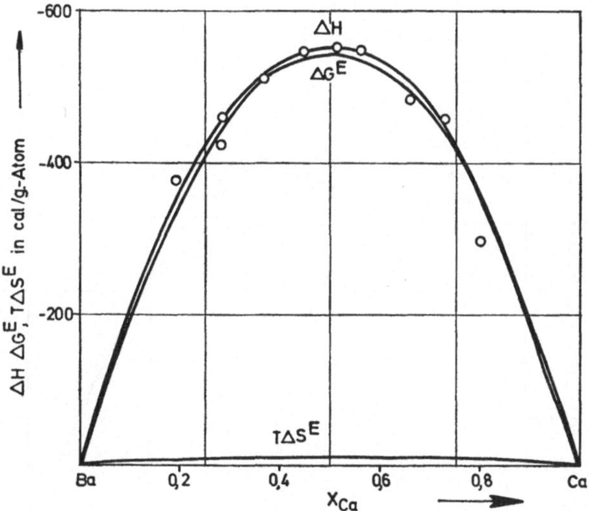

Fig. 10. Überschußfunktionen des Systems Ca-Ba bei $870 \pm 3\,°C$ mit $\delta_E = 0,025$

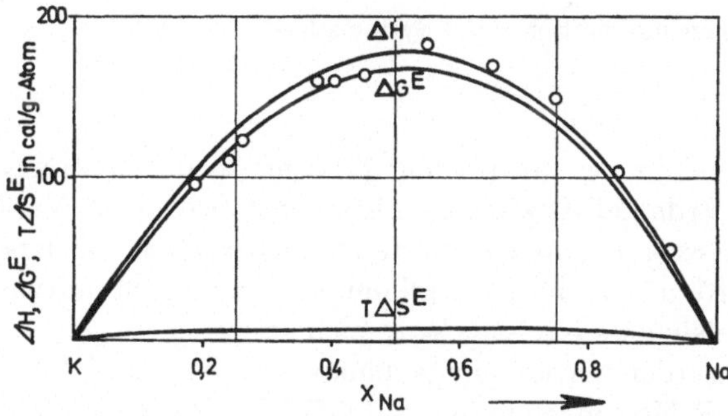

Fig. 11. Überschußfunktionen des Systems Na-K bei $111\,°C$ mit $\delta_E = -\,0,006$

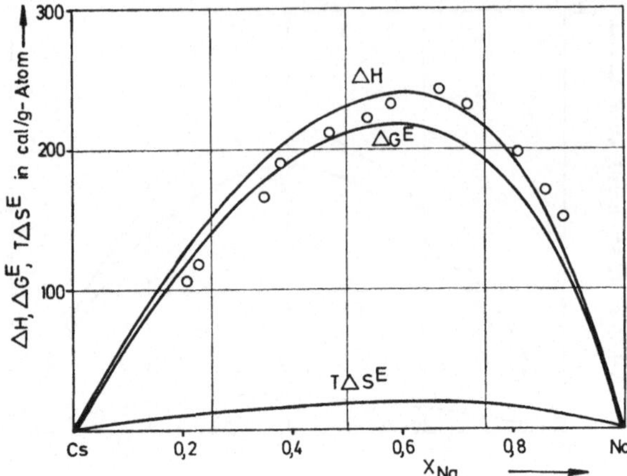

Fig. 12. Überschußfunktionen des Systems Na-Cs bei 111 °C mit $\delta_E = 0{,}007$

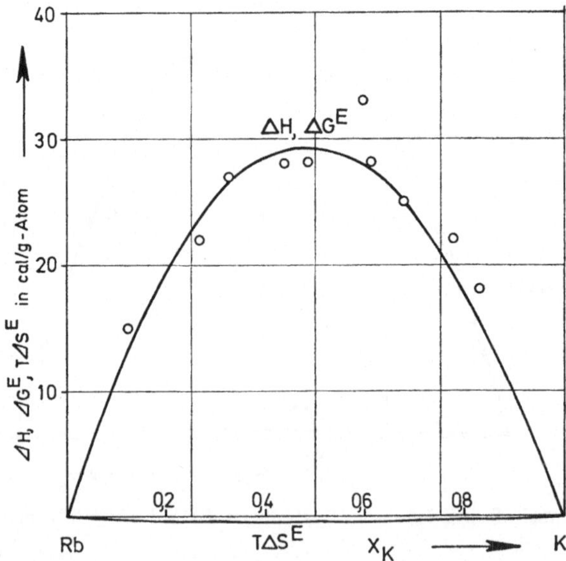

Fig. 13. Überschußfunktionen des Systems K-Rb bei 111 °C mit $\delta_E = -\,0{,}0035$

Über die Beziehungen (5), (6) und (7) wurden die Mischungsenthalpien für die Alkali- und Erdalkali-Legierungen berechnet, wobei die Variation von δ_E so erfolgte, daß die experimentellen Werte möglichst genau wiedergegeben wurden. Die Fig. 8 bis 10 zeigen die so bestimmten Überschußfunktionen. Die experimentellen Werte sind als Kreise gekennzeichnet.

Ergänzend wurden ferner einige binäre Systeme der Alkalimetalle ausgewertet, für die ΔH-Messungen vorlagen [9]. Die Ergebnisse sind in den Fig. 11—16 wiedergegeben.

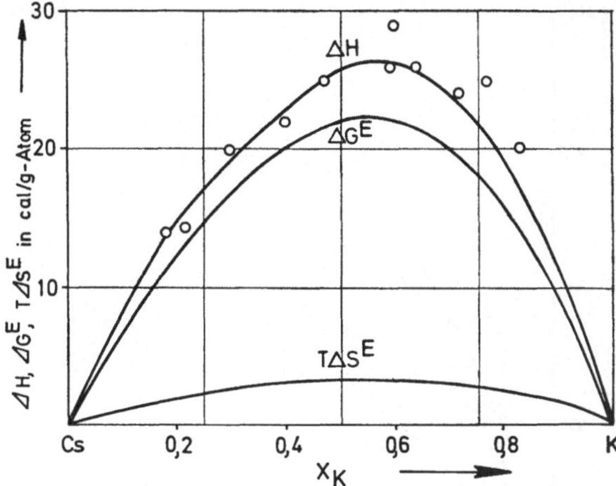

Fig. 14. Überschußfunktionen des Systems K-Cs bei 111 °C mit $\delta_E = 0{,}003$

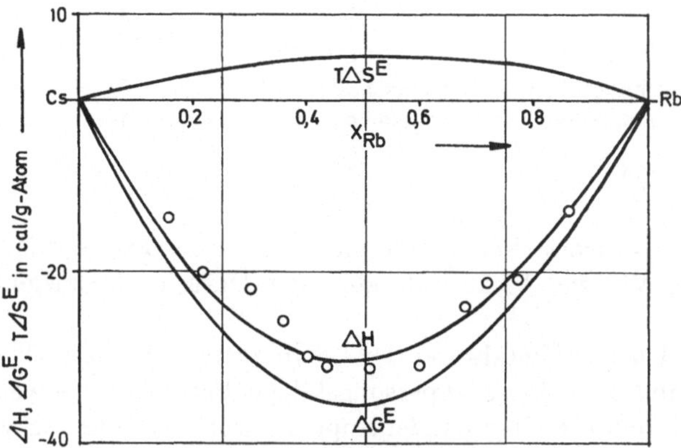

Fig. 15. Überschußfunktionen des Systems Rb-Cs bei 111 °C mit $\delta_E = 0{,}007$

Diskussion

Die von der „Significant-Structure"-Theorie gelieferte Konzentrationsabhängigkeit der Mischungsenthalpie stimmt mit den vorliegenden experimentellen ΔH-Werten zufriedenstellend überein. Am Beispiel des Systems Na-Rb deutet sich an, daß große Asymmetrien im Gang der ΔH-x-Kurve mit unserem Ansatz für E_S nicht wiedergegeben werden können. Da die beim Mischungsvorgang auftretenden Energieänderungen in der Mischungsenthalpie angezeigt werden, wurde von der additiv zusammengesetzten Sublimationsenergie das experimentell ermittelte ΔH abgezogen. Mit diesem Ansatz für E_S erfolgte über Gl. (4) und (5) die Bestimmung der Zustandsfunktionen. Das hiermit gewonnene Resultat für die Mischungsenthalpie stimmt mit den experimentellen Werten überein (gestrichelt gezeichneter Verlauf in Fig. 16). Dies ist ein Beweis dafür, daß auf diesem Wege eine wirklichkeitsnahe Zustandsfunktion gewonnen wird. Dieser Tatbestand ist naturgemäß dann von Bedeutung, wenn man mit Hilfe der „Signi-

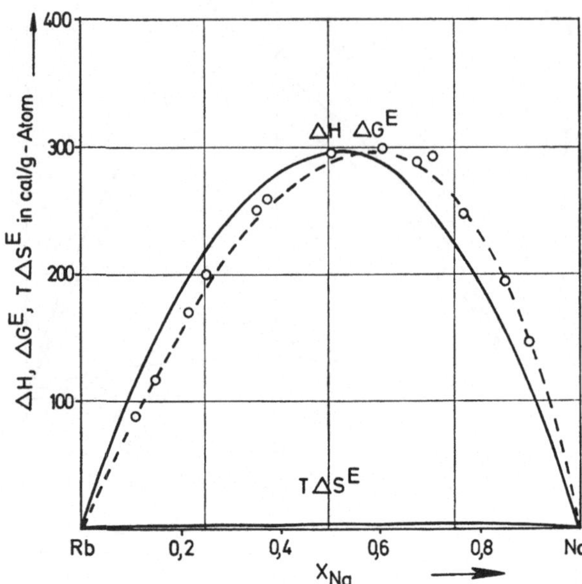

Fig. 16. —— Überschußgrößen des Systems Na-Rb bei 111 °C mit $\delta_E = 0{,}018$; ——— mit experimentellem ΔH berechnet; ooo experimentelle Werte [9]

ficant-Structure"-Theorie weitere thermodynamische Größen wie Kompressibilität, spezifische Wärme, Viskosität oder den Diffusionskoeffizienten berechnen will.

Wie schon bei der Einführung der „Significant-Structure"-Theorie erwähnt, kann man n und a auch als anpassungsfähige Parameter verwenden. Es kann gezeigt werden, daß die Überschußentropien durch Veränderung von n, a und T beeinflußt werden, nicht aber ΔH. Die Variation des Ansatzes für E_s Gl. (11) wirkt sich demgegenüber praktisch nur auf ΔH aus. Dieser Tatbestand erweist die innere Konsistenz des Modells, da n und a die Abhängigkeit von Koordinationszahl und Überschußvolumen wiedergeben. Um unsere Rechnungen zu überprüfen, verwandten wir neben den in Tabelle 5 aufgeführten Werten für n und a die von Vilcu und Misdolea [10] und von Hsu et al. [7] benutzten Werte. Die genannten Autoren erhielten diese Werte einmal durch Anpassung an die Schmelzentropie, zum anderen durch Anpassung an andere thermodynamische Größen reiner Metalle am Schmelzpunkt. Wir erhielten jeweils für gleiche δ_E gleiche ΔH, nur ΔS^E und damit ΔG^E veränderten sich etwas. Wir können jedoch als gesichert ansehen, daß, wie Fig. 8 bis 16 zeigen, die Überschußentropien der flüssigen Alkali- und Erdalkalimetall-Legierungen gering sind.

Verschiedentlich wird versucht, neben einer modellmäßigen Behandlung der thermodynamischen Eigenschaften flüssiger Legierungen die Mischungsfunktionen quantenmechanisch zu erschließen. Eine Möglichkeit besteht in der Anwendung der Pseudopotentialtheorie. Dabei kann von der Bestimmung der Bindungsenergie flüssiger Metalle ausgegangen werden. Unter Verwendung eines von Heine und Abarenkov [11] eingeführten Modellpotentials in einer von Appapilai

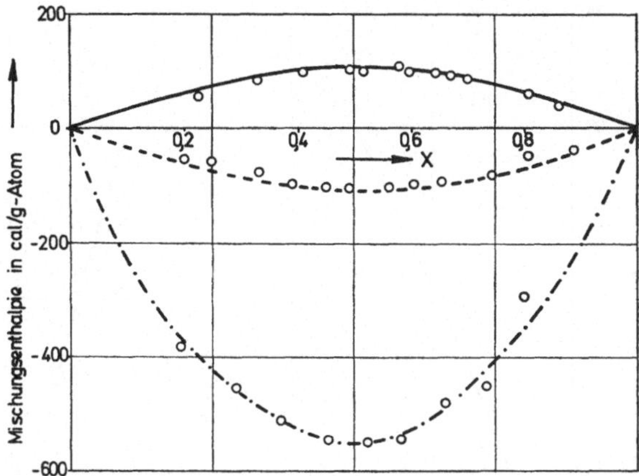

Fig. 17. Mischungsenthalpie für das System —— Ca-Sr, — — — Ba-Sr, –·—·– Ba-Ca

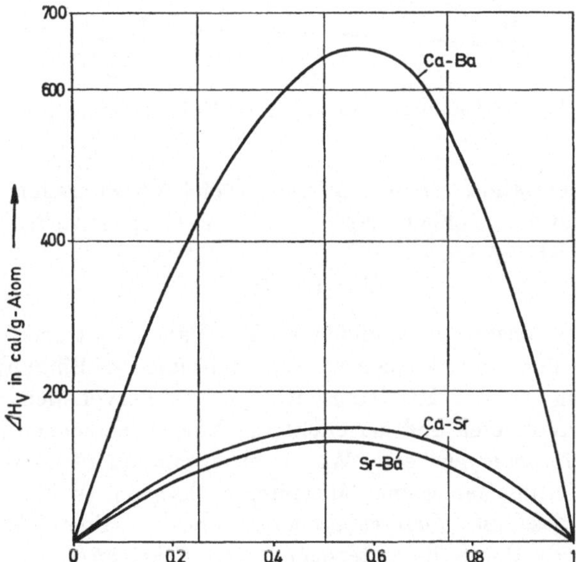

Fig. 18. Verzerrungsanteil der Mischungsenthalpie

und Heine [12] verbesserten Form erhält man für die vom Volumen abhängende
Energie pro Atom folgende Beziehung:

$$U = \frac{1{,}105\,z^{*5/3}}{r^2} - \frac{0{,}9\,z^{*2}}{r} + \frac{1{,}5\,z^{*2}\,R_{\mathrm{M}}^2}{r^3} - \frac{A_0\,z^*\,R_{\mathrm{M}}^3}{r^3} - \frac{0{,}458\,z^{*4/3}}{r}$$
$$- \,0{,}0575\,z^*$$

z^* = effektive Valenz,

R_{M} = Radius des Atomrumpfes,

A_0 = Potentialparameter.

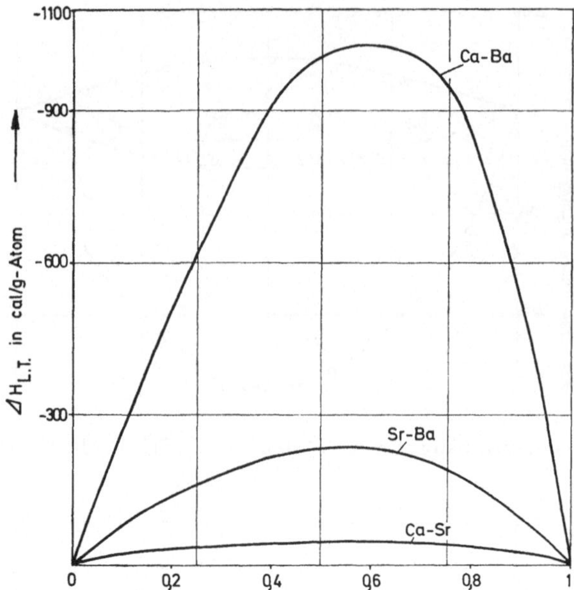

Fig. 19. Ladungstransferanteil der Mischungsenthalpie

Durch Differentiation erhält man den Gleichgewichtsradius R_0, der dem mittleren Abstand der Atome entspricht, und die Gleichgewichtsenergie U_0. Die Bindungsenergie ist

$$B = I - U_0 \,,$$

wobei I die Ionisationsenergie angibt, um alle z Valenzelektronen des Ions abzulösen. Mit dieser Beziehung können wir die Änderung der Bindungsenergie beim Mischungsvorgang — die Mischungsenthalpie —, hervorrufen durch Unterschiede der Atomvolumina und der effektiven Valenz, berechnen [5].

Die in Fig. 17 wiedergegebenen Werte zeigen eine quantitative Übereinstimmung mit den von uns gemessenen Mischungsenthalpien. In Fig. 18 und 19 sind für diese Ergebnisse die Verzerrungsenergie und die Energieänderung durch Ladungstransfer, die durch die Unterschiede der effektiven Valenz hervorgerufen wird, getrennt aufgeführt. Es sei bemerkt, daß die hier dargelegte Kombination von statistisch-thermodynamischer und quantenmechanischer Behandlung der Mischungseigenschaften auch allgemein sich als fruchtbare Möglichkeit zur umfassenden Erschließung der thermodynamischen Eigenschaften flüssiger Legierungen erweisen dürfte.

Dank. Dem Landesamt für Forschung des Landes Nordrhein-Westfalen, der Deutschen Forschungsgemeinschaft sowie dem Fonds der Chemischen Industrie danken wir für die Bereitstellung von Forschungsmitteln.

Literatur

1. Hultgren, R., Orr, R. L., Anderson, P. D., Kelley, K. K.: Selected Values of Thermodynamic Properties of Metals and Alloys. New York: John Wiley 1963
2. Wittig, F. E., Müller, E.: Z. Metallkde. **51**, 226 (1960)

3. Ehrlich, K.: Dissertation, Naturwissenschaftliche Fakultät der Ludwig-Maximilian-Universität in München (1965)
4. Hansen, M., Anderko, K.: Constitution of Binary Alloys. New York: McGraw-Hill 1958
5. Sommer, F.: Acta Metallurgica **21**, 1289 (1973)
6. Eyring, H.: Significant Liquid Structures. New York: John Wiley 1969
7. Hsu, Chen C., Macknight, Allen K., Eyring, Henry: J. phys. Chem. **76**, 1612 (1972)
8. Kubaschewski, O., Evans, E. L.: Metallurgical Thermo-Chemistry (Pergamon Press, New York 1958). — Culpin, M. F.: Proc. Phys. Soc. **70**, 1669 (1957)
9. Yohokama, T., Kleppa, O.: J. chem. Phys. **40**, 46 (1964)
10. Vilcu, R., Misdolea, C.: J. chem. Phys. **49**, 3179 (1968)
11. Heine, V., Abarenkov, I. V., Phil. Mag. **12**, 529 (1965)
12. Appapilai, M., Heine, V.: Technical Report No. 5, University of Cambridge Cavendish Laboratory (1972)

Prof. Dr. Bruno Predel
Dr. Ferdinand Sommer
Institut für Metallkunde der Universität Stuttgart
und Max-Planck-Institut für Metallforschung
D-7000 Stuttgart
Seestraße 75
Bundesrepublik Deutschland

Phys. cond. Matter 17, 267—291 (1974)

Magnetische und kalorische Eigenschaften von Alkali-Hyperoxid-Kristallen

A. Zumsteg *, M. Ziegler, W. Känzig und M. Bösch

Laboratorium für Festkörperphysik der Eidgenössischen Technischen Hochschule Zürich

Eingegangen am 10. Januar 1974

Magnetic and Caloric Properties of Alkali-Hyperoxide Crystals

The phase transitions of Alkali-Hyperoxide crystals. (NaO_2, KO_2, RbO_2, and CsO_2) grown in liquid ammonia have been investigated by means of the following measurements:
a) magnetic susceptibility,
b) differential magnetic susceptibility as magnetic field,
c) magnetization curve in static and pulsed fields,
d) specific heat.
The anomalies of the specific heat could be correlated with the magnetic properties and structural changes. Several new phase transitions were found. The magnetic behaviour of NaO_2 indicates magnetic order (of as yet unknown nature) at low temperatures. The magnetic and caloric behaviour of KO_2 at low temperatures is compatible with a Néel point at 7 K. A metamagnetic transition can be induced at temperatures below 12 K with fields of about 70 kOe. This transition is connected with structural changes. $RbsO_2$ and CO_2 are probably antiferromagnetic with Néel temperatures of 15 K and 9.6 K, respectively.

Die Phasenumwandlungen von im flüssigen Ammoniak gezüchteten Alkali-Hyperoxid-Kristallen (NaO_2, KO_2, RbO_2 und CsO_2) wurden mit Hilfe der folgenden Messungen untersucht:
a) magnetische Suszeptibilität,
b) differentielle magnetische Suszeptibilität als Funktion des Magnetfeldes,
c) Magnetisierung in statischen und gepulsten Feldern,
d) spezifische Wärme.
Die Anomalien der spezifischen Wärme konnten mit den magnetischen Eigenschaften und strukturellen Änderungen korreliert werden. Mehrere neue Phasenumwandlungen wurden gefunden. Das magnetische Verhalten von NaO_2 weist auf einen geordneten Zustand bei tiefen Temperaturen hin; doch ist dessen Natur noch nicht abgeklärt. Das kalorische und magnetische Verhalten von KO_2 bei tiefen Temperaturen ist kompatibel mit antiferromagnetischer Ordnung unterhalb 7 K. Eine metamagnetische Umwandlung kann induziert werden unterhalb 12 K mit Feldstärken von ca. 70 kOe. Sie ist mit einer Strukturänderung verknüpft. RbO_2 und CsO_2 sind sehr wahrscheinlich antiferromagnetisch mit Néel-Temperaturen von 15 K bzw. 9,6 K.

1. Einleitung

1.1. Magnetische Ordnung in p-Elektronen Systemen

Es gibt eine sehr große Zahl von festen Körpern, in denen ungepaarte d- oder f-Elektronen Anlaß geben zu magnetischen Ordnungserscheinungen. Im Gegen-

---

* *Jetzige Adresse:* Société Suisse pour l'Industrie Horlogère Dépt.: Recherche horlogère électronique, CH-2500 Biel/Schweiz.

18 *

satz dazu kennt man bis heute nur wenige Systeme, bei denen ungepaarte p-Elektronen magnetisch ordnen. Das bekannteste Beispiel dürfte wohl fester molekularer Sauerstoff sein. Der elektronische Grundzustand des O_2-Moleküls ist $^3\Sigma_g^-$, und fester Sauerstoff geht beim Unterschreiten einer Temperatur $T_N = 23{,}7$ K von einer paramagnetischen in eine antiferromagnetische Phase über [1]. Ein weiteres interessantes paramagnetisches Molekül ist das Stickoxid NO. Im Grundzustand $^2\Pi_{1/2}$ kompensieren sich zwar Bahn- und Spinmoment; aber 120 cm^{-1} über dem Grundzustand liegt der $^2\Pi_{3/2}$-Zustand, in welchem Bahn- und Spinmoment parallel sind. Magnetische Ordnung in diesem System ist kaum zu erwarten, umsomehr als sich schon bei der Kondensation des Gases Dimere mit gepaarten Spins bilden [2].

NO_2 hat wohl einen paramagnetischen Grundzustand; aber magnetische Ordnung kann auch hier im kondensierten Zustand nicht auftreten wegen der Bildung von N_2O_4.

Das freie Molekül-Ion O_2^+ hat denselben Grundzustand wie NO. Im festen Körper kann aber das Kristallfeld das Bahnmoment löschen, so daß Spin- und Bahnmoment sich nicht kompensieren. Möglicherweise tritt dies in der Verbindung O_2AsF_6 auf. Magnetische Ordnung ist indessen noch nicht festgestellt worden, wenigstens oberhalb $4{,}2$ K [3].

Das paramagnetische O_2^--Molekül-Ion (Hyperoxid-Ion) tritt als Anion auf in den *Hyperoxiden*, und diese zeigen nun magnetische Ordnungserscheinungen: KO_2 ist mit Sicherheit antiferromagnetisch mit einer Néel-Temperatur von 7 K [39]. RbO_2 und CsO_2 sind sehr wahrscheinlich antiferromagnetisch, und NaO_2 hat eine magnetische Ordnung, deren Natur noch nicht abgeklärt ist. Andere Hyperoxide, wie $Ca(O_2)_2$ und $N(CH_3)_4O_2$ [4] existieren auch; doch sind die bisher hergestellten Präparate noch nicht rein genug, als daß man aus den magnetischen Messungen definitive Schlüsse ziehen könnte.

Weitere Beispiele für magnetische Ordnung in p-Elektronen-Systemen könnte man unter den stabilen freien Radikalen der organischen Chemie suchen. Meist handelt es sich um ein ungepaartes p-Elektron, dessen Bahnmoment weitgehend gelöscht ist, so daß fast reiner Spinmagnetismus resultiert. Das wohlbekannte 1,1-diphenyl-2-pikrylhydrazyl (DPPH) zeigt bis hinunter auf $1{,}2$ K keine magnetische Ordnung. Hingegen scheint das freie Radikal p-Cl-BDPA unterhalb $3{,}25$ K antiferromagnetische Fernordnung zu zeigen [5].

Im in letzter Zeit aktuell gewordenen N-methylphenazin-tetracyanoquinodimethan (NMP-TCNQ) ist ein ungepaartes Elektron am ebenen $(TCNQ)^-$-Molekül-Ion lokalisiert. Der Nachweis einer langreichweitigen magnetischen Ordnung in dieser Verbindung, die stark eindimensionalen Charakter hat, ist noch nicht erbracht [6].

1.2. Die Alkali-Hyperoxide

Die Alkali-Hyperoxide sind neben dem festen molekularen Sauerstoff wohl die einfachsten Verbindungen, in welchen magnetische Ordnung von p-Elektronen beobachtet wurde. Trotzdem sind sie bis anhin nur wenig untersucht worden, was nicht zuletzt mit den Schwierigkeiten der Herstellung definierter Präparate zusammenhängen dürfte. Das Interesse unserer Forschungsgruppe an diesen Verbindungen wurde geweckt durch die spektroskopischen Arbeiten über das

O_2^--Molekül-Ion, welches als Verunreinigung in Alkalihalogeniden auftritt, wo es ein Halogen-Ion ersetzen kann. Über die elektronische Struktur und den Einfluß des Kristallfeldes ist man gut informiert [7]: Das ungepaarte p-Elektron ist im Grundzustand in einem π-Orbital. Die Bindung an die umgebenden Alkali-Ionen hat einen kovalenten Anteil, der sich in der Superhyperfeinstruktur und dem g-Faktor äußert. Die magnetische Wechselwirkung zwischen zwei benachbarten O_2^--Molekül-Ionen in stark dotierten Alkali-Halogeniden ist ebenfalls im Detail studiert worden [8]. Je nach gegenseitiger Anordnung ist der Austausch vorwiegend ferromagnetisch oder antiferromagnetisch. Die Austauschwechselwirkung entspricht in den bisher untersuchten Fällen einer Temperatur von der Größenordnung 1 K und darunter.

Tabelle 1. Strukturen und Umwandlungen der Alkalihyperoxide
Zahlen in [ ] beziehen sich auf das Literaturverzeichnis, * bedeutet vorliegende Arbeit, + mit Differentialthermoanalyse an Pulver festgestellt.

Phase	NaO_2	KO_2	RbO_2	CsO_2
I	NaCl, kubisch; O_2^- gehinderte Rotation [26—28]	Wie NaO_2 I [28, 38]	Wie NaO_2 I ?	Wie NaO_2 I [42]
	231 K↑ 1. Ordn. [10, *] 230 K↓	395 K ? [+, *]	420 K ? [+, *]	378 K ? [+, *]
II	Pyrit, kubisch; O_2^--Molekülachsen parallel verschiedene [111]-Richtungen [27, 28]	CaC_2, tetragonal O_2^--Molekülachse parallel [001] [32, 34, 35, 37]	Wie KO_2 II [34, 35, 42]	Wie KO_2 II [34, 35, 42]
	207 K↑ 1. Ordn. [10, *] 200 K↓	231 K [10, *]	194 K [11, 12, *]	190 K [*]
III	Markasit (Verzwillingung) [27, 29—31]	?	?	?
	43,3 K 1. Ordn. [*]	196,6 K 1. Ordn. [10, 33, 36, *]	15,1 K Néel-Punkt ? [11, 12, *]	9,6 K Néel-Punkt ? [*]
IV	Markasit mit magnetischer Ordnung ? [*]	?	Antiferromag. ? [40, *]	Antiferromag. ? [*]
		12,1 K↑ 1. Ordn. 10,6 K↓ [39, *]	14,7 K Néel-Punkt ? [11, 12, *]	
V		?	Antiferromag. ? [40, *]	—
		7,1 K Néel-Punkt [39, *]		
VI		Antiferromag. [39]	—	—

In der vorliegenden Arbeit werden unsere magnetischen und kalorischen Messungen an NaO_2, KO_2, RbO_2 und CsO_2 zusammengefaßt. Frühere Untersuchungen sind fast ausschließlich an Pulverpräparaten durchgeführt worden, wobei bemerkt wurde, daß insbesondere die magnetischen Eigenschaften stark

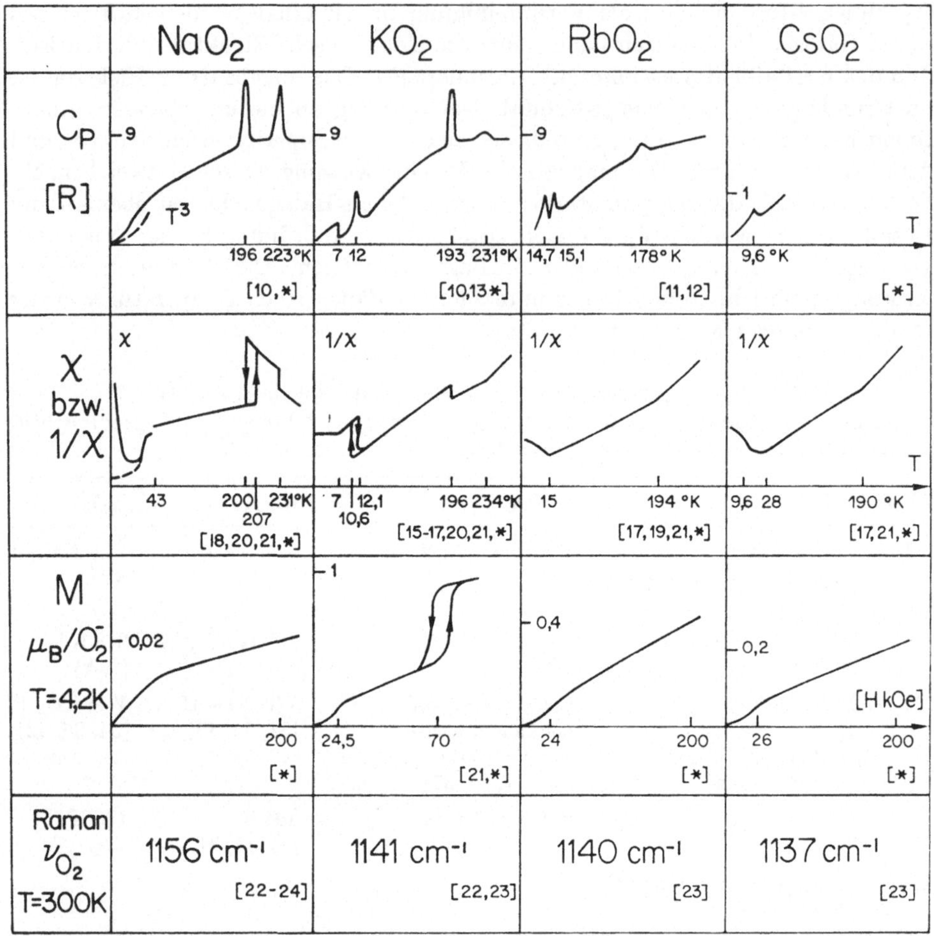

Fig. 1. Kalorische, magnetische und spektroskopische Eigenschaften der Alkali-Hyperoxide. Zahlen in [ ] beziehen sich auf das Literaturverzeichnis, * bedeutet vorliegende Arbeit

vom Herstellungsverfahren abhängen. Die Chemie der Darstellung von Pulvern ist in den Arbeiten von Vannerberg und von Volnov enthalten [9]. Unsere Messungen wurden an Kristallen durchgeführt, die durch Rekristallisation von Hyperoxidpulvern in flüssigem Ammoniak gewonnen wurden. Es wurden NaO$_2$-Einkristalle mit Kantenlängen bis zu 10 mm gezüchtet. Bei KO$_2$, RbO$_2$ und CsO$_2$ erhält man selten Einkristalle, deren Größe 1 mm übersteigt. Die Kristallzuchttechnik soll in einer späteren Publikation beschrieben werden.

Es ist eine Eigenart der Alkali-Hyperoxide, daß sie mehrere Phasenumwandlungen durchlaufen, bei denen sich insbesondere die Orientierung der O$_2^-$-Molekül-Ionen ändert. Einige dieser Umwandlungen können vermutlich mit dem Jahn-Teller-Effekt erklärt werden. Bei den Umwandlungen bei tiefen Temperaturen scheinen auch magnetische Wechselwirkungen eine wichtige Rolle zu spielen.

Die Unterscheidung zwischen „kristallographischen" und „magnetischen" Phasenumwandlungen dürfte in diesen p-Elektronen-Systemen in einigen Fällen sinnlos sein.

Die Umwandlungstemperaturen wurden anhand der Anomalien der spezifischen Wärme und der magnetischen Eigenschaften bestimmt. Für jede Substanz wurden die Phasen in der Reihenfolge ihres Auftretens mit sinkender Temperatur bezeichnet mit I, II, III .... Fig. 1 gibt eine Übersicht über die kalorischen, magnetischen und einige spektroskopische Eigenschaften, wie sie sich aus den Messungen anderer Autoren und den eigenen Messungen ergeben hat. In Tabelle 1 sind die Strukturdaten und die Umwandlungstemperaturen aufgeführt. Tabelle 2 enthält die Parameter des Curie-Weiss-Gesetzes, welches die magnetische Suszeptibilität der verschiedenen Phasen beschreibt.

2. NaO_2

2.1. Suszeptibilität

Die Suszeptibilität wurde nach der Faraday-Methode bei einem Feld von 8 kG gemessen. Die Apparatur wurde mit flüssigem Hg ($\chi = -\,0{,}1665 \cdot 10^{-6}$ cm³/g) geeicht. Die Auflösung der Kraftmessung war immer besser als $2 \cdot 10^{-3}$. Die Temperatur wurde mit einem Cu-Konstantan-Thermoelement gemessen. Die Eichtabelle von Powell et al. [43] wurde bei 89,66 K und zwischen 14 K und 20 K kontrolliert. Bei 89,66 K betrug die Abweichung 0,3 K und unterhalb 20 K ist sie kleiner als 0,1 K.

Die Proben bestehen aus vielen kleinen Kristallen mit einer typischen Kantenlänge von 0,5 mm und wiegen typisch 50 mg. Da die Kristalle lose in einem Quarzröhrchen liegen, ist die Packungsdichte inhomogen und verändert sich beim Ein- und Ausbau des Präparates. Verschiedene Messungen an der gleichen Probe haben deshalb Unterschiede bis zu 1,5% gezeigt. Dies ist die wichtigste Beeinträchtigung der Reproduzierbarkeit. Eine massenspektrometrische Analyse der Kristalle hat folgende Verunreinigung gezeigt: C^{12}: 10^4 ppm (Reaktion mit CO_2 beim Einbau des Präparates in das Massenspektrometer), H^1, Cl^{35} und K^{39} je 10^3 ppm. Übrige Elemente sind mit 300 ppm oder weniger vorhanden.

Die gemessene Suszeptibilität ist in Fig. 2 dargestellt. Die Phasenumwandlungen manifestieren sich durch sprunghafte Änderung der Magnetisierung als Funktion der Temperatur. Es handelt sich somit um Phasenübergänge 1. Ordnung. Bei den Phasenübergängen I/II und II/III wird eine thermische Hysterese von 1 K bzw. 7 K beobachtet. Die entsprechenden Anomalien der spezifischen Wärme wurden von Todd gemessen [10]. Die in dieser Arbeit gefundenen Umwandlungstemperaturen liegen etwa um 8 K höher, was wahrscheinlich darauf zurückzuführen ist, daß Todd feines kommerzielles Pulver und nicht grobkörniges rekristallisiertes Material verwendete. Beim Übergang III/IV läßt sich eine Hysterese nicht eindeutig feststellen; denn ein kleines Nachhinken der Probentemperatur (0,2 K) ist nicht ausgeschlossen.

In den Phasen I und II gehorcht die Suszeptibilität einem Curie-Weiss-Gesetz $\chi = C/(T - \theta)$.

Eine effektive Magnetonenzahl μ_{eff} wurde aus den Messungen gewonnen mit Hilfe der Beziehung $C = N\mu_B^2\,\mu_{\mathrm{eff}}^2/3\,\mathrm{k}$. Die angepaßten Werte von μ_{eff} und θ

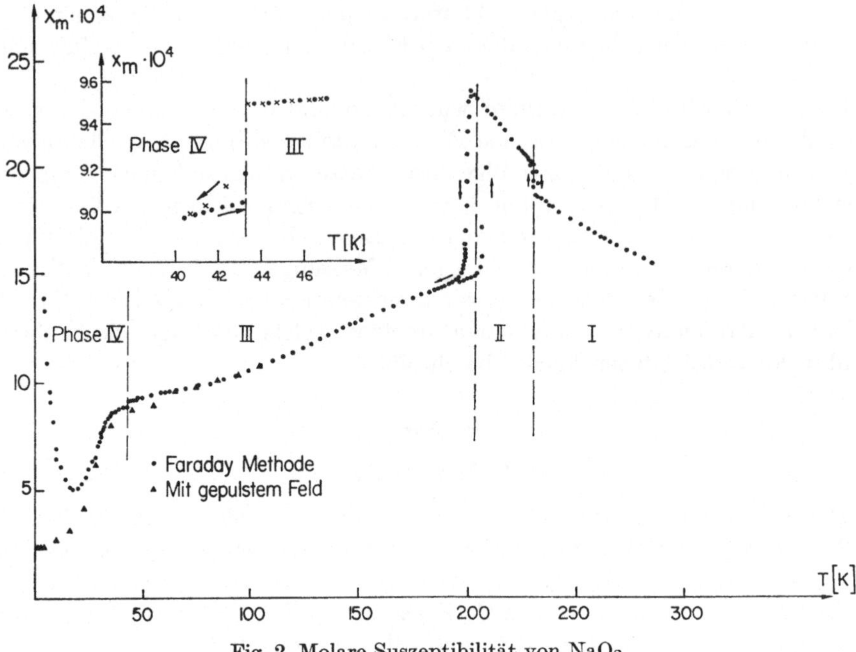

Fig. 2. Molare Suszeptibilität von NaO₂

betragen 1,97 und -31 K bzw. 1,79 und $+33$ K. Die Abnahme von μ_{eff} beim Übergang von der ungeordneten in die geordnete Pyritstruktur ist *unerwartet*. Um dies einzusehen, muß der g-Faktor des O_2^--Molekül-Ions im Kristallfeld näher betrachtet werden [7]. Der minimale Wert der effektiven Magnetonenzahl wird erhalten bei vollständig gelöschtem Bahnmoment, nämlich $\mu_{\text{eff}} = \sqrt{3}$. Der maximale Wert wird erhalten, wenn die Kristallfeldaufspaltung der Niveaus $^2\Pi_g$ in $^2\Pi_g^x$ und $^2\Pi_g^y$ gegen null geht. Der g-Tensor wird dann axial mit den Werten $g_{xx} = g_{yy} = 0$, $g_{zz} = 4$. Die Mittelung über die Molekülorientierungen ergibt dann $\mu_{\text{eff}} = 2$. In der Phase I ist die lokale Symmetrie des Kristallfeldes wegen der Unordnung der Orientierungen der O_2^--Molekülachsen genügend tief, daß eine teilweise Löschung des Bahnmomentes zu erwarten ist.

In der Phase II sind die Molekülachsen aber so geordnet, daß das Kristallfeld trigonal ist. Die Aufspaltung der Niveaus $^2\Pi_g$ sollte verschwinden. Beim Phasenübergang I → II erwartet man also eine Zunahme des effektiven magnetischen Momentes.

Die beobachtete Abnahme könnte damit zusammenhängen, daß die geordnete Pyritstruktur nur die *mittlere* Struktur ist. Tatsächlich erwartet man wegen des Jahn-Teller-Effektes eine Erniedrigung der lokalen Symmetrie [41].

Das Auftreten positiver und negativer Werte von θ deutet an, daß in NaO₂ der Austausch vorwiegend ferro- oder antiferromagnetisch sein kann, je nach Struktur. Dies ist nicht erstaunlich; denn die magnetischen Wechselwirkungen von Paaren von O_2^- in Alkali-Halogeniden sind ferro- oder antiferromagnetisch, je nach gegenseitiger Anordnung [8].

Die Temperaturabhängigkeit der Suszeptibilität und die Magnetisierungs-
kurven in der Phase III weisen eher auf Antiferromagnetismus hin. Die Neutronen-
streuungsexperimente an Pulver aus rekristallisiertem NaO_2 und an einem Einkristall können diese Hypothese nicht ausschließen [30, 31]. Es ist auch möglich,
daß das magnetische Verhalten durch kurzreichweitige Ordnung bestimmt wird.
Ein Indiz dafür wäre die Abnahme der Intensität der ESR-Absorption beim
Abkühlen.

Die Abnahme der Suszeptibilität in der Phase IV unterhalb 43,3 K könnte ein
Hinweis für die Existenz einer langreichweitigen Ordnung sein. Unterhalb 18 K
nimmt die Suszeptibilität, gemessen mit der Faraday-Methode, wieder zu und
zwar etwa wie $1/T$. Es scheint ein paramagnetischer Anteil vorhanden zu sein.
Wenn man paramagnetische Ionen mit einem magnetischen Moment von 1 μ_B
annimmt ($g = 2$, $J = 1/2$), ergibt sich aus der Curiekonstanten eine Konzentration, die 1% der totalen O_2^--Konzentration beträgt. (Im kommerziellen Pulver
ist dieser paramagnetische Anteil viel größer). Durch mehrfache Rekristallisation
konnte der paramagnetische Anstieg nicht weiter vermindert werden. Dieser
paramagnetische Anteil läßt sich in hohen Feldern sättigen. Die verbleibende
Suszeptibilität ist in Fig. 2 eingezeichnet. Bemerkenswert ist die sehr kleine Suszeptibilität für $T \to 0$.

2.2. Magnetisierung

Die Messungen der Magnetisierung in gepulsten Feldern wurden mit der Apparatur von Herrn Dr. H. Rohrer im IBM-Forschungslaboratorium Rüschlikon,
Schweiz, vorgenommen. Das hohe magnetische Feld wird durch Entladen einer
Kondensatorbatterie über eine im flüssigen He eingetauchte Feldspule erzeugt.
Der einer Sinus-Halbwelle entsprechende Feldpuls dauert 4 msec, und die maximale Feldstärke beträgt 225 kOe zwischen 1,3 K und 4,2 K bzw. 150 kOe oberhalb 4,2 K. In der Feldspule befinden sich 2 Pick-up-Spulen nebeneinander, von
denen die eine die Probe und die andere ein Thermoelement enthält. Die elektronische Integration der Differenz der Signale beider Pick-up-Spulen ist ein Maß
für die Magnetisierung. Geeicht wurde mit reinem Eisen [44—46]. Für sämtliche
Temperaturen ist die Magnetisierung eine stetige Funktion des Feldes und kann
wie folgt beschrieben werden:

$$M = \chi(T) \cdot H + M_0 B_J(x), \qquad x = g\,\mu_\mathrm{B}\,JH/kT,$$

wobei M_0 eine Konstante und $B_J(x)$ die Brillouinfunktion bedeuten. Der erste
Term beschreibt die nicht sättigbare Magnetisierung. χ ist innerhalb der Meßgenauigkeit (10% von χ) unabhängig von H bis zu den maximalen Feldern. Die
entsprechende Suszeptibilität ist in Fig. 2 eingetragen. Der zweite Term entspricht
dem paramagnetischen Anteil. Die beschränkte Meßgenauigkeit erlaubt es nicht,
den Drehimpuls J aus dem Verlauf der Magnetisierung als Funktion des Feldes
zu bestimmen. Wenn man ein magnetisches Moment von 1 μ_B ($g = 2$, $J = 1/2$)
annimmt, ergibt sich aus dem experimentellen Wert von M_0 eine Konzentration
von etwa 1% in Übereinstimmung mit der Faraday-Methode und den Spin-
Resonanz-Experimenten (cf. Abschnitt 2.4).

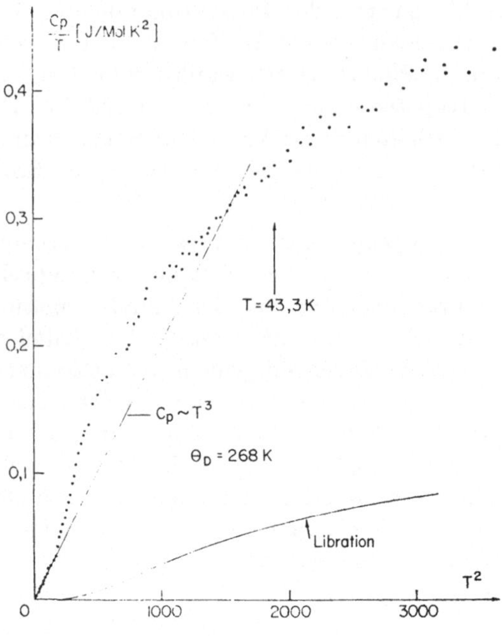

Fig. 3. Molwärme von NaO_2

2.3. Spezifische Wärme

Die spezifische Wärme wurde zwischen 1,8 K und 70 K mit einem adiabatischen Kalorimeter gemessen [14]. Die Kristalle wurden in einer Cu-Patrone bei Zimmertemperatur eingeschlossen. Zur Herstellung des Wärmekontaktes wurde 70 Torr He-Gas eingefüllt. Der Wärmekontakt darf nicht durch Zusammenpressen der Kristalle erzwungen werden, da dadurch das Gitter dermaßen beschädigt wird, daß ganz andere Meßwerte resultieren können. (Bei KO_2 verschwindet z.B. die Anomalie der spezifischen Wärme bei 12 K.)

Die Meßergebnisse sind in Fig. 3 dargestellt. Es sind keine ausgeprägten Anomalien vorhanden. Zwischen 1,8 K und 8 K gehorcht die spezifische Wärme einem T^3-Gesetz. Berücksichtigt man nur die translatorischen Freiheitsgrade von Na^+ und O_2^-, erhält man eine Debye-Temperatur von 268 K. In diesem Temperaturintervall sind die Vibrations- und Librationsfreiheitsgrade des O_2^- noch eingefroren; denn die Streckfrequenz des Molekül-Ions beträgt 1164 cm^{-1} und die Librationsfrequenzen sind von der Größenordnung 100 cm^{-1} [23, 24].

Ab 8 K wächst die spezifische Wärme schneller an als einem T^3-Gesetz entspricht. Zu welchen Anteilen diese Abweichung durch translatorische Gitterschwingungen, Auftauen der libratorischen Freiheitsgrade und durch Änderung der magnetischen Ordnung bedingt ist, entzieht sich unserer Kenntnis. Die auf Grund der spektroskopisch bestimmten Librationsfrequenzen (126 cm^{-1} und 217 cm^{-1}) berechnete Librationswärme ist in Fig. 3 eingezeichnet. Der Suszeptibilitätssprung bei 43,3 K manifestiert sich nicht messbar in den kalorischen

Messungen. Als obere Grenze einer Diskontinuität der spezifischen Wärme oder einer latenten Wärme kann 0,5 J/Mol K bzw. 1 J/Mol abgeschätzt werden.

2.4. Elektronen-Spin-Resonanz (ESR)

Im Jahre 1955 haben Benett, Ingram und Schonland [48, 49] ESR-Messungen ausgeführt an polykristallinem NaO_2, das als Verunreinigung in Na_2O_2 vorkommt, und an KO_2, das durch Oxydation des erhitzten Metalls gewonnen wurde. Bei 90 K war die Absorption stark asymmetrisch und wurde in beiden Alkali-Hyperoxiden interpretiert mit einem g-Tensor mit den Werten $g_\perp = 2,002$, $g_\parallel = 2,175$. Dieselbe Absorption beobachten wir an kommerziellem NaO_2-Pulver.

Das Verhalten des *rekristallisierten* NaO_2 ist aber völlig anders. Sowohl eine polykristalline Probe als auch ein Einkristall zeigen bei 77 K eine nahezu symmetrische Linie, deren g-Faktor 2,11 beträgt. Die Linienbreite (Abstand zwischen den Extrema der Ableitung der Absorptionskurve) ist beim Einkristall von der Orientierung des Magnetfeldes abhängig. Bei 77 K beträgt die minimale Linienbreite für H_0 in einer (100)-Ebene 920 Gauß im Q-Band und 560 Gauß im X-Band. Diese Linie ist somit inhomogen verbreitert.

An einem Einkristall wurde weiter der g-Faktor, die Linienbreite und die absolute Intensität als Funktion der Temperatur gemessen. Als Maß für die Intensität wird in Fig. 4 der Bruchteil x der Anzahl Spins in der Probe, die zur ESR-Absorption beitragen, angegeben. Die Bestimmung der Anzahl Spins, die zur ESR-Absorption beitragen, erfolgte durch direkten Vergleich mit einer geeichten DPPH-Probe. Anisotropie-Effekte der Größenordnung $\Delta g/g^2$ wurden nicht korrigiert [50]. Die Linie, die bei 77 K sehr glatt ist, weist unterhalb 30 K Struktur auf, und unterhalb 12 K kann man mindestens 20 aufgelöste Linien unterscheiden, deren g-Faktoren zwischen 1,98 und 2,32 liegen. Diese Werte sind für O_2^--Molekül-Ionen [7] und O^--Ionen [62] in Alkalihalogeniden typisch. Die Summe der Intensitäten aller dieser aufgelösten Linien bleibt unterhalb 20 K konstant und entspricht dem Wert $x = (1 \pm 0,3) \cdot 10^{-2}$. Dieser Wert stimmt überein mit dem aus Suszeptibilität und Magnetisierung bestimmten paramagnetischen Anteil.

Die Interpretation des ESR-Spektrums ist durch die Verzwillingung des NaO_2 in der Markasitphase erschwert. Röntgenographische Untersuchungen haben gezeigt, daß unterhalb 200 K 12 verschiedene Orientierungen der Zwillinge vorkommen [31].

Sehr wahrscheinlich ist die Zahl der Linien im ESR-Spektrum 24. Die Verschiebungen der Linien, die beobachtet werden bei einer Drehung des Kristalls im Felde H_0, zeigen tatsächlich, daß sie von paramagnetischen Ionen herrühren, deren g-Tensoren eine feste Beziehung zur Orientierung der Zwillinge haben, und zwar kommen in jedem Zwilling die Ionen in zwei verschiedenen Orientierungen vor. Da die O_2^--Molekül-Ionen in der Markasitstruktur in zwei verschiedenen Orientierungen vorkommen, ist es naheliegend anzunehmen, daß das paramagnetische Ion, welches Anlaß gibt zum beobachteten ESR-Spektrum, den Platz eines O_2^--Molekül-Ions einnimmt. Wahrscheinlich handelt es sich um O^-, d.h. die Zusammensetzung der im flüssigen Ammoniak gezüchteten NaO_2-Einkristalle könnte um ca. 1% von der stöchiometrischen Zusammensetzung abweichen.

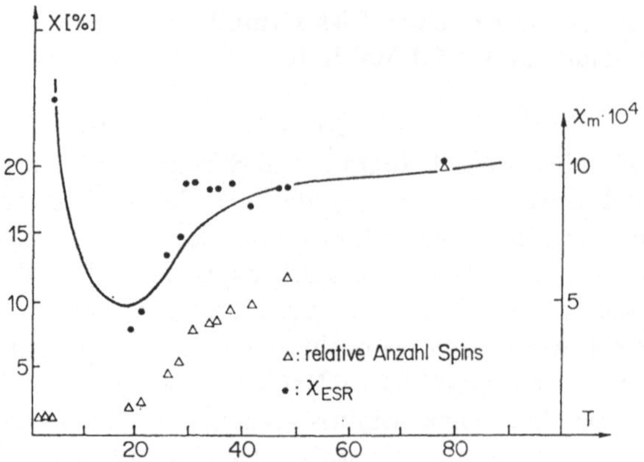

Fig. 4. Intensität der ESR Absorption von NaO_2 und entsprechende Suszeptibilität. Die ausgezogene Kurve entspricht den Messungen nach der Faraday-Methode (cf. Fig. 2)

Beim Erwärmen über 20 K nimmt die Anzahl der Spins, die zur ESR-Absorption beitragen, sehr stark zu (Fig. 4). Die entsprechende Suszeptibilität χ_{ESR} stimmt innerhalb 20% mit der nach Faraday bestimmten Suszeptibilität überein. Diese starke Zunahme hat nun mit der Abweichung von der stöchiometrischen Zusammensetzung nichts zu tun, sondern sie ist den O_2^--Molekül-Ionen im NaO_2-Gitter zuzuschreiben. Da schon eine kurzreichweitige antiferromagnetische Ordnung die ESR-Suszeptibilität stark reduzieren kann [8], führt Fig. 4 zur Vermutung, daß der Anstieg der Suszeptibilität mit steigender Temperatur dem Aufbrechen einer lokalen antiferromagnetischen Ordnung zugeschrieben werden könnte [51].

Die Linienbreite bleibt zwischen 20 K und 60 K konstant, $\Delta H_0 = 660$ Gauß im Q-Band und steigt dann rasch an. Der g-Faktor bleibt zwischen 20 K und 60 K konstant, $g = 2{,}12$, und scheint dann abzunehmen. Seine Bestimmung bei höherer Temperatur ist infolge der Linienverbreiterung problematisch [52].

3. KO_2

3.1. Suszeptibilität (Faraday-Methode)

Die Proben bestanden aus vielen kleinen Kristallen mit einer typischen Kantenlänge von ca. 0,1 mm. Die Phase I, oberhalb 395 K wurde nicht untersucht. Eine massenspektrometrische Analyse von KO_2-Kristallen hat folgende Verunreinigungen gezeigt: $C^{12}: 2 \cdot 10^3$ ppm, Na^{23}, $Rb^{85} + Rb^{87}$, Cs^{133} je 100 ppm. Übrige Elemente sind mit weniger als 100 ppm vorhanden.

Die reziproke Suszeptibilität ist in Fig. 5 dargestellt. Die Phasenumwandlungen manifestieren sich auf verschiedene Arten. Beim Übergang II/III ändern sich μ_{eff} und θ (cf. Tabelle 2), eventuell ist noch ein kleiner Sprung der Suszeptibilität in der Größenordnung von 0,5% vorhanden. Beim Übergang III/IV findet ein Sprung der Suszeptibilität statt. Beide Übergänge weisen Anomalien der spezifischen Wärme auf, die von Todd [10] an kommerziellem Pulver gemessen wurden. Auch hier besteht eine kleine Diskrepanz in den Umwandlungs-

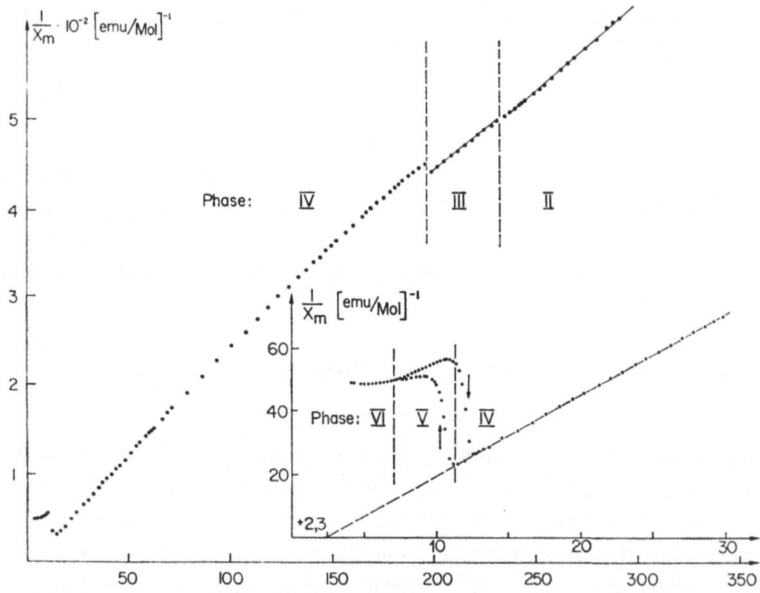

Fig. 5. Inverse molare Suszeptibilität von KO_2

temperaturen, die wahrscheinlich auf die verschiedenen Herstellungsverfahren des Materials zurückzuführen ist. Beim Übergang IV/V ändert sich die Suszeptibilität sprunghaft um einen Faktor 2, und eine thermische Hysterese von 1,6 K wird beobachtet. Der Übergang V/VI bei 7 K entspricht nach Smith *et al.* [39] dem Néel-Punkt. Unterhalb 7 K konnten diese Autoren Überstruktur-Interferenzen in den Neutronenstreuexperimenten beobachten. Bei diesem Übergang ist die Suszeptibilität stetig. Bei den zwei letzten Umwandlungen treten ausgeprägte Anomalien der spezifischen Wärme auf (cf. Fig. 11). Die Suszeptibilität der Phasen II und III kann durch ein Curie-Weiß-Gesetz beschrieben werden. Die angepaßten Werte von μ_{eff} und θ betragen 2,01 und -18 K bzw. 2,10 und -44 K. Die Temperaturabhängigkeit der reziproken Suszeptibilität der Phase IV weicht oberhalb 80 K von einer Geraden ab. Unterhalb 80 K gehorcht die Suszeptibilität wieder einem Curie-Weiß-Gesetz mit den Parametern $\mu_{eff} = 1,76$ und $\theta = +2,3$ K. Die Suszeptibilität der Phase V scheint auch einem Curie-Weiß-Gesetz mit Parametern $\mu_{eff} = 1,93$ und $\theta = -16$ K zu gehorchen. Unterhalb des Néel-Punktes nimmt die Suszeptibilität noch leicht zu und durchläuft ein sehr flaches Maximum bei 5 K. Präparate aus kommerziellem Pulver, oder aus Pulver, das durch Oxydation von Kalium im flüssigen Ammoniak gewonnen wurde, verhalten sich anders bei tiefen Temperaturen.

Die Zunahme des magnetischen Moments beim Übergang II $\rightarrow$ III von der tetragonalen Ca-C_2-Struktur zu einer Struktur mit tieferer Symmetrie [42] ist, ähnlich wie beim Übergang I/II des NaO_2, im Rahmen einer statischen Kristallfeldtheorie unverständlich. Halverson vermutet, daß ein dynamischer Jahn-Teller-Effekt die Röntgendaten von Abrahams *et al.* erklären konnte [41]. Die Werte des effektiven magnetischen Momentes sind unerwartet hoch. Daß sie größer als 2 sind, ist ein weiterer Hinweis für die Unzulänglichkeit einer statischen Kristallfeldtheorie. Nach Ausweis der Massenspektren kann eine Vergrößerung des magnetischen Momentes nicht durch eine chemische Verunreinigung erklärt werden.

Es ist denkbar, daß die Erhöhung der effektiven Magnetonenzahl dadurch zustande kommt, daß sich die magnetisch sehr anisotropen O_2^--Molekül-Ionen unter der Wirkung des Magnetfeldes etwas aus der ursprünglichen Gleichgewichtsorientierung herausdrehen. Allerdings könnte man auch Abweichungen von der Stöchiometrie in Betracht ziehen, z.B. O_2 assoziiert mit einer Kationenleerstelle.

3.2. Differentielle Suszeptibilität und Magnetisierung bei tiefer Temperatur

3.2.1. Apparatur

Die adiabatische differentielle Suszeptibilität wurde mit einem feldmodulierten Differentialmagnetometer gemessen. Die Probe besteht aus vielen kleinen Kristallen und wiegt 20 mg. Die Kristalle befinden sich dicht zusammengelegt (nicht gepreßt) in einem Plexiglasröhrchen. Dem statischen Magnetfeld wird ein kleines Wechselfeld überlagert. Die differentiellen Änderungen der Magnetisierung induzieren in einer Pick-up-Spule Signale, die mit Lock-in-Technik verstärkt werden. Die Kompensation des Signals der Pick-up-Spule ohne Probe erfolgt grob (10^{-3}) mit einer zweiten Pick-up-Spule, und fein $(3 \cdot 10^{-6})$ durch variable Spannungsteilung einer Kompensationsspule [47]. Die Fig. 6 zeigt die Anordnung der Pick-up-Spulen. Mit dieser Anordnung wird eine sehr gute Stabilität des Nullabgleichs für große Änderungen des Magnetfeldes und der Temperatur erreicht. Der Fehler der absoluten Messungen liegt unter 1%. Bei einer Pulverprobe ist die Genauigkeit durch die Homogenität der Packungsdichte begrenzt. Die Modulationsfeldstärke betrug 4 Oe_{eff} und die Modulationsfrequenz 116 Hz. Bei der verwendeten Modulationsfrequenz (116 Hz) mißt man die adiabatische Suszeptibilität. Um zu entscheiden, ob Spins und Gitter im thermodynamischen Gleichgewicht sind, wurden die ESR Daten herbeigezogen. Die Unmöglichkeit, die ESR-Linien mit der verfügbaren Leistung zu sättigen, setzt die obere Grenze der Spin-Gitter-Relaxationszeit bei $T_1 < 10^{-4}$ sec fest. In diesem Fall gilt folgende Beziehung zwischen der adiabatischen differentiellen Suszeptibilität

$$\chi_s = (\partial M / \partial H)_{ds=0}$$

und der isothermen differentiellen Suszeptibilität $\chi_T = (\partial M / \partial H)_{dT=0}$

$$\frac{1}{\chi_s} = \frac{1}{\chi_T} \left[1 + \frac{T}{C_M} \frac{\left(\frac{\partial M}{\partial T} \right)_H^2}{\chi_T} \right]. \tag{1}$$

C_M bedeutet die spezifische Wärme bei konstanter Magnetisierung. Sie dürfte sich nur wenig von der spezifischen Wärme des Gitters unterscheiden.

Im Bereich einer magnetischen Hysterese versagt die Feldmodulationstechnik (cf. Fig. 7 und 9). Um die differentielle Suszeptibilität zu ermitteln, ändert man in einem solchen Fall das Feld monoton in einer Richtung. Bei genügend langsamer Variation des Feldes ist dann die Spannung an der Pick-up-Spule pro-

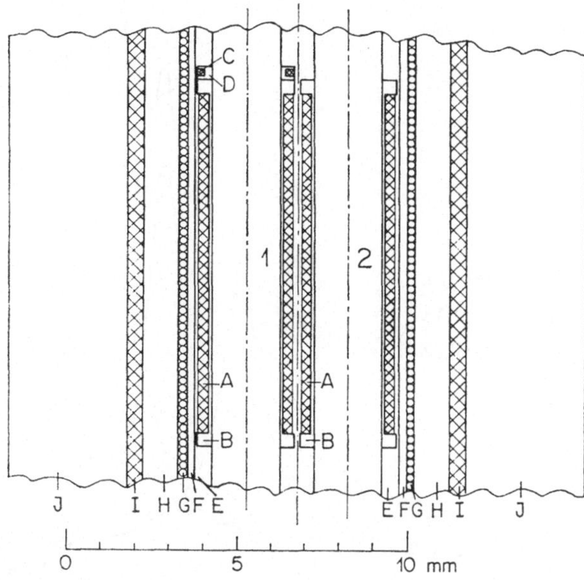

Fig. 6. Pick-up-Spulen mit Heizung und Modulation. A, B: Pick-up-Spulen, 2000 Windungen, Cu $\varnothing$ 0,03 mm; C, D: Kompensationsspule, 4 Windungen, Cu $\varnothing$ 0,03 mm; E: Halter mit eingegossenen Pick-up-Spulen und Kompensationsspule; F, G: Heizung, Wicklung: 1 Lage, bifilar, Konstantan $\varnothing$ 0,1 mm; H, I: Modulationsspule, Länge 50 mm, Wicklung: 4 Lagen, Cu $\varnothing$ 0,1 mm; J: Thermische Isolation. — Alle Spulenkörper, sowie der Halter sind aus Plexiglas. Die Bohrung 1 enthält das Thermometer, Bohrung 2 das Präparat

portional zur isothermen differentiellen Suszeptibilität. Im Experiment wurde das Feld mit einer Geschwindigkeit von 60 Oe/sec geändert. Die induzierte Spannung betrug dann 70 Nanovolt außerhalb der metamagnetischen Umwandlung. Der Nullabgleich ist wegen den unvermeidlichen Thermospannungen nur auf etwa 15 Nanovolt stabil. Bei der metamagnetischen Phasenumwandlung induzieren die sprunghaften Änderungen der Magnetisierung (cf. Abschnitt 3.2.2) Spannungsspitzen, die von einem Gleichspannungs-Verstärker mit Unterbrecher-eingang nicht richtig verarbeitet werden. Diese Fehlerquellen beschränken die Genauigkeit dieser Messungen auf 20%.

3.2.2. Ergebnisse

Die Fig. 7 zeigt die isotherme differentielle Suszeptibilität und die Fig. 8 die entsprechenden Integralkurven. Ganz deutlich kommt eine metamagnetische Phasenumwandlung mit einer großen statischen magnetischen Hysterese zum Ausdruck. Bei den meisten Metamagneten beobachtet man bei langsamer Feld-variation keine Hysterese, z. B. $FeCl_2$, $FeBr_2$, $CoCl_2\,2\,H_2O$, DyAl-Granat [53—59]. In gepulsten Feldern hingegen wird oft eine Hysterese beobachtet, z. B. $FeCl_2$, $CoCl_2\,2\,H_2O$ [54, 57, 58]. Besonders merkwürdig ist, daß die differentielle Suszep-tibilität als Funktion des Feldes noch weit oberhalb der Néel-Temperatur (bei steigender Temperatur bis zu 12 K $= 1,7\ T_N$) ein Maximum durchläuft (Wende-punkt in der Magnetisierungskurve). Ein ähnliches Verhalten wurde dem $FeBr_2$

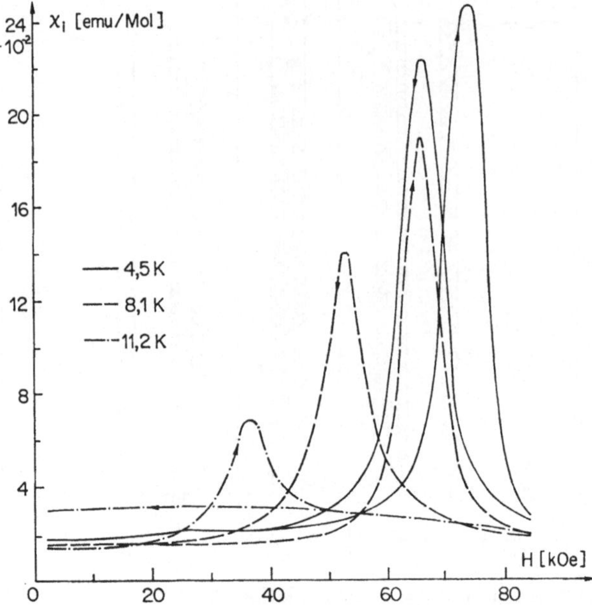

Fig. 7. Molare isotherme differentielle Suszeptibilität von KO_2 als Funktion der magnetischen Feldstärke

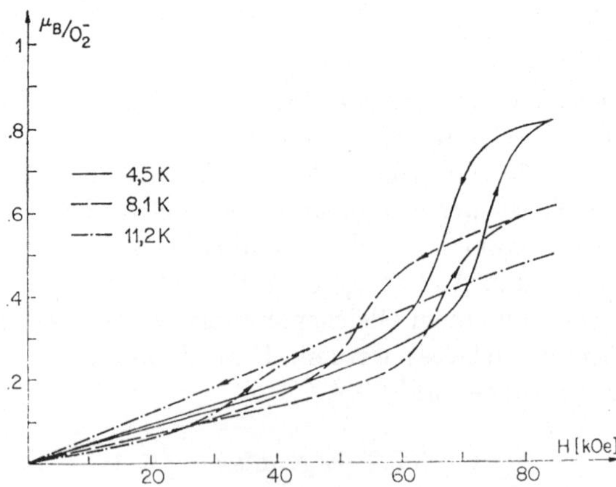

Fig. 8. Magnetisierung von KO_2 als Funktion der magnetischen Feldstärke

ünd dem $FeCl_2$ zugeschrieben, aber beim $FeBr_2$ beruhte es auf einer falschen Festlegung der Néel-Temperatur und beim $FeCl_2$ ist es widerrufen worden [53, 55, 56]. Ein Wendepunkt bei einer Temperatur, die nur einige Prozent oberhalb T_N liegt, wurde bei DyAl-Granat beobachtet. Dieses Verhalten entspräche übrigens der exakten Lösung eines 2-dimensionalen dekorierten Ising-Modells [61].

Im Gebiet der metamagnetischen Umwandlung ist die differentielle Suszeptibilität mit starkem „Rauschen" behaftet, das durch sprunghafte Änderungen

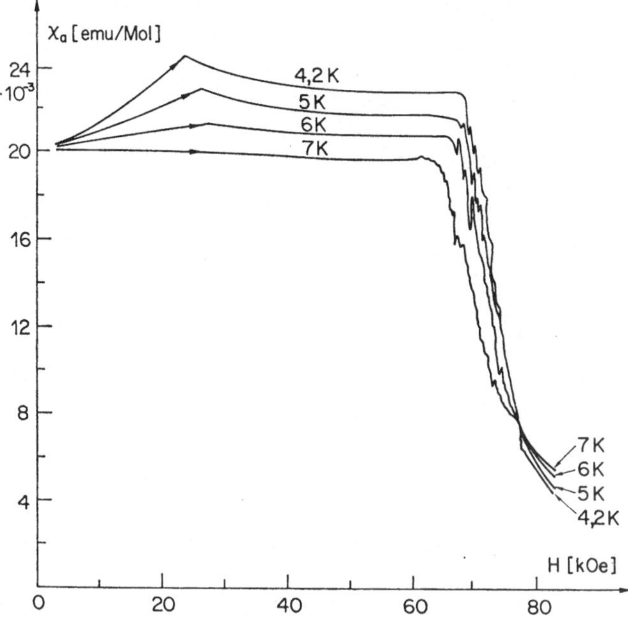

Fig. 9. Molare adiabatische differentielle Suszeptibilität von KO_2 als Funktion der magnetischen Feldstärke

der Magnetisierung verursacht wird. Bei einer metamagnetischen Umwandlung 1. Ordnung ohne statische Hysterese kann sich die Magnetisierung infolge des Entmagnetisierungsfeldes nicht sprunghaft ändern. Ist aber die statische magnetische Hysterese größer als das maximale Entmagnetisierungsfeld, so wird ein Magnetisierungssprung möglich. Dies trifft beim KO_2 bei 4,2 K zu. Die magnetische Hysterese beträgt 7,5 kOe, während das Entmagnetisierungsfeld im schlimmsten Fall 2,1 kOe betragen kann. Die Sprünge der Magnetisierung bei einer langsamen Änderung des äußeren Magnetfeldes konnten durch breitbandiges Verstärken des Signals der Pick-up-Spulen direkt beobachtet werden.

Eine Messung im gepulsten Feld zeigt, daß die Magnetisierung bei 4,2 K zwischen 100 kOe und 225 kOe innerhalb der Meßgenauigkeit (3%) konstant bleibt und einem magnetischen Moment von $0,92 \pm 0,1 \mu_B$ pro O_2^- entspricht.

Die adiabatische differentielle Suszeptibilität χ_s ist in Fig. 9 dargestellt. Für $H = 0$ erhält man innerhalb 1% dieselben Werte der Suszeptibilität wie mit der Faraday-Methode (für $H = 0$ gilt $\chi_T = \chi_s$ im para- oder antiferromagnetischen Zustand, cf. Gleichung (1).) Bei dieser Messung zeigt sich eine weitere Anomalie (Unstetigkeit in $\partial\chi_s/\partial H$) unterhalb 7 K bei Magnetfeldern von etwa 25 kOe. Es fragt sich nun, wie sich χ_T verhält. Leider sind die Messungen der differentiellen Suszeptibilität ohne Feldmodulation nicht genau genug zur quantitativen Beantwortung dieser Frage. Sie zeigen aber, daß χ_T immer positiv und beschränkt bleibt. Aus dieser Tatsache und mit Hilfe der Gleichung (1) folgt, daß sich die Feldabhängigkeiten von χ_T und χ_s qualitativ nicht unterscheiden. Somit findet hier eine Phasenumwandlung 3. Ordnung statt, indem $\partial\chi_s/\partial H$ springt. Die

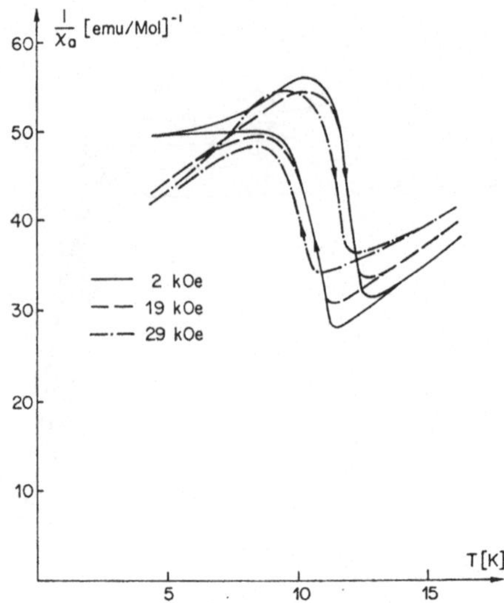

Fig. 10. Inverse molare adiabatische differentielle Suszeptibilität von KO_2 als Funktion der Temperatur

Temperaturabhängigkeit des Feldes H, bei welchem $\partial\chi_s/\partial H$ eine Diskontinuität aufweist, ist im Phasendiagramm Fig. 12 eingezeichnet. Die Diskontinuität verschwindet bei 7 K (Néel-Punkt). Die Fig. 10 zeigt die reziproke adiabatische differentielle Suszeptibilität als Funktion der Temperatur für verschiedene Magnetfelder. Für kleine Magnetfelder stimmt sie überein mit der nach der Faraday-Methode gemessenen isothermen Suszeptibilität, wie es Gleichung (1) fordert.

3.3. Spezifische Wärme

Die Probe wiegt 400 mg und die Qualität der Kristalle ist dieselbe wie bei den Suszeptibilitätsmessungen. Die Fig. 11 zeigt die Temperaturabhängigkeit der molaren spezifischen Wärme. Die ausgeprägte Anomalie bei $7{,}1 \pm 0{,}1$ K zeigt keine Divergenz und ist dem Néel-Punkt zuzuordnen [39]. Die Spitze bei 12,3 K entspricht der Phasenumwandlung IV/V, die mit einer kristallographischen Strukturänderung verknüpft ist [39]. Aus der thermischen Hysterese der Suszeptibilität und der Andeutung latenter Wärme schließt man auf eine Umwandlung erster Ordnung.

Die Meßwerte lassen sich zwischen 14 K und 20 K durch folgende Beziehung approximieren:

$$C_p = a\,T^3 + b\,T^{-2}\,,$$

wobei $a = 4{,}3 \cdot 10^{-4}$ J/Mol K^4 und $b = 120$ JK/Mol. Man kann versuchen, den ersten Term als Debye-Wärme zu interpretieren. Im betrachteten Temperaturintervall sind die Streckschwingungen ($\nu = 1150$ cm^{-1}) und die Librationsschwingungen ($\nu \cong 100$ cm^{-1}) der O_2^--Moleküle noch eingefroren, so daß nur die translatorischen Freiheitsgrade berücksichtigt werden müssen. Man erhält eine

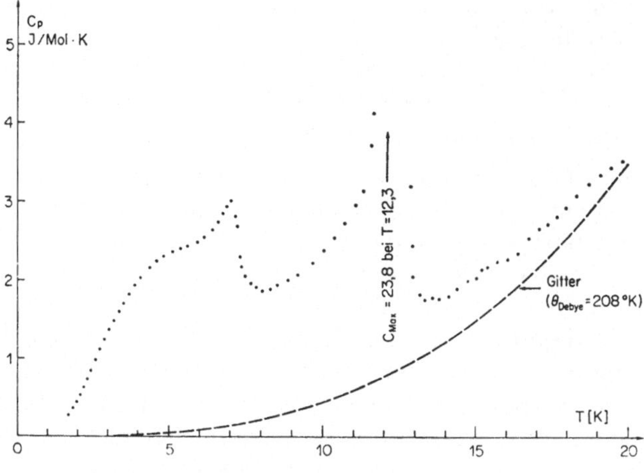

Fig. 11. Molwärme von KO_2

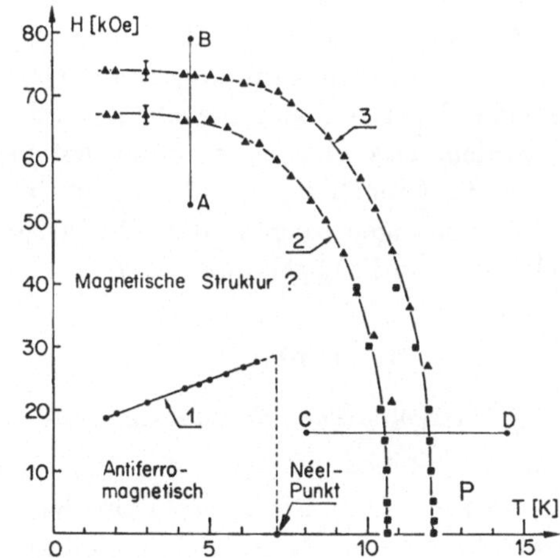

Fig. 12. Phasendiagramm von KO_2

Debye-Temperatur von 208 K. Der zweite Term könnte von der Änderung einer kurzreichweitigen Ordnung herrühren.

Die totale Entropiezunahme zwischen 1,7 K und 7,1 K beträgt 0,40 $R \ln 2$. Die totale Entropie bei 1,7 K wird auf 0,01 $R \ln 2$ geschätzt. Wenn das System bei tiefer Temperatur sein thermodynamisches Gleichgewicht tatsächlich erreicht, müßte man schließen, daß die Phase oberhalb des Néel-Punktes noch eine beachtliche magnetische Ordnung besitzt; darauf weisen auch die metamagnetischen Eigenschaften dieser Phase hin. Der Entropiezuwachs zwischen 9,9 K und 13,7 K beträgt 0,37 $R \ln 2$, und bei 20 K beträgt die totale Entropie 1,10 $R \ln 2$. Eine Aufteilung der Umwandlungsentropie in einen magnetischen Anteil, in Vibrationsentropie und einen Anteil aus einer kristallographischen Umwandlung

19 *

ist wahrscheinlich nicht sinnvoll, da die magnetischen, elastischen und strukturellen Eigenschaften stark gekoppelt sind.

3.4. Phasendiagramm bei tiefer Temperatur

Die Phasengrenzen (Grenzlinien 1, 2 und 3) des H-T-Phasendiagramms (Fig. 12) wurden durch Messung der differentiellen Suszeptibilität als Funktion des Magnetfeldes bei konstanter Temperatur und als Funktion der Temperatur bei konstantem Magnetfeld bestimmt. Im Gebiet, wo beide Methoden anwendbar sind, liefern sie dieselben Phasengrenzen. Die Grenzlinie 1 entspricht der Unstetigkeit von $\partial \chi_s / \partial H$ (cf. Fig. 9). Die Grenzlinie 3 wird wie folgt erhalten: Der Schnittpunkt des Weges AB mit der Grenzlinie 3 entspricht dem Maximum der isothermen differentiellen Suszeptibilität (Wendepunkt der Magnetisierungskurve) bei zunehmendem Magnetfeld (cf. Fig. 7). Der Schnittpunkt des Weges CD mit der Grenzlinie 3 entspricht der annähernd sprunghaften Änderung der adiabatischen differentiellen Suszeptibilität bei zunehmender Temperatur (cf. Fig. 10). Läßt man nun das Magnetfeld bzw. die Temperatur abnehmen, erhält man nach dem entsprechenden Verfahren die Grenzlinie 2.

Die beiden so konstruierten Grenzlinien berühren oder kreuzen sich nicht. Unabhängig von der Richtung, in welcher die Grenzlinie überschritten wird, tritt Hysterese auf, was auf eine Umwandlung erster Ordnung schließen läßt. Bei $H = 0$ ist mit Hilfe der Neutronenbeugung eine kristallographische Strukturänderung festgestellt worden. Das Phasendiagramm legt nahe, daß eine solche auch auftritt, wenn man die Grenzlinien (2 oder 3) in einem beliebigen Punkt überschreitet. Diese Erscheinungen zeigen, daß die kristallographischen und magnetischen Eigenschaften von KO_2 gekoppelt sind.

4. RbO$_2$

4.1. Suszeptibilität (Faraday-Methode)

Die Probe wiegt 35 mg und besteht aus vielen kleinen Kristallen mit einer typischen Kantenlänge von 0,1 mm. Die Phase I oberhalb 420 K wurde nicht untersucht. Die reziproke Suszeptibilität ist in Fig. 13 dargestellt. Zwei Phasenumwandlungen lassen sich eindeutig erkennen. Beim Übergang II/III ($T = 194 \pm 10$ K) ändert sich μ_{eff} und θ (cf. Tabelle 2). Dieser Übergang entspricht einer kristallographischen Phasenumwandlung, die röntgenographisch [42] und mit Neutronenstreuung [40] festgestellt wurde. Die entsprechende Anomalie der

Tabelle 2. Effektive magnetische Momente und Curie-Weiss-Temperaturen der Alkalihyperoxide. Zahl links: μ_{eff} (gemessen in μ_{Bohr}), Zahl rechts: θ [°K]

Phase	NaO$_2$		KO$_2$		RbO$_2$		CsO$_2$	
I	1,97,	-31	—		—		—	
II	1,79,	$+33$	2,01,	-18	2,05,	-12	1,92,	-7
III	—		2,10,	-44	2,11,	-26	2,00,	-28
IV	—		1,76,	$+2,3$	—		—	
V	—		1,93,	-16	—		—	

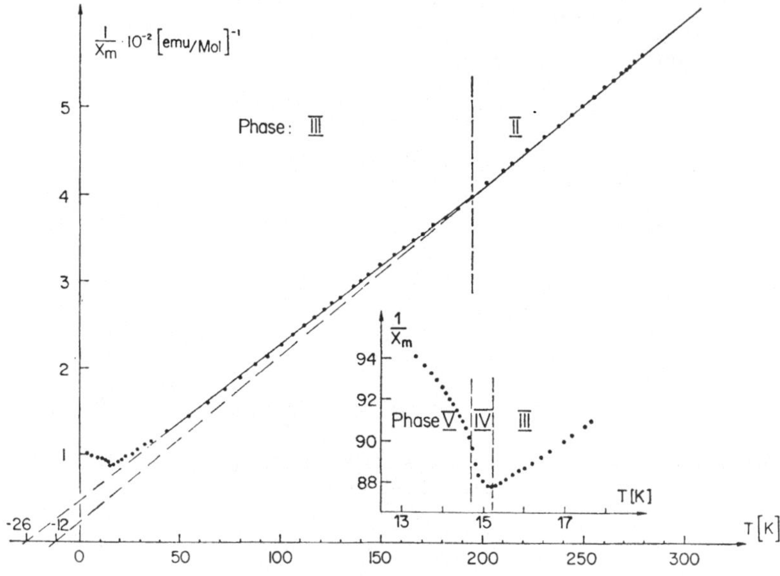

Fig. 13. Inverse molare Suszeptibilität von RbO_2

spezifischen Wärme (kleine Spitze in C_p bei 178,3 K) wurde von Pankov *et al.*
[11, 12] an Pulver gemessen, das durch Verbrennen von Rb in einem Gasgemisch
von Sauerstoff und Argon gewonnen wurde. Wiederum scheint die Umwandlungs-
temperatur des rekristallisierten Materials höher als beim Pulver zu liegen. In
der Gegend von 15 K zeigen die Messungen der spezifischen Wärme eine ausgeprägte Anomalie, die aus zwei Spitzen besteht: Die eine bei 15,1 K (Übergang
III/IV), die andere bei 14,7 K (Übergang IV/V) [11, 12]. Das scharfe Minimum
der reziproken Suszeptibilität bei 15,2 ± 0,2 K entspricht dem Übergang III/IV.
Der Übergang IV/V ist in der Suszeptibilität kaum feststellbar. (In der $1/\chi(T)$-
Kurve könnte man eventuell eine kleine Änderung der Neigung herauslesen.)
Smith *et al.* [40] haben mit Neutronenstreuung eine Phasenumwandlung, die
möglicherweise magnetischer Natur ist, zwischen 78 K und 4,2 K festgestellt.
Der Verlauf der Suszeptibilität als Funktion der Temperatur deutet auf Antiferromagnetismus unterhalb 15 K hin. Ob ein eventueller Néel-Punkt bei 15,1 K
oder 14,7 K liegt, bleibt offen.

Die Suszeptibilität der Phase II und der Phase III oberhalb 50 K kann durch
ein Curie-Weiß-Gesetz beschrieben werden. Die angepaßten Werte von μ_{eff} und θ
betragen 2,05 und -12 K bzw. 2,11 und -26 K. Die Zunahme des effektiven
magnetischen Momentes beim Übergang von der tetragonalen CaC_2-Struktur
(Phase II) zu einer Struktur mit tieferer Symmetrie (Phase III), und auch die
Werte des effektiven magnetischen Momentes, die höher als 2 sind, sind im
Rahmen einer statischen Kristallfeldtheorie wiederum unverständlich.

4.2. Adiabatische differentielle Suszeptibilität und Magnetisierung

Die Fig. 14 zeigt den Verlauf von χ_s als Funktion des Magnetfeldes für verschiedene Temperaturen. Bei Feldern von etwa 26 kOe tritt eine erste Anomalie

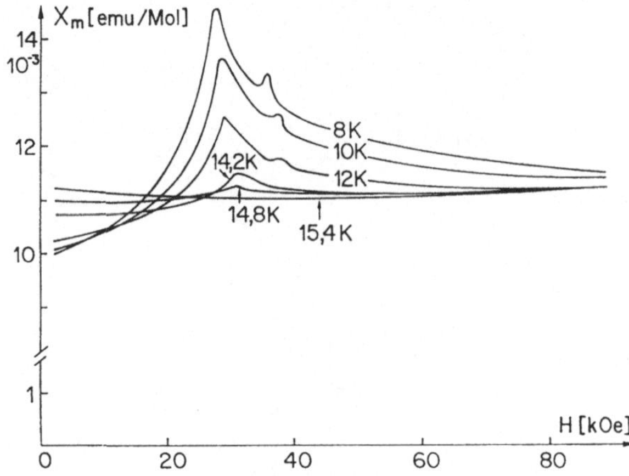

Fig. 14. Molare adiabatische differentielle Suszeptibilicät von RbO_2 als Funktion der magnetischen Feldstärke

(Maximum) auf, die als Phasenumwandlung 3. Ordnung (Sprung in $\partial\chi_s/\partial H$) klassifiziert werden kann (cf. Abschnitt 3). Diese Anomalie verschwindet zwischen 14,9 K und 15,4 K. Eine weitere kleine Anomalie (lokales Maximum) tritt bei Feldern von etwa 34 kOe auf. Diese zweite Anomalie können wir bei 14,5 K nicht mehr feststellen. Die Temperaturabhängigkeit der Felder, bei welchen χ_s Anomalien aufweist, ist linear; für die erste Anomalie variiert das Feld von 24 kOe bei 4,2 K zu 29,4 kOe bei 14 K und für die zweite von 32 kOe bei 4,2 K zu 36,5 kOe bei 12 K. Es scheint, daß diese Anomalien von χ_s und eine eventuelle antiferromagnetische Ordnung bei nahezu derselben Temperatur verschwinden, wie es beim KO_2 der Fall ist.

Messungen in gepulsten Feldern bis zu 225 kOe zwischen 1,4 K und 4,2 K und bis zu 150 kOe oberhalb von 4,2 K haben gezeigt, daß innerhalb der Meßgenauigkeit (3%) die Magnetisierung eine lineare Funktion des Feldes ist. Im erfaßten Feld- und Temperaturbereich konnte kein Anzeichen einer metamagnetischen Umwandlung beobachtet werden.

5. CsO_2

5.1. Suszeptibilität (Faraday-Methode)

Die Probe wiegt 23 mg und besteht aus vielen kleinen Kristallen mit einer typischen Kantenlänge von 0,15 mm. Die Phase I oberhalb 378 K wurde nicht untersucht. Die reziproke Suszeptibilität ist in Fig. 15 dargestellt. Bei $T = 190 \pm 10$ K läßt sich eine Phasenumwandlung (Übergang II/III) erkennen, in dem sich μ_{eff} und θ ändern (cf. Tabelle 2). Dieser Übergang entspricht einer kristallographischen Phasenumwandlung, die auch röntgenographisch festgestellt wurde [42]. Bei tiefer Temperatur gehorcht die Suszeptibilität keinem Curie-Weiß-Gesetz mehr. Bei 28 K durchläuft die Suszeptibilität ein flaches Minimum und bei $9,5 \pm 0,5$ K ist ein Wendepunkt erkennbar.

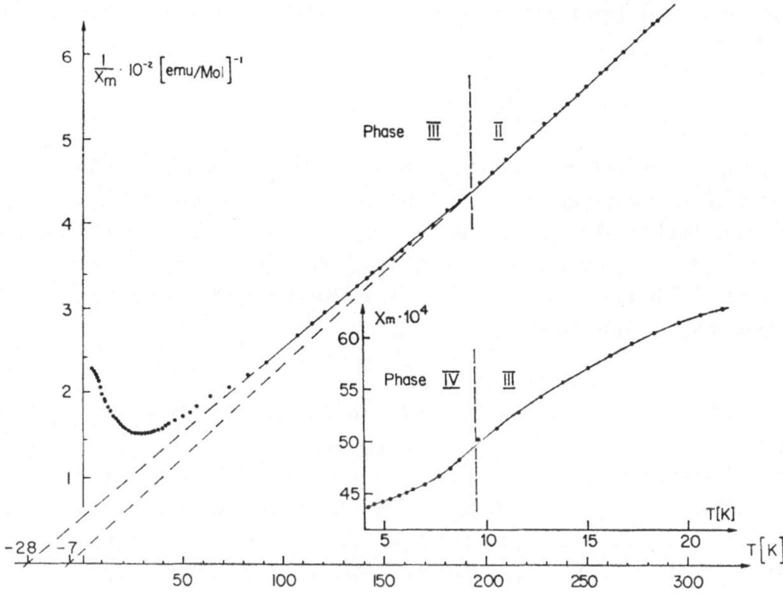

Fig. 15. Inverse molare Suszeptibilität von CsO_2

Die Suszeptibilität der Phasen II und III oberhalb 100 K kann durch ein Curie-Weiß-Gesetz beschrieben werden. Die angepaßten Werte von μ_{eff} und θ betragen 1,92 und — 7 K, bzw. 2,00 und — 28 K. Die Zunahme des effektiven magnetischen Momentes beim Übergang von der tetragonalen CaC_2-Struktur (Phase II) zu einer Struktur mit tieferer Symmetrie (Phase III), ist im Rahmen einer statischen Kristallfeldtheorie wiederum unverständlich.

5.2. Adiabatische differentielle Suszeptibilität und Magnetisierung

Der Verlauf der adiabatischen differentiellen Suszeptibilität von CsO_2 als Funktion des Magnetfeldes bei konstanter Temperatur ist demjenigen bei RbO_2 sehr ähnlich (cf. Fig. 14). Es tritt indessen nur ein Maximum auf bei Feldern von etwa 28 kOe. Die Temperaturabhängigkeit des Magnetfeldes, bei welchem das Maximum auftritt, ist zwischen 25,8 kOe bei 4,2 K und 31 kOe bei 9 K linear. Das Maximum verschwindet bei 9,5 $\pm$ 0,3 K, was sich mit der Temperatur, bei welcher die isotherme Suszeptibilität (Faraday) einen Wendepunkt aufweist, korrelieren läßt.

Bei KO_2 und wahrscheinlich auch bei RbO_2 verschwinden die langreichweitige Ordnung und das Maximum der adiabatischen differentiellen Suszeptibilität bei derselben Temperatur; wir vermuten, daß dies auch für das CsO_2 zutrifft. Der Néel-Punkt von CsO_2 wäre demnach bei 9,5 $\pm$ 0,3 K. Die Anomalie der spezifischen Wärme bei 9,58 K (cf. Fig. 16) scheint diese Vermutung zu stützen.

Messungen in gepulsten Feldern bis zu 225 kOe zwischen 1,4 K und 4,2 K und bis zu 150 kOe oberhalb von 4,2 K haben gezeigt, daß innerhalb der Meßgenauigkeit (3%) die Magnetisierung eine lineare Funktion des Feldes ist. Im erfaßten Feld- und Temperaturbereich konnte kein Anzeichen einer metamagnetischen Umwandlung beobachtet werden.

5.3. Spezifische Wärme

Die Probe wiegt 800 mg und die Qualität der Kristalle ist dieselbe wie bei den Suszeptibilitätsmessungen. Die Fig. 16 zeigt die Temperaturabhängigkeit der molaren spezifischen Wärme. Die Anomalie bei 9,58 K zeigt keine Divergenz und entspricht wahrscheinlich dem Néel-Punkt (Übergang III/IV).

Zwischen 1,7 K und 4 K lassen sich die Meßwerte innerhalb 3% durch folgende Beziehungen approximieren:

$$C = a \cdot T^3 + b \cdot T,$$

wobei $a = 3,7 \cdot 10^{-3}$ J/Mol K^4, $b = 8,0 \cdot 10^{-3}$ J/Mol K^2. Man kann versuchen, den ersten Term als Debye-Wärme zu interpretieren. Falls die Streckschwingungen und die Librationsschwingungen des O_2^- im betrachteten Temperaturintervall noch eingefroren sind, wie es beim NaO_2 und KO_2 der Fall ist, entspricht der erste Term einer Debye-Temperatur von 102 K. Der zweite Term, linear in der Temperatur, könnte dem magnetischen Anteil zugeschrieben werden.

Die totale Entropiezunahme zwischen 1,7 K und 9,58 K beträgt 0,25 $R \ln 2$. Die totale Entropie bei 1,7 K wird auf 0,003 $R \ln 2$ geschätzt. Bei 20 K beträgt die totale Entropie 1,20 $R \ln 2$. Wenn der Néel-Punkt tatsächlich bei 9,58 K liegt, so besitzt die Phase oberhalb des Néel-Punktes noch eine beachtliche magnetische Ordnung; die allmähliche Auflösung dieser Ordnung mit steigender Temperatur liefert eine Erklärung für die Zunahme der Suszeptibilität zwischen 9,58 K und 28 K. Ein ähnliches Verhalten der Suszeptibilität ist z.B. für $CuCl_2 2 H_2O$ bekannt und wurde von Nagai diskutiert [60].

6. Schlußbemerkungen

Wie diese Arbeit zeigt, ist das Verhalten der Alkalihyperoxidkristalle sehr komplex. Dies hängt mit folgenden Tatsachen zusammen:

1. Die O_2^--Molekül-Ionen können verschiedene Orientierungen einnehmen.

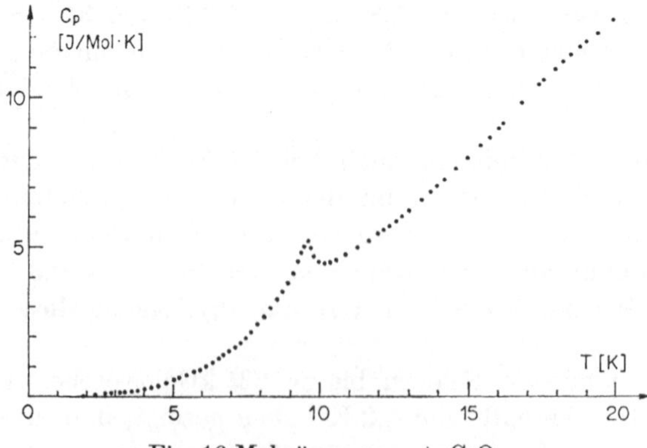

Fig. 16 Molwärme von ± CsO_2

2. Im trigonalen und tetragonalen Kristallfeld hat das O_2^--Molekül-Ion eine Bahnentartung.

3. Die O_2^--Molekül-Ionen sind paramagnetisch, und es bestehen magnetische Wechselwirkungen zwischen ihnen.

Jeder dieser drei Punkte könnte schon für sich allein zu einem kooperativen Phänomen Anlaß geben. Es ist schon möglich, daß es unter den vielen beobachteten Phasenumwandlungen in den Alkalihyperoxiden solche gibt, welche primär auf eine der drei Tatsachen zurückführbar sind. Man darf aber nicht vergessen, daß die drei Effekte nicht unabhängig sind. So ist z. B. die Zeeman-Aufspaltung des Grundzustandes von O_2^- einerseits von der Orientierung der Molekülachse und andererseits von der Kristallfeldaufspaltung abhängig. Es ist deshalb möglich, daß mit der Magnetisierung dieser Substanzen eine Orientierungsänderung der Molekülachsen verbunden ist. Es sind insbesondere zwei Beobachtungen, welche zu dieser Vermutung führen:

1. Die effektiven Magnetonenzahlen sind in gewissen Phasen größer als diejenigen, die bei fester Molekülachse auftreten können.

2. Bei KO_2 ist mit der metamagnetischen Umwandlung eine Strukturänderung verknüpft, bei welcher die Moleküle ihre Orientierung sehr wahrscheinlich ändern.

Über die Strukturuntersuchungen an Einkristallen mit Hilfe der Streuung von Röntgenstrahlen und Neutronen, sowie über optische und spektroskopische Untersuchungen soll in einer späteren Publikation berichtet werden.

Verdankung. Diese Arbeit wäre nicht möglich gewesen ohne die tatkräftige Hilfe, die uns von vielen Physikern gewährt wurde, und die wir hier verdanken:

Dr. H. Rohrer vom IBM-Laboratorium in Rüschlikon führte uns in die Meßtechnik der hohen gepulsten Magnetfelder ein und stellte seine Apparatur zur Verfügung.

Im Laboratorium für Festkörperphysik der ETH führte Dr. J. Muheim die massenspektrometrischen Analysen durch, Dr. W. Stutius machte die ersten Messungen der spezifischen Wärmen und stellte dann seine Apparatur zur Verfügung; die Messungen der Suszeptibilität nach der Faraday-Methode wurden mit der Apparatur von Dr. L. Schlapbach durchgeführt. Dr. H. Arend stellte seine wertvollen Kenntnisse in der Differentialthermoanalyse, sowie seine Apparatur zur Verfügung.

Besonderer Dank gebührt auch H. R. Aeschbach und M. Wächter für stete Hilfe in technischen Belangen und in der Kristallzucht.

Die Arbeit wurde vom Schweizerischen Nationalfonds unterstützt.

Literatur

1. Alikhanov, R. A.: Soviet Physics JETP 18, 556 (1964); Cox, D. E., Samuelsen E. J., and Beckurts, K. H.: Phys. Rev. B 7, 3102 (1973)
2. Lips, E.: Helv. Phys. Acta 8, 247 (1935); Herzog, T., Schwab, G. M.: Z. Phys. Chemie 66, 190 (1969)
3. Grill, A., Schieber, M., Shamir, J.: Phys. Rev. Letters 25, 747 (1970)
4. Ehrlich, P.: Z. anorg. Chemie 252, 370 (1944); Evdokimova, A. D., Cherkasov, E. N.: Izv. Akad. Nauk SSSR, Ser. Khim. 1969, 2692
5. Van Itterbeek, Labro, M.: Physica 30, 157 (1960); Jun Yamauchi *et al.*: J. Phys. Soc. Japan 35, 443 (1973)
6. Epstein, A. J., Etemad, S., Garito, A. F., Heeger, A. J.: Phys. Rev. B 5, 952 (1972)
7. Zeller, H. R., Känzig, W.: Helv. Phys. Acta 40, 845 (1967)
8. Baumann, R., Beyeler, H. U., Känzig, W.: ibid 44, 252 (1971)
9. Vannerberg, N. G.: Progress Inorganic Chemistry 4, 125 (1962); Volnov, I. I.: Monographs in Inorganic Chemistry, Plenum Press, New York (1966)

10. Todd, S. S.: J. Amer. Chem. Soc. **75**, 1229 (1953)
11. Pankov, I. E., Rakhmenkulov, F. S., Dobrolyubova, M. S., Tsentsiper, A. B.: Izv. Akad. Nauk SSSR, Ser. Khim. **9**, 2135 (1970)
12. Pankov, E., Rakhemenkulov, F. S.: Soviet Physics Solid State **13**, 1846 (1972)
13. Zumsteg, A., Ziegler, M., Bösch, M., Känzig, W.: Helv. Phys. Acta **46**, 15 (1973)
14. Stutius W.: Phys. kondens. Materie **10**, 152 (1969)
15. Neuman, E. W.: J. Chem. Phys. **2**, 31 (1934)
16. Klemm, W., Sodomann, H.: Z. anorg. allg. Chemie **225**, 273 (1935)
17. Helms, A., Klemm, W.: ibid **241**, 97 (1939)
18. Stephanou, S. E., Schechter, W. H., Argersinger, W. J., Kleinberg, J.: J. Amer. Chem. Soc. **71**, 1819 (1949)
19. Evdokimova, A. D., Dobrolyubova, U. S., Tsentsiper, A. B.: Izv. Akad. Nauk SSSR, 446 (1971)
20. Sparks, J. T., Komoto, T.: J. appl. Phys. **37**, 1040 (1966)
21. Zumsteg, A., Bösch, M., Känzig, W., Rohrer, H.: Helv. Phys. Acta **45**, 28 (1972)
22. Creighton, J. A., Lippincott, E. R.: J. Chem. Phys. **40**, 1779 (1964)
23. Bates, J. B., Brooker, M. H., Boyd, G. E.: Chem. Phys. Letters **16**, 391 (1972)
24. Bösch, M., Känzig, W., Steigmeier, E. F.: Phys. kondens. Materie **16**, 107 (1973)
25. Neiding, A. B., Kazarnovskii, I. A.: Zhur. Fiz. Khim. **24**, 1407 (1950)
26. Templeton, D. H., Dauben, C. H.: J. Amer. Chem. Soc. **72**, 2251 (1950)
27. Carter, G. F., Templeton, D. H.: ibid. **75**, 5247 (1953)
28. Zhdanov, G. S., Zvonkova, Z. V.: Doklady Akad. Nauk SSSR **82**, 743 (1952)
29. Fischer, P.: Progress Report, EIR, Würenlingen, Schweiz, AF-SSP-2 (1966)
30. Ziegler, M., Fischer, P.: ibid. AF-SSP-**61**, 22 (1972)
31. Ziegler, M., Fischer, P.: ibid. AF-SSP-**65**, 25 (1972)
32. Kassatochkin, W., Kotow, W.: J. Chem. Phys. **4**, 458 (1936)
33. Rode, T. V.: Doklady Akad. Nauk SSSR **90**, 1075 (1953)
34. Helms, A.: Angew. Chemie **29**, 498 (1938)
35. Helms, A., Klemm, W.: Z. anorg. allg. Chem. **241**, 97 (1939)
36. Neuman, E. W.: J. Chem. Phys. **3**, 254 (1935)
37. Abrahams, S. C., Kalnas, J.: Acta Cryst. 8, 503 (1955)
38. Carter, G. E., Margrave, J. L., Templeton, D. H.: ibid. **5**, 851 (1952).
39. Smith, H. G., Niclow, R. M., Raubenheimer, L. J., Wilkinson, M. K.: J. appl. Phys. **37**, 1047 (1966)
40. Smith, H. G., Niclow, R. M., Raubenheimer, L. J., Wilkinson, M. K.: Acta Cryst. **21**, A 210 (1966)
41. Halverson, F.: J. Phys. Chem. Solids **23**, 207 (1962)
42. Ziegler, M.: nicht publiziert.
43. Powell, R. L., Bunch, M. D., Corruccini, R. J.: Cryogenics 1, 139 (1961)
44. Olsen, J. L.: Helv. Phys. Acta **26**, 798 (1953)
45. Cotti, P.: Z. angew. Math. Phys. 11, 17 (1960)
46. Blazey, K. W., Rohrer, H.: Phys. Rev. **173**, 574 (1968)
47. Goldstein, A., Williamson, S. J., Foner, S.: Rev. Sci. Instr. **36**, 1356 (1965)
48. Bennett, J. E., Ingram, D. J., Symons, M. C.: Phil. mag. **46**, 443 (1955)
49. Bennett, J. E., Ingram, D. J., Schonland, D.: Proc. Phys. Soc. A **69**, 556 (1956)
50. Bleaney, B.: Proc. Phys. Soc. **75**, 621 (1960)
51. Wangsness, R. K.: Phys. Rev. **89**, 142 (1953)
52. Hedgcock, F. T., Raudorf, T. W.: Can. J. Phys. **50**, 579 (1972)
53. Bizette, H., Terrier, C., Tsai, B.: Compt. Rend. **243**, 895 (1956); ibid. **261**, 653 (1965)
54. Jacobs, I. S., Lawrence, P. E.: Phys. Rev. **164**, 866 (1967)
55. Fert, A. R., Carrara, P., Lanusse, M. C., Mischler, G., Redoules, J. P.: J. Phys. Chem. Solids **34**, 223 (1973)
56. Jacobs, I. S., Lawrence, P. E.: J. appl. Phys. **35**, 996 (1964)
57. Motokawa, M., Date, M.: J. Phys. Soc. Japan **20**, 465 (1965)
58. Kuramitsu, Y., Amaya, K., Haseda, T.: ibid. **33**, 83 (1972)
59. Ball, M., Wolf, W. P., Wyatt, A. F. G.: Phys. Letters **10**, 7 (1964)

60. Nagai, O.: J. Phys. Soc. Japan **18**, 510 (1963)
61. Fisher, M. E.: Proc. Roy. Soc. (London) A **254**, 66 (1960)
62. Beyeler, H. U.: Diplomarbeit ETH, unveröffentlicht

Prof. Dr. W. Känzig
Laboratorium für Festkörperphysik
Eidgen. Technische Hochschule
Zürich, Hönggerberg
CH-8049 Zürich, Schweiz

Phys. cond. Matter 17, 293—305 (1974)

Analytische Näherung für elastische Wellen in anisotropen kubischen Kristallen

Walter Orth

Institut für Theoretische Physik der Universität des Saarlandes, Saarbrücken

Eingegangen am 29. Juni 1973/Revidierte Form: 6. Juli 1973

Analytic Approximation for Elastic Waves in Anisotropic Cubic Crystals

It is shown that the frequencies $\omega_\alpha(k)$ and the polarisation vectors $e_\alpha(k)$ ($\alpha = 1, 2, 3$) of the elastic waves in anisotropic cubic crystals can be described exactly as Taylor series in the parameter $\delta = \dfrac{c_{11} - c_{12} - 2\,c_{44}}{c_{12} + c_{44}}$ for all wave number vectors k. As the expansion functions of these series include no elastic constants, δ is taken as the proper anisotropy parameter. The series are converging very fast for almost all substances and may be broken off after the third expansion term.

Es wird gezeigt, daß die Frequenzen $\omega_\alpha(k)$ und die Polarisationsvektoren $e_\alpha(k)$ ($\alpha = 1, 2, 3$) der elastischen Wellen in anisotropen kubischen Kristallen sich exakt als Taylor-Reihen in dem Parameter $\delta = \dfrac{c_{11} - c_{12} - 2\,c_{44}}{c_{12} + c_{44}}$ für alle Wellenzahlvektoren k darstellen lassen. Da die Entwicklungsfunktionen dieser Reihen keine elastischen Konstanten enthalten, wird δ als der wahre Anisotropieparameter gedeutet. Die Reihen konvergieren für fast alle Substanzen sehr rasch und können dann nach dem dritten Entwicklungsglied abgebrochen werden.

I. Einführung

Die analytische Darstellung der Frequenzen und Polarisationsvektoren in einem elastischen Medium der Dichte ϱ mit kubischer Symmetrie bereitete lange Zeit große Schwierigkeiten. Während Goens (1937) [1] für Wellenzahlvektoren k ausgezeichneter Ebenen die Säkulargleichung des anisotropen Tensors $T_{il}(k)$ exakt gelöst hat — die analytische Gestalt der Lösungen erweist sich schon in diesen ausgezeichneten Ebenen als recht kompliziert —, haben Born und Karman (1913) [2] die Säkulargleichung für $|\gamma| = |\varepsilon/c_{11} - c_{44}| \ll 1$ untersucht. Die hierbei erzielten Ergebnisse sind auf Grund einer unzulässigen Vernachlässigung eines Terms der Säkulargleichung fehlerhaft und werden heute noch in der Literatur [3, 4] als die Schallgeschwindigkeiten für nahezu isotrope Materialien geführt (der Fehler in den Schallgeschwindigkeiten ist von der Größenordnung γ und verschwindet für den isotropen Fall $\varepsilon = 0$). Ein für alle Kristallsysteme und Kristallrichtungen gültiges Verfahren wurde von Fedorow (1968) [5] angegeben. Das Verfahren be-

---

1 Die Fedorowsche Minimalisierungsvorschrift lautet:

$$\sum_{k,l} (T'_{kl})^2 = \mathrm{Sp}.\,(T'^2) \overset{!}{=} \mathrm{Min}.$$

d.h. die Summe der Quadrate aller Komponenten von T' soll minimal sein.

ruht auf der Minimalisierung[1] der Differenz $T'(k) = T(k) - \hat{T}(k)$, wobei der „isotrope" Tensor $\hat{T}(k)$ die Form $\hat{T}(k) = \hat{a}(k)\, k^2 + \hat{b}(k)\, \boldsymbol{k}\boldsymbol{k}$ hat (nur für konstante $\hat{a}$ und $\hat{b}$ beschreibt $\hat{T}(k)$ ein isotropes Medium). Es zeigt für kubische Kristalle eine rasche Konvergenz, die analytische Gestalt dieser Lösungen ist jedoch kompliziert, so daß der praktische Umgang mit diesen Lösungen Schwierigkeiten bereitet.

In der vorliegenden Arbeit wird ein einfaches Verfahren aufgestellt, das die Lösungen der Säkulargleichung als Reihenentwicklung nach einem Parameter $\delta = \dfrac{\varepsilon}{c_{12} - c_{44}}$, der Abweichung von der Isotropie, liefert. Die Reihen sind konvergent für $\delta > -1$, und sie konvergieren für fast alle kubischen Kristalle sehr schnell. Da die bei dem Parameter δ stehenden Entwicklungsfunktionen keine elastischen Konstanten enthalten werden, kann δ als der wahre Anisotropieparameter angesehen werden, dessen Absolutwert die Stärke der Anisotropie wiedergibt. Mit Hilfe der nach δ entwickelten Eigenwerte von $T(k)$ werden dann durch Lösung der Eigenwertgleichung die Polarisationsvektoren ebenfalls als Entwicklung nach δ angegeben.

II. Theoretische Herleitung des Tensors $T(k)$

Die elastischen Wellen sind Lösungen der Bewegungsgleichung für ein elastisches Medium der Dichte ϱ [6, 7]

$$\varrho\, \ddot{s}_i = \sum_{k,l,m} C_{iklm}\, \frac{\partial^2 s_l}{\partial X_k\, \partial X_m}, \tag{1}$$

wobei $C_{iklm} = \dfrac{\partial^2 F}{\partial \varepsilon_{ik}\, \partial \varepsilon_{lm}}$ die Entwicklungskoeffizienten der freien Energie sind und elastische Konstanten genannt werden:

$$F(T, \varepsilon_{ik}) = F_0(T) + \tfrac{1}{2} \sum_{i,k,l,m} C_{iklm}(T)\, \varepsilon_{ik}\, \varepsilon_{lm} + \cdots. \tag{2}$$

$\varepsilon_{ik} = \dfrac{1}{2}\left(\dfrac{\partial s_i}{\partial X_k} + \dfrac{\partial s_k}{\partial X_i} \right)$ stellt den symmetrischen Verzerrungstensor dar und

$s(\boldsymbol{R}) = (s_1, s_2, s_3)$ die Auslenkung des Punktes $\boldsymbol{R}$. Die Lösungen der partiellen Differentialgleichung mit konstanten Koeffizienten sind

$$s_{\alpha i} = e_{\alpha i} \exp[i(\boldsymbol{k}\boldsymbol{R} - \omega_\alpha t)] \quad (\alpha = 1, 2, 3). \tag{3}$$

Man erhält mit diesem Ansatz aus der obigen Wellengleichung folgende Eigenwertgleichung:

$$\sum_l T_{il}(\boldsymbol{k})\, e_{\alpha l}(\boldsymbol{k}) = \omega_\alpha^2(\boldsymbol{k})\, e_{\alpha i}(\boldsymbol{k}) \quad \text{mit} \quad T_{il}(\boldsymbol{k}) = 1/\varrho \sum_{k,m} C_{iklm}\, k_k\, k_m \tag{4}$$

als symmetrischem Tensor einer positiv definit quadratischen Form, den Quadraten der Frequenzen $\omega_\alpha^2(\boldsymbol{k})$ als Eigenwerten und den Polarisationsvektoren $e_\alpha(\boldsymbol{k})$ als Eigenvektoren. Da letztere Eigenvektoren eines symmetrischen Tensors sind, müssen sie orthogonal sein:

$$e_\alpha(\boldsymbol{k})\, e_{\alpha'}(\boldsymbol{k}) = \delta_{\alpha\alpha'}. \tag{5}$$

Der Tensor der Elastizitätsmoduln C_{iklm} besitzt auf Grund folgender Symmetriebeziehungen — sie ergeben sich aus der Vertauschbarkeit der Differentiation und der Symmetrie von ε_{ik} — statt 81 nur 21 verschiedene Komponenten

$$C_{iklm} = \frac{\partial^2 F}{\partial \varepsilon_{ik}\,\partial \varepsilon_{lm}} = C_{lmik} = C_{kilm} = C_{ikml}\,. \tag{6}$$

Die Symmetrie des Kristalls führt zu Relationen zwischen den einzelnen Komponenten des Tensors C_{iklm}, so daß die Anzahl seiner unabhängigen Komponenten kleiner als 21 ist:

$$C_{iklm} = \sum_{i',k',l',m'} D_{ii'}\,D_{kk'}\,D_{mm'}\,D_{ll'}\,C_{i'k'l'm'}\,, \tag{7}$$

wobei die Operationen D Symmetrieelemente des Kristalls sind.

Für den kubischen Kristall — die Symmetrieelemente sind die Operationen $D \in \mathrm{Oh}^2$ — erhält man deshalb nur noch 3 unabhängige Komponenten:

$$C_{1111} = c_{11}\,, \qquad C_{1122} = c_{12}\,, \qquad C_{1212} = c_{44}\,.$$

Der Tensor $T_{il}(\boldsymbol{k})$ nimmt für den kubischen Kristall dann folgende Gestalt an:

$$T_{il}(\boldsymbol{k}) = \frac{1}{\varrho}\left[(c_{12} + c_{44})\,k_i\,k_l + c_{44}\,k^2\,\delta_{il} + \varepsilon\,k_i^2\,\delta_{il}\right]$$

$$= T_{il}^{(0)}(\boldsymbol{k}) + \frac{1}{\varrho}\,\varepsilon\,k_i^2\,\delta_{il} \qquad (\varepsilon = c_{11} - c_{12} - 2\,c_{44})\,, \tag{8}$$

wobei $T_{il}^{(0)}(\boldsymbol{k})$ der entsprechende Tensor des isotropen Mediums ist und aus $T_{il}(\boldsymbol{k})$ durch $\lim$ hervorgeht[3].
$\varepsilon \to 0$

Die Eigenwerte und Eigenvektoren des isotropen Tensors $T_{il}^{(0)}(\boldsymbol{k})$ sind einfach und bekannt, man erhält:

longitudinale Lösung:

$$\boldsymbol{e}_1(\boldsymbol{k}) = \frac{\boldsymbol{k}}{|\boldsymbol{k}|}\,, \qquad \omega_1^2(\boldsymbol{k}) = \frac{c_{11}}{\varrho}\,k^2 = c_l^2\,k^2 \tag{9a}$$

und 2 unabhängige transversale Lösungen:

$$\boldsymbol{e}_{2/3}(\boldsymbol{k}) \perp \boldsymbol{k}\,, \qquad \omega_{2/3}^2(\boldsymbol{k}) = \frac{c_{44}}{\varrho}\,k^2 = c_t^2\,k^2\,. \tag{9b}$$

(c_l ist hierbei die konstante Wellengeschwindigkeit oder auch Schallgeschwindigkeit der longitudinalen, c_t die der transversalen Wellen des isotropen Mediums.)

Wesentlich komplizierter sind die Frequenzen und Polarisationsvektoren eines anisotropen Mediums; aus Gleichung (4) ist ersichtlich (wenn man das Verhältnis

---

2 Die 48 Elemente D der Oh-Gruppe beschreiben eine beliebige Vertauschung der Koordinaten von $\boldsymbol{R} = (X_1, X_2, X_3)$ sowie die Einfügung eines beliebigen Vorzeichens:

$$\boldsymbol{D}\boldsymbol{R} = (\pm X_i, \pm X_\gamma, \pm X_k) \qquad (i \neq \gamma \neq k)\,.$$

3 Da im isotropen Medium alle Drehungen $\boldsymbol{D} \in \delta_3$ (δ_3 = unendliche Drehgruppe) Symmetrieelemente sind, ergibt sich aus

$$C'_{iklm} = C_{iklm} = \sum D_{ii'}\,D_{kk'}\,D_{ll'}\,D_{mm'}\,C_{i'k'l'm'}$$

eine weitere Beziehung zwischen den elastischen Konstanten:

$$\varepsilon = c_{11} - c_{12} - 2\,c_{44} = 0\,.$$

$\omega_\alpha(k)/k$ als unbekannte Größe einführt, hängen die Koeffizienten der Gleichung nicht mehr von k ab), daß die Frequenzen die Form

$$\omega_\alpha = c_\alpha(\varkappa)\, k \qquad (\alpha = 1, 2, 3) \tag{10}$$

— $\varkappa = (\varkappa_1, \varkappa_2, \varkappa_3)$ ist der Einheitsvektor im Wellenzahlraum — haben werden, wobei die Wellengeschwindigkeit $c_\alpha(\varkappa) = \omega_\alpha(k)/k$ nicht von k abhängt.

III. Berechnung der Frequenzen und Polarisationsvektoren

Mit den Abkürzungen

$$\varepsilon = c_{11} - c_{12} - 2\,c_{44}, \qquad B = c_{12} + c_{44} \tag{1}$$

lautet der Tensor $T(k)$ des anisotropen kubischen Kristalls

$$T_{il}(k) = \frac{1}{\varrho}\,[B\,k_i\,k_l + (c_{44}\,k^2 + \varepsilon\,k_i^2)\,\delta_{il}]. \tag{2}$$

Seine Eigenwerte $\omega_\alpha^2(k)$ $(\alpha = 1, 2, 3)$ erhält man durch Lösung der Säkulargleichung

$$\det[T(k) - \omega^2(k)\,E] = \det[\varrho\,T(k) - \lambda\,E] = 0$$

mit
$$\lambda = \varrho\,\omega^2(k)$$

die auf ein Polynom 3. Grades in λ führt:

$$\lambda^3 + c_1\,\lambda^2 + c_2\,\lambda + c_3 = 0 \tag{3}$$

mit
$$\begin{aligned}
c_1 &= -\,k^2[\varepsilon + B + 3\,c_{44}], \\
c_2 &= k^4[c_{44}(3\,c_{44} + 2\,(\varepsilon + B)) + \varepsilon(\varepsilon + 2\,B)\,\chi], \\
c_3 &= -\,k^6[c_{44}^2(c_{44} + \varepsilon + B) + \varepsilon(\varepsilon + 2\,B)\,c_{44}\,\chi + \varepsilon^2(\varepsilon + 3\,B)\,\Gamma], \\
\chi &= \varkappa_1^2\varkappa_2^2 + \varkappa_1^2\varkappa_3^2 + \varkappa_2^2\varkappa_3^2, \qquad \Gamma = \varkappa_1^2\varkappa_2^2\varkappa_3^2.
\end{aligned}$$

Die kubische Gleichung (3) kann jedoch durch die Transformation

$$\lambda = B\,k^2\,v - \frac{c_1}{3}$$

entscheidend vereinfacht werden; durch sie wird einerseits der quadratische Term der kubischen Gleichung eliminiert und andererseits, was sehr viel wichtiger ist, hängen die Wurzeln der Gleichung (4) nicht mehr von den 3 Parametern ε, B, c_{44} ab, sondern nur noch von einem einzigen Parameter δ.

$$v^3 + p(\varkappa, \delta)\,v + q(\varkappa, \delta) = 0 \tag{4}$$

mit
$$\begin{aligned}
p(\varkappa, \delta) &= -\,\tfrac{1}{3}(1 + \delta)^2 + \delta(2 + \delta)\,\chi, \\
q(\varkappa, \delta) &= -\,\tfrac{2}{27}(1 + \delta)^3 + \tfrac{1}{3}\delta(1 + \delta)(2 + \delta)\,\chi - \delta^2(3 + \delta)\,\Gamma, \\
\delta &= \varepsilon/B.
\end{aligned}$$

Es läßt sich zeigen (siehe Anhang), daß die Lösungen $v_\alpha(\varkappa, \delta)$ der Säkulargleichung für $\delta > -1$ in Taylor-Reihen nach δ entwickelt werden können.

$$v_\alpha(\varkappa, \delta) = \sum_{m=0}^{\infty} \delta^m\,v_{\alpha'm}(\varkappa), \qquad v_{\alpha'm}(\varkappa) = (1/m!)\,v_\alpha^{(m)}(\varkappa, 0). \tag{5}$$

Die Reihen konvergieren im allgemeinen (siehe Diskussion, IV) sehr rasch und können nach dem Glied δ^3 abgebrochen werden, denn für fast alle kubischen Kristalle liegt δ im Intervall

$$0 < |\delta| \leqq 0{,}55 \tag{6}$$

(siehe Tabelle 1).

Tabelle 1. Adiabatische Elastizitätsmoduln für kubische Kristalle bei tiefen Temperaturen und bei Zimmertemperatur (in 10^{12} dyn/cm^2-Einheiten) [8, 9]

Kristall	Temperatur°K	c_{11}	c_{12}	c_{44}	ε	δ
W	0	5,326	2,049	1,631	0,015	0,004
	300	5,233	2,045	1,607	$-$ 0,026	$-$ 0,007
Ta	0	2,663	1,582	0,874	$-$ 0,67	$-$ 0,27
	300	2,609	1,574	0,818	$-$ 0,601	$-$ 0,255
Cu	0	1,762	1,249	0,818	$-$ 1,123	$-$ 0,543
	300	1,684	1,214	0,754	$-$ 1,038	$-$ 0,527
Ag	0	1,315	0,973	0,511	$-$ 0,68	$-$ 0,458
	300	1,24	0,937	0,461	$-$ 0,619	$-$ 0,44
Au	0	2,016	1,697	0,454	$-$ 0,589	$-$ 0,273
	300	1,423	1,631	0,42	$-$ 0,548	$-$ 0,267
Al	0	1,413	0,619	0,316	$-$ 0,108	$-$ 0,115
	300	1,068	0,607	0,282	$-$ 0,103	$-$ 0,116
Pb	0	0,555	0,454	0,194	$-$ 0,287	$-$ 0,44
	300	0,495	0,423	0,141	$-$ 0,226	$-$ 0,4
Ni	0	2,612	1,508	1,317	$-$ 1,53	$-$ 0,541
	300	2,508	1,5	1,235	$-$ 1,462	$-$ 0,534
Pd	0	2,341	1,761	0,712	$-$ 0,844	$-$ 0,341
	300	2,271	1,761	0,717	$-$ 0,124	$-$ 0,372
V	0	2,324	1,194	0,46	0,21	0,126
	300	2,28	1,187	0,426	0,241	0,149
Li	300	0,135	0,114	0,088	$-$ 0,155	$-$ 0,767
KCl	4	0,483	0,054	0,066	0,297	2,474

Die Entwicklungskoeffizienten $v_{\alpha' m}(\varkappa)$ erhält man durch fortgesetzte Differentiation der Säkulargleichung, es ergibt sich so folgende Rekursionsformel:

$$v_\alpha^{(m)}(\varkappa, 0) = \left[\frac{- q^{(m)} - \binom{m}{1} p' v_\alpha^{(m-1)} - \binom{m}{2} p'' v_\alpha^{(m-2)} - (v_\alpha^3)^{(m)} + 3 v_\alpha^2 v_\alpha^{(m)}}{3 v_\alpha^2 + p} \right]_{\delta = 0} \tag{7}$$

mit $q^{(m)} = 0$ $(m > 3)$.

Für $\alpha = 2, 3$ ist die Rekursionsformel von der unbestimmten Form $\frac{0}{0}$, so daß die L'Hôpitalsche Regel angewandt werden muß. Man erhält für die gesuchten Frequenzen bis zur Ordnung δ^3 folgende Ausdrücke:

$$\omega_1^2(k) = k^2 \frac{B}{\varrho} \left[1 + \frac{c_{44}}{B} + \delta(1 - 2\chi) + \delta^2(3\Gamma + \chi - 4\chi^2) \right.$$

$$\left. - \delta^3(5\Gamma + \chi - 18\Gamma\chi - 8\chi^2 + 16\chi^3) + \delta^4 \cdots \right], \tag{8}$$

$$\omega_\alpha^2(\boldsymbol{k}) = k^2 \frac{B}{\varrho} \left[\frac{c_{44}}{B} + \delta K_\alpha - \delta^2 L_\alpha + \delta^3 M_\alpha + \delta^4 \cdots \right] \qquad (\alpha = 2, 3) \qquad (9)$$

mit

$$K_\alpha = \chi \mp [\chi^2 - 3\,\Gamma]^{1/2} \,,$$

$$L_\alpha = \frac{1}{2} \left[\chi - 4\chi^2 + 3\,\Gamma \pm \frac{4\chi^3 - \chi^2 - 9\,\Gamma\chi + 2\,\Gamma}{[\chi^2 - 3\,\Gamma]^{1/2}} \right] \,,$$

$$M_\alpha = \frac{1}{2} \left[16\chi^3 - 8\chi^2 - 18\,\Gamma\chi + 5\,\Gamma + \chi \right.$$
$$\mp \frac{(16\chi^6 - 8\chi^5 + \chi^4 - 90\,\Gamma\chi^4 + 41\,\Gamma\chi^3 - \frac{19}{4}\,\Gamma\chi^2}{(\chi^2 - 3\,\Gamma)^{3/2}}$$
$$\left. + \frac{135\,\Gamma^2\chi^2 - 99\,\Gamma^2\chi + 5\,\Gamma^2 - \frac{135}{4}\,\Gamma^3)}{[\chi^2 - 3\,\Gamma]^{3/2}} \right] \,.$$

Die entsprechenden Polarisationsvektoren $\boldsymbol{e}_\alpha(\boldsymbol{k})$ ermittelt man aus dem Gleichungssystem

$$[\boldsymbol{T}(\boldsymbol{k}) - \omega_\alpha^2(\boldsymbol{k})\,\boldsymbol{E}]\,\boldsymbol{e}_\alpha(\boldsymbol{k}) = 0 \,.$$

Nach äußerst aufwendigen Rechnungen erhält man für

$$e_{11}(\boldsymbol{k}) = \varkappa_1[1 - \delta(1 - \varkappa_1^2 - 2\chi) + \delta^2\{10(\chi^2 - 2\,\Gamma)$$
$$+ (1 - \varkappa_1^2 - 2\chi) - 4(1 - \varkappa_1^2)\chi + 16\,\Gamma + \tfrac{1}{2}(1 - \varkappa_1^2)(\varkappa_2^2\varkappa_3^2 - \varkappa_1^2)\}] \,, \qquad (10)$$
$$e_{12}(\boldsymbol{k}) = e_{11}(k_2, k_3, k_1) \,, \qquad e_{13}(\boldsymbol{k}) = e_{11}(k_3, k_1, k_2) \,.$$

Für $\boldsymbol{e}_\alpha(\boldsymbol{k})$ $(\alpha = 2, 3)$ werden die Formeln komplizierter:

$$e_{\alpha 1}(\boldsymbol{k}) = \frac{\varkappa_1 \varkappa_3 (\varkappa_2^2 - G_\alpha)}{F_\alpha^{1/2}} \,,$$

$$e_{\alpha 2}(\boldsymbol{k}) = \frac{\varkappa_2 \varkappa_3 (\varkappa_1^2 - G_\alpha)}{F_\alpha^{1/2}} \,, \qquad (11)$$

$$e_{\alpha 3}(\boldsymbol{k}) = - \frac{[\varkappa_1^2 (\varkappa_2^2 - G_\alpha) + \varkappa_2^2 (\varkappa_1^2 - G_\alpha) + \delta(\varkappa_1^2 - G_\alpha)(\varkappa_2^2 - G_\alpha)]}{F_\alpha^{1/2}}$$

mit

$$F_\alpha = \varkappa_1^2 \varkappa_3^2 (\varkappa_2^2 - G_\alpha)^2 + \varkappa_2^2 \varkappa_3^2 (\varkappa_1^2 - G_\alpha)^2$$
$$+ [\varkappa_1^2 (\varkappa_2^2 - G_\alpha) + \varkappa_2^2 (\varkappa_1^2 - G_\alpha) + \delta(\varkappa_1^2 - G_\alpha)(\varkappa_2^2 - G_\alpha)]^2 \,,$$
$$G_\alpha = K_\alpha - \delta L_\alpha + \delta^2 M_\alpha + \delta^3 \cdots \,.$$

IV. Diskussion der Ergebnisse

Die Frequenzen und Polarisationsvektoren konnten als Taylor-Reihen in dem Anisotropieparameter δ, der Abweichung von der Isotropie, für das Gebiet $\delta > -1$, das alle in der Praxis vorkommenden Substanzen enthält (siehe Fig. 5), dargestellt werden. Sie sind weiterhin analytisch in der Wellenzahlrichtung $\boldsymbol{\varkappa}$, wobei lediglich die Polarisationsvektoren $\boldsymbol{e}_{2/3}(\boldsymbol{\varkappa})$ in den speziellen Richtungen der Achsen- und Raumdiagonalen eine Ausnahme bilden. In diesen Richtungen wird die zweifache

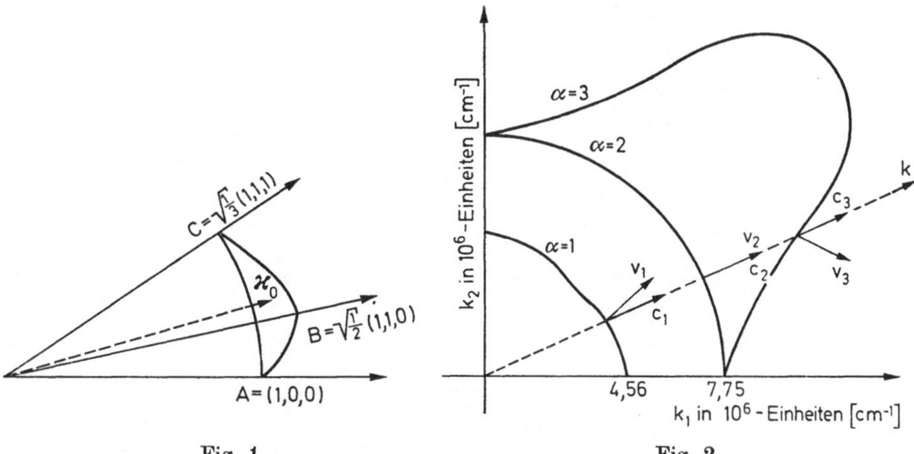

Fig. 1 Fig. 2

Fig. 1. Das sphärische Dreieck ABC stellt 1/48 der Einheitskugeloberfläche dar; es enthält alle nichtäquivalenten k-Werte. $\varkappa_0$ liegt im Zentrum von $\Delta\,ABC$:

$$\varkappa_0 = \tilde{\varkappa}_0/|\tilde{\varkappa}_0|, \quad \tilde{\varkappa}_0 = (1,\ \sqrt{2}-1,\ \sqrt{3}-\sqrt{2})$$

Fig. 2. Kurven konstanter Frequenz in der (k_1, k_2)-Ebene für Pb ($\delta < 0$) bei $T = 0\,°K$:

$$\omega_\alpha(k) = 10^{12}\,\mathrm{Hz}\quad (\alpha = 1, 2, 3)$$

Frequenzentartung des isotropen Mediums $\omega_2(k) = \omega_3(k)$ durch die Anisotropie ($\delta \neq 0$) des Mediums nicht aufgehoben, d.h. die Polarisationsvektoren $e_{2/3}(\varkappa)$ sind in diesen Richtungen unbestimmt.

Das Konvergenzverhalten der Reihe wird in der folgenden Tabelle 2 für die stark anisotropen (siehe Fig. 5) Substanzen Li (größter negativer Wert für δ: $-0{,}767$), KCl (größter positiver Wert für δ: $2{,}474$) und Pb (δ: $-0{,}44$), einem Metall mittlerer Anisotropie, veranschaulicht. Hierbei werden statt der Frequenz-

Tabelle 2. Normierte Schallgeschwindigkeiten in der Richtung $\varkappa_0$ für Pb, Li, KCl

δ	$-0{,}44$	$-0{,}767$	$2{,}474$
$c_1(\varkappa_0; 0)$	1,125	1,23	0,68
$c_1(\varkappa_0; \cdot\,\delta)$	0,99	0,97	0,96
$c_1(\varkappa_0; \cdot\cdot\,\delta^2)$	1,0	0,99	1,03
$c_1(\varkappa_0; \cdot\cdot\cdot\,\delta^3)$	—	1,0	0,97
$c_2(\varkappa_0; 0)$	1,08	1,1	0,85
$c_2(\varkappa_0; \cdot\,\delta)$	1,0	1,0	1,01
$c_2(\varkappa_0; \cdot\cdot\,\delta^2)$	—	—	1,001
$c_2(\varkappa_0; \cdot\cdot\cdot\,\delta^3)$	—	—	1,007
$c_3(\varkappa_0; 0)$	1,34	1,56	0,74
$c_3(\varkappa_0; \cdot\,\delta)$	1,05	1,15	1,1
$c_3(\varkappa_0; \cdot\cdot\,\delta^2)$	1,01	1,04	0,87
$c_3(\varkappa_0; \cdot\cdot\cdot\,\delta^3)$	1,0	1,0	1,09

4 $c_\alpha(\varkappa_{0i} \dots \delta^3)$ bedeutet die Schallgeschwindigkeit in der Richtung $\varkappa_0$ unter Berücksichtigung der 3 ersten Entwicklungsglieder; die angegebenen Werte sind auf den entsprechenden exakten Wert der Schallgeschwindigkeit normiert.

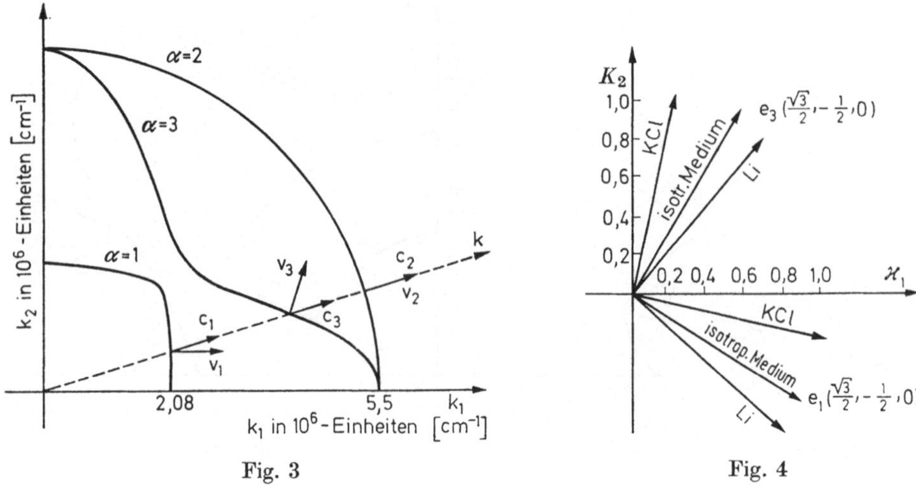

Fig. 3   Fig. 4

Fig. 3. Kurven konstanter Frequenz in der (k_1, k_2)-Ebene für KCl $(\delta > 0)$ bei $T = 0\,°\text{K}$:
$$\omega_\alpha(k) = 10^{12}\,\text{Hz} \quad (\alpha = 1, 2, 3)$$

Fig. 4. Anisotropieeffekt bei den Polarisationsvektoren: Die Drehung aus der isotropen Lage
$(e_1 \,\|\, k,\ e_{2/3} \perp k)$ ist bei verschiedenen Vorzeichen von δ entgegengesetzt

werte die Werte für die Schallgeschwindigkeiten[4] angegeben, und zwar für den
allgemeinen Wellenzahlvektor $\varkappa_0$ (siehe Fig. 1); für Wellenzahlvektoren, die in
ausgezeichneten Ebenen liegen, ist die Konvergenz noch besser.

Anhand dieser extrem gewählten Beispiele zeigt es sich, daß für fast alle
Materialien (siehe Fig. 5 u. Tab. 1) die exakten Frequenzwerte schon durch die
Reihe mit den 3 ersten Entwicklungsgliedern genau wiedergegeben werden. Für
$\delta > 2$ (KCl) konvergiert die Reihe langsam, hier müßten noch höhere Potenzen
in δ berücksichtigt werden. Approximativ könnte die Reihe schon nach dem
1. Entwicklungsglied abgebrochen werden, die Abweichung vom exakten Fre-
quenzwert wird dann für $\omega_3(k)$ um 10% liegen. Für die Frequenzen $\omega_1(k)$, $\omega_2(k)$
kann man generell die Reihe nach dem 1. Entwicklungsglied abbrechen, da die
vernachlässigten Korrekturen minimal sind.

Der Betrag des Anisotropieparameters gibt insofern die Stärke der Anisotropie
eines Materials an, als die Größe der Abweichung von der Isotropie in den Gleichungen für die Frequenzen und Polarisationsvektoren direkt proportional δ ist.
Da hierbei der Anisotropieparameter δ sowohl positive wie negative Werte annehmen kann — in der Fig. 5 sind die Materialien mit negativem δ-Wert von jenen
mit positivem δ-Wert durch die Gerade $\delta = 0$ getrennt —, sind die Frequenzen
zu einem Wellenvektor k im anisotropen Medium sowohl größer wie kleiner als die
zugehörigen isotropen Frequenzanteile (siehe Fig. 2 und 3). Entsprechend ist der
Anisotropieeffekt bei den Polarisationsvektoren (siehe Fig. 4): Je stärker die
Anisotropie, desto größer ist die Drehung aus der „isotropen" Lage, und zwar ist
die Drehung bei verschiedenen Vorzeichen von δ entgegengesetzt. In besonderen
symmetrischen Richtungen (Achsen, Flächen- und Raumdiagonalen) wird die
isotrope Lage beibehalten.

Die infolge der Anisotropie auftretende Dispersion $c_\alpha = $ const. (isotropes Medium) $\to c_\alpha(\varkappa)$ (anisotropes Medium) ist mit dem Anisotropieparameter δ eng verknüpft: Ein Maß für die Dispersion ist der Gradient der Schallgeschwindigkeit $\nabla_k c_\alpha(\varkappa)$, der infolge der Beziehung

$$v_\alpha(\varkappa) = \nabla_k \omega_\alpha(k) = c_\alpha(\varkappa) + k\,\nabla_k c_\alpha(\varkappa)$$

den Unterschied zwischen der Schallgeschwindigkeit $c_\alpha(\varkappa)$ und der Gruppengeschwindigkeit $v_\alpha(\varkappa)$ bestimmt. Mit Hilfe der abgeleiteten Formeln für die Frequenzen (bzw. Schallgeschwindigkeiten) wird der Dispersionsterm in 1. δ-Näherung

$$k\,\nabla_k c_\alpha(\varkappa) = \frac{B}{\varrho}\,\delta\,\frac{f_\alpha(\varkappa)}{c_\alpha(\varkappa)}\,.$$

($f_\alpha(\varkappa)$ enthält keine elastischen Konstanten mehr und ist nur von $\varkappa$ abhängig.)

Eine Materialgröße ist der dimensionslose Parameter

$$\sigma_\alpha = \frac{B}{\varrho}\,\delta\,\frac{r_\alpha}{c_{\alpha\,\mathrm{Max}}^2} \tag{12}$$

der ein Maß für die relative Dispersion $\dfrac{v_\alpha(\varkappa) - c_\alpha(\varkappa)}{c_{\alpha\,\mathrm{Max}}}$ ist und sich als die maximale relative Abweichung der Schallgeschwindigkeitsquadrate interpretieren läßt:

$$\sigma_\alpha = \frac{c_{\alpha\,\mathrm{Max}}^2 - c_{\alpha\,\mathrm{Min}}^2}{c_{\alpha\,\mathrm{Max}}^2}\,, \qquad r_\alpha = \begin{cases} \frac{2}{3}\,, & \alpha = 1\,; \\ \frac{1}{3}\,, & \alpha = 2\,; \\ \frac{1}{2}\,, & \alpha = 3\,. \end{cases} \tag{13}$$

Da die Minimum- und Maximumwerte der Schallgeschwindigkeiten in den Symmetrierichtungen angenommen werden, ist es leicht, die Parameter σ_α zu bilden. Der Vergleich der 3 Größen σ_α ($\alpha = 1, 2, 3$) zeigt (siehe Fig. 2 und 3), daß $\sigma_3 > \sigma_2 > \sigma_1$ gilt, d.h. die relative Dispersion des Mediums macht sich bei der Schallgeschwindigkeit $c_3(\varkappa)$ am stärksten bemerkbar. Der Parameter σ_3 erscheint dadurch als ein geeignetes Maß für die Stärke der relativen Disperison eines Mediums:

$$\sigma_3 = \begin{cases} -\dfrac{\varepsilon}{2\,c_{44}}\,, & \delta < 0\,; \\[2ex] \dfrac{\varepsilon}{2\,c_{44} + \varepsilon}\,, & \delta > 0\,; \end{cases} \tag{14}$$

Tabelle 3 zeigt, daß KCl trotz des 5—6fachen δ-Betrages von Pb eine kleinere relative Dispersion zeigt als Pb (siehe Fig. 2 und 3) und Li die stärkste von allen Materialien. Allgemein läßt sich somit sagen, daß einem stärker anisotropen Material (größerer δ-Wert) keineswegs auch eine stärkere relative Dispersion entsprechen muß.

Die in den Frequenzen auftretenden elastischen Konstanten können natürlich nicht völlig beliebige Werte annehmen. Die Forderung $\omega_\alpha(k) = c_\alpha(\varkappa)\,k > 0$ für alle $\varkappa$ ergibt folgende Einschränkungen für die elastischen Konstanten:

$$c_{11} > 0\,, \quad c_{44} > 0\,, \quad c_{11} - c_{12} > 0\,, \quad c_{11} + 2\,c_{12} + 4\,c_{44} > 0\,.^5 \tag{14}$$

---

5 Letztere Bedingung — eine Einschränkung für negative c_{12} — fehlt bei Dederichs und Leibfried [10].

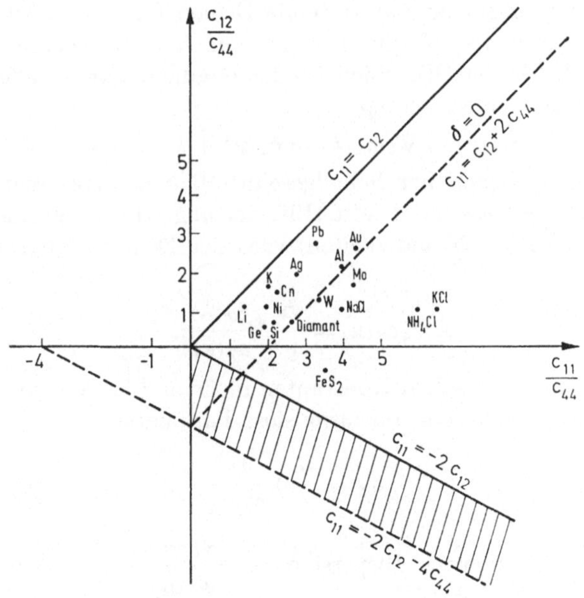

Fig. 5. c_{12}/c_{44} gegen c_{11}/c_{44} für verschiedene Kristalle. Die Gerade $\delta = 0$ trennt die Materialien entsprechend dem Scherungsverhältnis $\dfrac{c_{11} - c_{12}}{2\,c_{44}} \gtrless 1$. Der aus $\frac{1}{2}\,C_{iklm}\,\varepsilon_{ik}\,\varepsilon_{lm} > 0$ gewonnene Stabilitätsbereich wird durch die Geraden $c_{11} = c_{12}$ und $c_{11} = -2c_{12}$ begrenzt. Dynamisch stabil ($\omega_\alpha(k) > 0$) sind die Materialien auch noch in dem gestrichelten Gebiet

Tabelle 3

	Ag	Al	Au	Cu	KCl	Li	Ni	Pb
δ	$-0{,}46$	$-0{,}115$	$-0{,}27$	$-0{,}54$	$2{,}474$	$-0{,}767$	$-0{,}54$	$-0{,}44$
σ_3	$0{,}69$	$0{,}165$	$0{,}65$	$0{,}67$	$0{,}69$	$0{,}88$	$0{,}59$	$0{,}73$

Diese Bedingungen gewinnt man aus den Hauptminoren der dynamischen Matrix $\boldsymbol{T}(\boldsymbol{k})$, die wegen $\omega_\alpha(\boldsymbol{k}) > 0$ positiv sein müssen.

Die Bedingungen (14) sind weniger einschränkend (siehe Fig. 5) als jene, die sich für die elastischen Konstanten aus der Stabilitätsbedingung $\frac{1}{2}\,C_{iklm}\,\varepsilon_{ik}\,\varepsilon_{lm} > 0$ [11] ergeben:

$$c_{44} > 0, \quad c_{11} - c_{12} > 0, \quad c_{11} + 2c_{12} > 0 . \tag{15}$$

Da die dynamische Matrix $\boldsymbol{T}(\boldsymbol{k})$ nur für zeitabhängige Verschiebungs- bzw. Verzerrungsfelder, die sich aus den Eigenschwingungen $\omega_\alpha(\boldsymbol{k}) \neq 0$ des Mediums zusammensetzen lassen, definiert ist und somit statische Verzerrungen ($\omega_\alpha(\boldsymbol{k}) = 0$) wie die Kompression nicht berücksichtigt, müssen die aus $\boldsymbol{T}(\boldsymbol{k})$ gewonnenen Einschränkungen für die elastischen Konstanten schwächer sein als die der Stabilitätsbedingung, die Verzerrungen jeglicher Art zuläßt. So erlaubt die schwächere Bedingung $c_{11} + 2c_{12} + 4c_{44} > 0$, daß der Kompressionsmodul $K = \frac{1}{3}\,(c_{11} + 2c_{12})$ negativ werden kann.

V. Anhang

Im folgenden soll die Behauptung aus Abschnitt III bewiesen werden, daß die Lösungen $v_\alpha(\varkappa, \delta)$ ($\alpha = 1, 2, 3$) der Säkulargleichung $v^3 + p(\varkappa, \delta)\, v + q(\varkappa, \delta) = 0$ für $\delta > -1$ in Taylor-Reihen nach dem Anisotropieparameter δ entwickelt werden können.

Beweis:

Entwickelt man die Funktion

$$F(v, \delta) = v^3 + p(\varkappa, \delta)\, v + q(\varkappa, \delta) \tag{1}$$

um die Werte v_0 und δ_0 mit $F(v_0, \delta_0) = 0$, so gilt:

$$\begin{aligned}
F(v, \delta) = {}& [3\,v_0^2 + p(\varkappa, \delta_0)] \cdot (v - v_0) + [p'(\varkappa, \delta_0)\,v_0 + q'(\varkappa, \delta_0)] \cdot (\delta - \delta_0) \\
& + 3\,v_0(v - v_0)^2 + \tfrac{1}{2}\,[p''(\varkappa, \delta_0)\,v_0 + q''(\varkappa, \delta_0)]\,(\delta - \delta_0)^2 + p'(\varkappa, \delta_0) \\
& \cdot (v - v_0)\,(\delta - \delta_0) + (v - v_0)^3 + \tfrac{1}{6}\,q'''(\varkappa, \delta_0)\,(\delta - \delta_0)^3 + \tfrac{1}{2}\,p''(\varkappa, \delta_0) \\
& \cdot (v - v_0)\,(\delta - \delta_0)^2 \,.
\end{aligned} \tag{2}$$

Ist $F(v, \delta)$ die Säkulargleichung, d.h. $F(v, \delta) = 0$, und setzt man $\delta_0 = 0$, so folgt für v_0 aus $F(v_0, \delta_0) = 0$:

$$v_0 = v_\alpha(\varkappa, 0) = \begin{cases} \tfrac{2}{3} & \alpha = 1 \\ -\tfrac{1}{3} & \alpha = 2, 3 \,. \end{cases}$$

Aus Gleichung (2) ergeben sich dann für die Lösungen $v_\alpha(\varkappa, \delta)$ der Säkulargleichung die folgenden Darstellungen in Abhängigkeit von δ:

$$\begin{aligned}
w_1 = {}& v_1(\varkappa, \delta) - v_1(\varkappa, 0) \\
= {}& 2(\chi - \tfrac{1}{3})\,\delta + (\tfrac{4}{9} - \tfrac{5}{3}\chi + 3\,\Gamma)\,\delta^2 - 2\,w_1^2 + 2(\tfrac{1}{3} - \chi)\,w_1\,\delta - w_1^3 \\
& + (\tfrac{2}{27} - \tfrac{1}{3}\chi + \Gamma)\,\delta^3 + (\tfrac{1}{3} - \chi)\,\delta^2\,w_1
\end{aligned} \tag{3}$$

und für $\alpha = 2, 3$:

$$\begin{aligned}
w_\alpha^2 = {}& [v_\alpha(\varkappa, \delta) - v_\alpha(\varkappa, 0)]^2 \\
= {}& (-\tfrac{1}{9} + \tfrac{2}{3}\chi - 3\,\Gamma)\,\delta^2 + (-\tfrac{2}{3} + 2\,\chi)\,w_\alpha\,\delta + w_\alpha^3 + (-\tfrac{2}{27} + \tfrac{1}{3}\chi - \Gamma)\,\delta^3 + \\
& + (-\tfrac{1}{3} + \chi)\,\delta^2\,w_\alpha \,.
\end{aligned} \tag{4}$$

Da $p(\varkappa, \delta)$ und $q(\varkappa, \delta)$ Polynome in δ sind, können die Lösungen $v_\alpha(\varkappa, \delta)$ keine Unendlichkeitsstellen besitzen; der Term v^3 der Säkulargleichung würde dort stärker gegen Unendlich streben als ihre übrigen Terme, d.h. in der Summe könnten die Terme der Säkulargleichung an einem Unendlichkeitspunkt von $v_\alpha(\varkappa, \delta)$ nicht Null ergeben. Da weiterhin $\delta = 0$ kein Verzweigungspunkt sein kann, denn die Darstellungen

$$w_\alpha(\varkappa, \delta) = \sum_{\nu=1}^{\infty} \delta^{\nu/m}\, w_{\alpha, \nu}(\varkappa, 0)\,, \quad \alpha = 1, 2, 3\,.$$

($m \in \mathscr{N}$, $m \geqq 2$ und $w_{\alpha, 0}(\varkappa, 0) = w_\alpha(\varkappa, 0) = 0$ nach Definition) sind mit den Darstellungen (3, 4) nicht verträglich, müssen $w_\alpha(\varkappa, \delta)$ und demnach $v_\alpha(\varkappa, \delta)$ bei $\delta = 0$ analytisch und dort in Taylor-Reihen nach δ darstellbar sein:

$$v_\alpha(\varkappa, \delta) = \sum_{\nu=0}^{\infty} \frac{\delta^\nu}{\nu!}\, v_\alpha^{(\nu)}(\varkappa, 0)\,, \quad \alpha = 1, 2, 3\,. \tag{5}$$

Den Gültigkeitsbereich der Taylor-Reihen entnimmt man dem folgenden Satz [12]:

Es sei $F(v, \delta)$ eine in der Umgebung der Werte v_0 und δ_0 analytische Funktion mit $F(v_0, \delta_0) = 0$. Ist $\dfrac{\partial F(v_0, \delta_0)}{\partial v} \neq 0$, so existiert eine Funktion $v = f(\delta)$ mit $v_0 = f(\delta_0)$, die der Gleichung $F(v, \delta) = 0$ genügt und bei δ_0 analytisch ist.

(Für $\delta_0 = 0$ und $v_0 = v_1(\varkappa, 0)$ ist $\dfrac{\partial F(v_0, \delta_0)}{\partial v} \neq 0$, so daß die Taylor-Reihe für $v_1(\varkappa, \delta)$ aus diesem Satz unmittelbar folgt; dagegen können die Reihendarstellungen für $\alpha = 2, 3$ aus dem Satz nicht gefolgert werden, da für $\delta_0 = 0$ und $v_0 = v_2(\varkappa, 0) = v_3(\varkappa, 0)$ die Ableitung $\dfrac{\partial F(v_0, \delta_0)}{\partial v}$ verschwindet.)

Den Konvergenzbereich der Taylor-Reihen von $v_\alpha(\varkappa, \delta)$ um $\delta = 0$ findet man über die δ-Werte, die den Gleichungen

$$F(v, \delta) = v^3 + p(\varkappa, \delta)\, v + q(\varkappa, \delta) = 0\,, \tag{6}$$

$$\frac{\partial F(v, \delta)}{\partial v} = 3\,v^2 + p(\varkappa, \delta) = 0$$

genügen. Aus dem angegebenen Satz folgt dann die Gültigkeit der Reihen in den Gebieten, welche diese δ-Werte nicht enthalten. Andererseits können die Reihen selbst für diese δ-Werte existieren, wie der Fall $\delta = 0$ zeigte. Durch Kombination der Gleichungen (6) erhält man als Bestimmungsgleichung für die gesuchten δ-Werte:

$$\frac{q^2(\varkappa, \delta)}{4} + \frac{p^3(\varkappa, \delta)}{27} = 0 \tag{7}$$

bzw. $\qquad\qquad \delta^2(a_4\,\delta^4 + a_3\,\delta^3 + a_2\,\delta^2 + a_1\,\delta + a_0) = 0$

mit
$$\begin{aligned}
a_0 &= 4\,\chi^2 - 12\,\Gamma\,, \\
a_1 &= 12\,\chi^2 - 32\,\chi^3 - 40\,\Gamma + 108\,\chi\,\Gamma\,, \\
a_2 &= 13\,\chi^2 - 48\,\chi^3 - 48\,\Gamma + 198\,\chi\,\Gamma - 243\,\Gamma^2\,, \\
a_3 &= 6\,(\chi^2 - 4\,\chi^3 - 4\,\Gamma + 18\,\chi\,\Gamma - 27\,\Gamma^2) = 6\,a_4\,.
\end{aligned}$$

Die δ-Werte, die neben $\delta = 0$ der Gleichung (7) genügen, sind Lösungen der Gleichung 4. Grades $a_4\,\delta^4 + a_3\,\delta^3 + a_2\,\delta^2 + a_1\,\delta + a_0 = 0$. Durch Substitution $\delta = \bar{\delta} - 1$ kann gezeigt werden, daß für diese Lösungen $\delta_\nu \leqq -1$ ($\nu = 1, 2, 3, 4$) gelten muß:

$$b_4\,\bar{\delta}^4 + b_3\,\bar{\delta}^3 + b_2\,\bar{\delta}^2 + b_1\,\bar{\delta} + b_0 = 0$$

mit
$$\begin{aligned}
b_0 &= 4\,\chi^3 - 108\,\Gamma^2\,, \\
b_1 &= 8\,\chi^3 - 36\,\chi\,\Gamma + 108\,\Gamma^2\,, \\
b_2 &= \chi^2 - 18\,\chi\,\Gamma + 81\,\Gamma^2\,, \\
b_3 &= 2\,a_4\,, \qquad b_4 = a_4\,.
\end{aligned}$$

Da die Koeffizienten $b_\nu > 0$ sind (lediglich in den Richtungen der Achsen und Raumdiagonalen verschwinden sie), folgt:

$$\bar{\delta}_\nu \leqq 0 \qquad \text{bzw.} \qquad \delta_\nu \leqq -1\,, \tag{8}$$

d.h. die Taylor-Reihen der Lösungen $v_\alpha(\varkappa, \delta)$ der Säkulargleichung existieren mindestens für $\delta > -1$, wie behauptet wurde. (Die Richtigkeit der Reihen in den Richtungen der Achsen und Raumdiagonalen kann leicht durch Bestimmung der Eigenwerte des Tensors $T(k)$ mit dem Wellenzahlvektor k in diesen speziellen Richtungen gezeigt werden.)

Literatur

1. Goens, E.: Ann. d. Phys. **29**, 279 (1937)
2. Born, M., Karman, T.: Z. f. Phys. **14**, 18 (1913)
3. Blackman, M.: Hdb. d. Phys. VII/1, 344 (1955)
4. Mason, W. P.: Physical Acoustics, Vol. III, B, 31 (1965)
5. Fedorow, A. F. J.: Theory of Elastic Waves in Crystals, S. 245. New York: Plenum Press 1968
6. Landau, L. D., Lifschitz, E. M.: Elastizitätstheorie (Lehrbuch d. theor. Phys. Bd. VII), S. 188. Berlin: Akademie-Verlag 1966
7. Leibfried, G.: Hdb. d. Phys. VII/1, 119 (1955)
8. Hearmon, R. F. S.: Rev. mod. Phys. **18**, 409 (1946)
9. Huntington, H. B.: Solid State Physics (London) **7**, 274 (1958)
10. Dederichs, P. H. ,Leibfried, G.: Phys. Rev. **188**, 1175 (1969)
11. Nye, J. F.: Physical Properties of Crystals. Oxford: Clarendon Press 1957
12. Golubew, W.: Differentialgl. im Komplexen, S. 30. Berlin: Deutscher Verlag der Wissenschaften 1958

Dr. Walter Orth
D-6602 Saarbrücken-Dudweiler
Dürerstraße 1
Bundesrepublik Deutschland

Phys. cond. Matter 17, 307—315 (1974)

Photoemissionsmessungen an Europiumchalkogeniden

P. Cotti und P. Munz

Laboratorium für Festkörperphysik, ETHZ, Hönggerberg, Zürich

Eingegangen am 26. Oktober 1973

Photoemission-Measurements on Europium-Chalcogenides

From new photoemission measurements from EuO we show that the existing discrepancy between photoelectric- and optical results can be eliminated. The measurements on cleaved single crystal surfaces show sharper structure than measurements on evaporated films. The energy difference between the center of the $4f$-levels and the maximum of the valence band density of states can be read directly from the energy distribution curves. This energy difference decreases with increasing lattice constant from 2.8 eV for EuO to 1.6 eV for EuSe. However the two substances exhibit similar structure in the valence band.

1. Einleitung

Die magnetischen Isolatoren EuO, EuS, EuSe und EuTe können wegen ihrer magnetischen, kristallographischen und chemischen Einfachheit als Modellsubstanzen für das Studium magnetischer Wechselwirkungen betrachtet werden. Hieraus wie aus der Großzahl neuer magnetischer, magneto-optischer und elektrischer Effekte erklärt sich das anhaltende Interesse, das diese Substanzen gefunden haben [1, 11].

Diese vielfältigen Eigenschaften werden durch die elektronische Struktur bestimmt, die am direktesten mit Hilfe optischer — und photoelektrischer Methoden untersucht werden kann. Eine umfassende Darstellung der aus optischen Messungen gewonnenen Information ist in [1] zu finden. Über Messungen der Photoemission hingegen ist nur vereinzelt berichtet worden [2, 3, 15, 16]. Übereinstimmend ergeben die Photoemissionsmessungen, daß die das magnetische Moment tragenden, lokalisierten $4f$-Elektronen des Europium-Ions die höchsten besetzten Zustände bilden. Die zweiwertigen Europiumchalkogenide zeigen also eine einzigartige elektronische Struktur, da sie inhärente, lokalisierte Zustände besitzen, die in der Energielücke liegen. Allerdings ergaben bisherige Photoemissionsmessungen im Bereich des Vakuumultravioletten keine Übereinstimmung mit optischen Messungen in der relativen Lage der verschiedenen Energiezustände.

Durch die vorliegenden Messungen kann erstens diese Diskrepanz beseitigt werden und damit die Tauglichkeit der Photoemission für die Untersuchung der elektronischen Struktur der Europiumchalkogenide nachgewiesen werden. Zweitens sind die nun aus der Quantenausbeute erhaltenen Werte der Ionisierungsenergie konsistent mit den aus den Energieverteilungskurven bestimmten Werten.

2. Theorie

Die Photoemissions-Spektroskopie stellt eine neuerdings viel gebrauchte und viel diskutierte Methode zur Untersuchung der elektronischen Struktur von Fest-

körpern und Flüssigkeiten dar. Dabei wird meist die Energieverteilung der durch
Einstrahlen von monochromatischem Licht emittierten Elektronen analysiert.
(Im folgenden abgekürzt als EDC bezeichnet.)

Die allgemeinste theoretische Behandlung der Photoemission aus kondensierter
Materie findet man bei Ashcroft und Schaich [5]. Aber auch der auf die Näherung
unabhängiger Teilchen vereinfachte Ausdruck in obiger Arbeit ist bis jetzt nur in
einfachen Fällen explizit berechnet worden, nämlich für das Modell der freien
Elektronen, das Kronig Penney Modell und das Goodwin Modell [5]. Diese Be-
handlungen zeigen, daß insbesondere in Metallen Oberflächenanregungen möglicherweise nicht vernachlässigbar sind.

In den hier untersuchten Europiumchalkogeniden dürften diese Effekte von
geringerer Bedeutung sein, weil in Halbleitern und Isolatoren größere Austritts-
tiefen zu erwarten sind als in Metallen.

Eine vereinfachte Behandlungsweise der Photoemission, die einen direkten
Zusammenhang zwischen den gemessenen Energieverteilungskurven und der elektronischen Bandstruktur vermittelt, wurde von Spicer [6] und Kane [7] diskutiert
und an verschiedenen Substanzen erfolgreich angewendet. In diesem Drei-Stufen-
Modell der Photoemission werden die optische Anregung, der Transport zur Oberfläche und der Austritt ins Vakuum als unabhängige Prozesse behandelt. Dabei
ergibt sich folgender Ausdruck für die Energieverteilung $N(E)$ der bei Einstrahlung
von monochromatischem Licht der Energie $h\nu_0$ emittierten Elektronen:

$$N(E, h\nu_0) = C(\nu_0, E) \sum_{i,f} \int_{\mathrm{BZ}} dk^3 \, |M_{if}|^2 \, \delta(E_f(k) - E_i(k) - h\nu_0) \, \delta(E - E_f(k)). \quad (1)$$

Die Summe erstreckt sich über alle besetzten Bänder i und unbesetzten Bänder f.
Die Integration ist über die ganze Brillouinzone auszuführen. M_{if} ist das Matrix-
element für den optisch induzierten Übergang zwischen dem Anfangszustand $E_i(k)$
und dem Endzustand $E_f(k)$. Der Faktor $C(\nu_0, E)$ enthält die Wahrscheinlich-
keiten für den Transport eines angeregten Elektrons zur Oberfläche und für den
Austritt durch die Oberfläche. Berechnungen zeigen, daß $C(\nu_0, E)$ für Endenergien,
die mehr als 1 bis 2 eV über dem Vakuumniveau liegen, nur schwach energie-
abhängig ist. Dies ist auf die gegenläufige Wirkung von Transport- und Austritts-
wahrscheinlichkeit zurückzuführen. Der Ausdruck (1) kann in der folgenden Form
besser interpretiert werden:

$$N(E, h\nu_0) = C(\nu_0, E) \sum_{i,f} \int \frac{dl| \ |M_{if}|^2}{|\nabla_k E_f(k) \times \nabla_k E_i(k)|}. \quad (2)$$

Das Linienintegral erstreckt sich über die Schnittlinie der beiden Flächen

$$E_f(k) - E_i(k) = h\nu_0 \quad \text{und} \quad E_f(k) = E.$$

Maxima in der Energieverteilung sind, abgesehen von den Einflüssen durch
die Matrixelemente, unter den folgenden Bedingungen zu erwarten (7):

1. Endzustand gehört zu optischem kritischem Punkt:

$$\nabla_k(E_f(k) - E_i(k)) = 0.$$

2. Endzustand gehört zu zweidimensionalem kritischem Punkt:

$$\nabla_k(E_f(k) - aE_i(k)) = 0 \quad a \text{ Konstante}.$$

3. Anfangs- oder Endzustand zeigen kritischen Punkt in Zustandsdichte

$$\nabla_k E_i(k) = 0 \quad \text{oder} \quad \nabla_k E_f(k) = 0\,.$$

Der Typ einer experimentellen Struktur kann oft anhand der Abhängigkeit von der Photonenergie bestimmt werden [6. 7]. Zum Beispiel ist für ein Maximum der EDC, das auf Grund eines Maximums in der Zustandsdichte der Anfangszustände $\nabla_k E(k) = 0$ entsteht, zu erwarten, daß sich diese Struktur in der EDC bei Änderung der Photonenergie um $\Delta h\nu$ um denselben Energiebetrag ΔE verschiebt. Diese Zuordnung ist um so eindeutiger, je größer der Photonenergiebereich ist, in dem diese Gesetzmäßigkeit nachgewiesen werden kann. Wird keine k-Erhaltung vorausgesetzt, so sind in den Energieverteilungskurven nur noch Maxima des Typs 3) zu erwarten.

3. Experiment

Die vorliegenden Messungen wurden im Ultraviolett- und Vakuum-Ultraviolett-Gebiet durchgeführt (allgemein als UPS bezeichnet).

Als Lichtquelle dient das H_2 Spektrum einer modifizierten Hintereggerlampe, welche durch einen McPherson Monochromator des Seya Namioka Typs spektral zerlegt wird. Um in der Meßzelle Ultrahochvakuum zu erreichen, wurde ein LiF Fenster im Strahlengang eingebaut. Dadurch wird die maximale Photoenergie auf 11,0 eV begrenzt.

Im Unterschied zur konventionellen Technik wurde in unserem System ein zusätzlicher, toroidaler Spiegel eingebaut, der den Austrittspalt des Monochromators auf die Probe abbildet und dort einen relativ kleinen Lichtfleck erzeugt. Damit können auch Messungen an kleineren Einkristall-Proben vorgenommen werden, als bei der sonst üblichen direkten Einstrahlung des den Monochromator verlassenden Lichtes.

Vier Einkristall-Proben sind in einem drehbaren Halter eingespannt und können an den verschiedenen Präparations- und Meßstationen vorbeigeführt werden. Nach dem Spalten im Ultrahochvakuum kann die Austrittsarbeit nach der Kelvin-Methode bestimmt werden. Ferner erlaubt eine Cs-Quelle, monoatomare Schichten kontrollierter Dicke aufzutragen.

Die Energieverteilung wird nach einer modifizierten Retardierungsmethode, der Methode des abgeschirmten Emitters, gemessen [8]. In dieser Anordnung ist der Kollektor als Halbkugel ausgebildet, so daß über alle Emissionsrichtungen integriert wird. Dies vereinfacht die Interpretation insofern, als keine weiteren Auswahlregeln berücksichtigt werden müssen, wie dies bei der Messung unter einem kleineren Raumwinkel der Fall sein kann. Neue Messungen an Gold [9] haben gezeigt, daß sogar an polykristallinen Proben von Gold eine starke Anisotropie in der Winkelverteilung der emittierten Elektronen auftritt.

Das Auflösungsvermögen wurde durch Messungen an einem metallischen Emitter geprüft. Für die Breite der Fermikante zwischen 10% und 90% der Gesamthöhe wurde experimentell ein Energieintervall von 0,18 eV beobachtet. Dies ist etwa 0,08 eV mehr als die thermodynamische Breite. Daraus ergibt sich im betrachteten Energiebereich ein energetisches Auflösungsvermögen von ca. 0,1 eV.

Eine fundamentale Bedingung für Untersuchungen der Photoemission ist die atomare Reinheit der emittierenden Probenoberfläche. Deshalb wurden die Einkristall-Proben im Vakuum von $1 \cdot 10^{-10}$ Torr gespalten und gemessen. Beim Spalten der Kristalle wird eine sehr geringe Gasmenge frei, die zu einem starken Druckanstieg, verursacht durch den „memory effect" [21] der Ionengetterpumpe, führen kann. Durch spezielle Vorkehrungen konnte diese sekundäre Druckzunahme vermieden werden, so daß der Druck auch kurzzeitig nach dem Spalten nie über den 10^{-10} Torr Bereich steigt. Massenspektroskopische Untersuchungen ergaben, daß die Restgase sich zu 98,5% aus Wasserstoff und Edelgasen zusammensetzen. Der Partialdruck von O_2 und H_2O liegt unterhalb 10^{-13} Torr. Innerhalb eines Tages nach der Spaltung konnten keine Veränderungen festgestellt werden, die auf eine Kontamination der Probenoberfläche hinweisen.

Die Energieverteilungskurven wurden durch elektronische Differenzierung [10] der Strom-Spannungskurve erhalten und direkt mit einem x-y Schreiber aufgezeichnet.

4. Ergebnisse an EuO und EuSe und Diskussion

a) Quantenausbeute

In der ersten Figur ist der Verlauf der Quantenausbeute (Anzahl emittierter Elektronen pro eingestrahltem Photon) als Funktion der Photonenergie für EuO und EuSe dargestellt. Der starke Anstieg der Kurve für EuO in der Umgebung von 2 eV ist typisch für das Verhalten von Halbleitern und Isolatoren knapp oberhalb der Schwellenenergie. Er befolgt das hier übliche Gesetz:

$$y = c(h\nu - E_1)^3 \quad (c \text{ Konstante}). \tag{3}$$

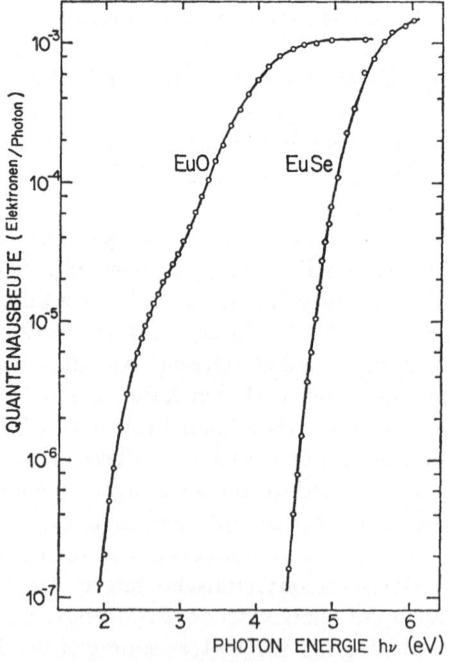

Fig. 1. Spektrale Abhängigkeit der Quantenausbeute

Für die Ionisierungsenergie E_1 erhält man an EuO $1{,}7 \pm 0{,}1$ eV. Die Messung der Austrittsarbeit nach Kelvin liefert innerhalb der Meßgenauigkeit etwa denselben Wert. Ein weiterer starker Anstieg der Kurve zeigt sich in der Umgebung von 3 eV. Versuchsweise kann hier eine zweite Ionisierungsenergie E_2 bestimmt werden, unter der Annahme, daß die Beziehung gilt:

$$y = c\,(h\nu - E_1)^3 + c'(h\nu - E_2)^3 \qquad (c' \text{ Konstante})\,.$$

Der erste Summand wird aus dem niederenergetischen Teil der Quantenausbeutekurve erhalten. Er ist klein und spielt deshalb bei der Bestimmung von E_2 keine große Rolle. Die auf diese Art erhaltene zweite Ionisierungsenergie beträgt etwa 2,7 eV. Die Bedeutung der beiden Ionisierungsenergien kann im Zusammenhang mit den Energieverteilungskurven interpretiert werden (im nächsten Abschnitt). Es scheint, daß Eastman [3] nur diese zweite Ionierungsenergie bestimmt hat.

Die zweite Kurve in Fig. 1 zeigt für EuSe einen einfacheren Verlauf. Die Ionisierungsenergie nach Gleichung (3) bestimmt, beträgt: $4{,}5 \pm 0{,}05$ eV. In der Arbeit [3] wurde dafür ein Wert von $2{,}8 \pm 0{,}3$ eV erhalten. Im Unterschied zur vorliegenden Arbeit wurde jener Wert an Filmen, die Emission aus Verunreinigungszuständen zeigen, gemessen. Die Überlegenheit von gespaltenen Einkristall-Oberflächen gegenüber diesen Filmen zeigt sich auch in der Schärfe der Strukturen in den Energieverteilungskurven.

b) Energieverteilungskurven

In der zweiten und dritten Figur sind gemessene Energieverteilungskurven für EuO und EuSe für verschiedene Photonenergien als Parameter aufgezeichnet. Die Anzahl emittierter Elektronen (willkürliche Einheiten) ist hier als Funktion ihrer Energie aufgetragen. Angegeben ist nicht die kinetische Energie E der emittierten Elektronen, die direkt gemessen wurde, sondern die Energie der Elektronen vor ihrer optischen Anregung. Als Nullpunkt wurde das Ferminiveau gewählt.

Die Lage des Ferminiveaus wurde wie folgt bestimmt: An einem metallischen Emitter, der in elektrischem Kontakt mit der Probe steht, wird gleichzeitig mit der Messung an der Probe (bei gleicher Photonenergie) eine Energieverteilungskurve aufgenommen. Die Energie der am metallischen Emitter beobachteten Fermikante (bei 50% der Gesamthöhe) wurde sodann als Referenz gewählt. Diese Lage ist wegen der experimentellen Verbreiterung der Kante durch endliches Auflösungsvermögen mit einem Fehler von etwa 0,1 eV behaftet.

Die Normierung der Energieverteilungskurven bezüglich der Energie der ursprünglichen Zustände erweist sich als zweckmäßig, da — wie sich im folgenden zeigen wird — die Strukturen in den Energieverteilungskurven im wesentlichen durch entsprechende Strukturen in den besetzten, also noch nicht angeregten Zuständen, bestimmt werden. Die Energieverteilungskurven erstrecken sich mit zunehmender Photonenergie nach immer tieferen Energien, da mit höheren Photonenergien auch energetisch tiefer gelegene Zustände zur Photoemission beitragen können.

Die Energieverteilungskurven in Fig. 2 und Fig. 3 zeigen bis zu drei Maxima. Auffallend ist die von der Photonenergie unabhängige Lage der beiden hochenergetischen Maxima von Fig. 2. Diese Strukturen werden den $4f$- bzw. Valenzzuständen zugeordnet [2], [3]. Das Maximum niedrigster Energie in Fig. 2 und

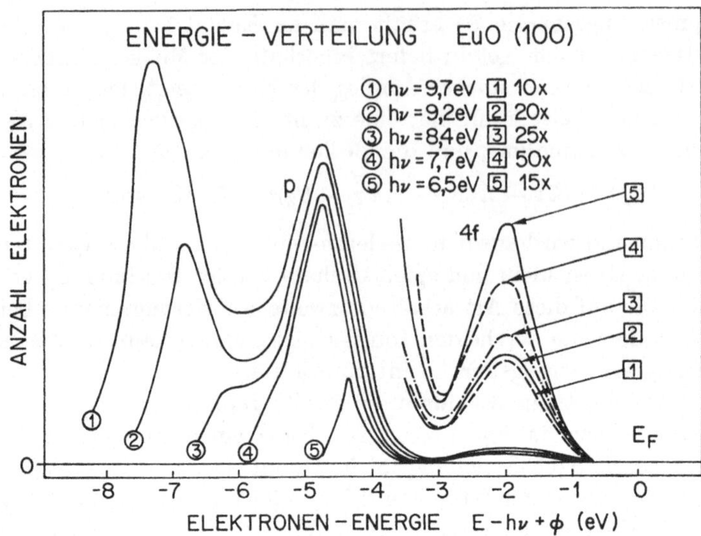

Fig. 2. Energieverteilungskurven gemessen an EuO für verschiedene Photonenergien als Parameter. Die Anzahl der emittierten Elektronen ist in willkürlichen Einheiten aufgetragen. Die vergrößerten Kurven sind durch quadratisch eingerahmte Ziffern gekennzeichnet

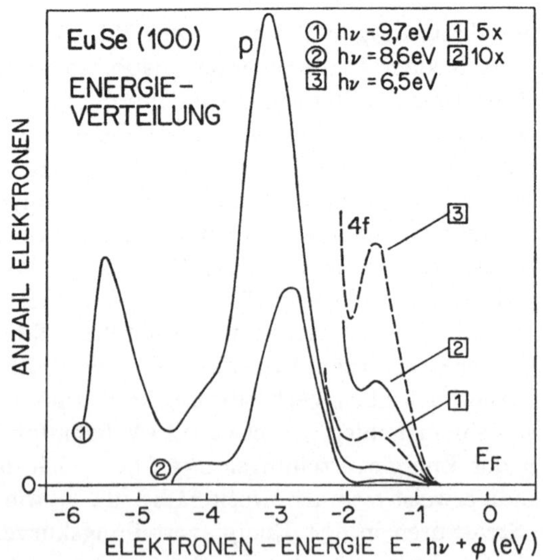

Fig. 3. Energieverteilungskurven gemessen an EuSe

Fig. 3 ist erst bei Photonenergien oberhalb 9 eV deutlich ausgeprägt. Ob es sich auch hier um eine von der Photonenergie unabhängige Struktur handelt, kann mit den zur Verfügung stehenden Photonenergien nicht abgeklärt werden. Außerdem können gestreute Elektronen in diesem Energiebereich einen wesentlichen Beitrag zur EDC liefern.

c) 4f-Zustände

Das bei der größten Energie liegende Maximum in Fig. 2 und Fig. 3 bei
$-2,0$ eV bzw. $-1,6$ eV wird der Emission aus $4f$-Zuständen zugeschrieben. Es
ist relativ schwach ausgeprägt, in Übereinstimmung mit Erwartungen für die
Wahrscheinlichkeit optischer Anregungen der $4f$-Zustände. Deshalb wurde diese
Struktur noch mit vergrößerter Empfindlichkeit gemessen und dargestellt. Es
zeigt sich, daß die in einer früheren Arbeit [2] gemessene Stationarität dieses
Maximums in EuO auch im Bereich der Photonenergie von 6,5 eV bis 10,5 eV
weiterbesteht. Die Breite dieser Struktur bleibt über den ganzen Bereich der
Photonenergie ungefähr konstant. Dies zeigt, daß für das Entstehen dieser
Struktur die Anfangszustände verantwortlich sind. Im Einelektronenbild ist dieses
Verhalten gemäß Gleichung (3) aus Abschnitt II als Maximum in der Zustandsdichte der besetzten Zustände zu interpretieren. Der Schwerpunkt der $4f$-Zustände
liegt für EuO 2,0 eV und für EuSe 1,6 eV unterhalb des Ferminiveaus.

Die $4f$-Zustände werden aber als sehr lokalisiert betrachtet, denn ihre Wellen-
funktionen befinden sich innerhalb des Xe-Rumpfes des Eu-Ions [11]. Die Photonenergien der Vakuumultraviolett-Photonemissionsspektroskopie genügen im allgemeinen nicht, um Emission aus lokalisierten Atomniveaus zu beobachten, wie
dies zum Beispiel mit der ESCA-Technik möglich ist. Die $4f$-Elektronen zeigen
als Besonderheit eine geringe Bindungsenergie wegen der starken Coulomb-Ab-
stoßung zwischen den Elektronen dieser Schale [11]. Da die $4f$ Elektronen unter-
einander stark in Wechselwirkung stehen, darf deren Verhalten nicht bedenkenlos
im Einelektronenbild interpretiert werden. Es muß zumindest die Totalenergie ε
der $4f$-Elektronen betrachtet werden. Es gilt dann

$$E_K - h\nu + \varphi = \varepsilon_{^8S_{7/2}} - \varepsilon_{^7F_J} + \varphi,$$

wobei E_K die kinetische Energie des emittierten Elektrons und $\varepsilon_{^8S_{7/2}}$ und $\varepsilon_{^7F_J}$
Anfangs- bzw. Endzustand der $4f$-Elektronen bezeichnen. Auch wenn $\varepsilon_{^8S_{7/2}}$ als
scharfe Energie zu betrachten wäre, kann E_K wegen der Spin Bahn-Kopplung
des ^7F_J Multipletts in einem gewissen Energiebereich variieren. $E_K - h\nu + \varphi$
ist bei der Emission aus $4f$-Zuständen keine eigentliche Einelektronen-Energie
der $4f$-Zustände, kann aber beim Vergleich mit optischen Messungen als quasi
Einelektronen-Energie betrachtet werden. $E_K - h\nu$ ist die erste Ionisierungsenergie für diese Zustände.

Die $4f$-Struktur zeigt bei EuO eine Halbwertsbreite von etwa 1,1 eV. Das
^7F_J Multiplett besitzt im Atom eine Breite von 0,63 eV [12]. Neben dem Multiplett
und der experimentellen Verbreiterung müssen noch andere Effekte, wie z.B.
Gitterrelaxationen um das $4f$-Loch, zur Breite der $4f$-Struktur beitragen. Auch
an Metallen zeigen Photoemissions- [13] und ESCA-Messungen [14] breite $4f$-
Strukturen. Eine quantitative Erklärung dieser Tatsache ist vorläufig noch nicht
bekannt.

In letzter Zeit sind verschiedene weitere Arbeiten erschienen, die nachweisen,
daß die obersten besetzten Zustände im reinen EuO, EuS und EuSe $4f$-Charakter
besitzen:

Einen direkten Beweis liefert die neue Methode der Messung der Spinpolarisa-
tion von Photoelektronen von Siegmann *et al.* [15].

Eastman [16] hat über die Energieabhängigkeit der Matrixelemente den $4f$-Charakter dieser Zustände nachgewiesen.

Eine OPW Bandberechnung wurde von Lendi [17] durchgeführt. Sie zeigt, daß die $4f$-Zustände in der Energielücke liegen müssen.

Neueste Messungen der optischen Konstanten von EuS bei tiefen Temperaturen von Güntherodt [20] zeigen im ersten Absorptionspeak sechs Strukturen, die dem ^7F_J Multiplett zugeordnet werden können.

d) Valenzzustände

Die zweite, mit p bezeichnete Spitze in den Energieverteilungskurven der Fig. 2 und 3 wird den Valenzzuständen zugeschrieben. Im Gegensatz zu anderen Messungen [3] erscheint diese Struktur in EuO im betrachteten Photonenergiebereich an der gleichen Stelle. Dieser Unterschied scheint daher zu rühren, daß in der Messung [3] vor allem bei niedrigen Photonenergien starke Verschmierungen vorliegen, was sich vor allem an der zu großen totalen Breite der Energieverteilungskurve zeigt. (Breite der EDC größer als $h\nu - E_1$, E_1 Schwellenenergie.) Nach der in Abschnitt II dargelegten Theorie scheint diese Struktur p in Fig. 2 und 3 durch eine entsprechende Spitze in der Zustandsdichte der Valenzbänder verursacht zu sein. Diese Interpretation steht in Einklang mit optischen Messungen [18]. Diese liefern energetische Abstände von 2,9 eV und 1,5 eV zwischen dem Maximum in der Zustandsdichte des Valenzbandes und dem Schwerpunkt der $4f$-Zustände.

Aus den Energieverteilungskurven in Fig. 2 und Fig. 3 lassen sich die entsprechenden Werte von 2,8 bzw. 1,6 eV herauslesen. Für beide Substanzen findet man also eine Übereinstimmung zwischen der Photoemission und optischen Messungen innerhalb 0,1 eV. Der Abstand zwischen den $4f$-Zuständen und dem Maximum in der Zustandsdichte des Valenzbandes nimmt dementsprechend von EuO nach EuSe, also mit zunehmender Gitterkonstanten, ab. Beide Bandberechnungen, nach der APW Methode [19] und nach der OPW Methode [17], zeigen eine solche erhöhte Zustandsdichte am oberen Valenzbandrand.

Interessant ist der Vergleich von Energieverteilungskurven, gemessen an EuSe zwischen Schichten Fig. 1 in [3] und an gespaltenen Einkristall-Oberflächen in vorliegender Arbeit: Die Einkristallmessungen zeigen deutlich schärfere Struktur. Die $4f$-Spitze ist hier innerhalb des gesamten Photonenergiebereichs sichtbar, während im Fall der Filme [3] für Photonenergien größer als 6,5 eV diese Struktur höchstens noch als Schulter feststellbar ist.

Anhand der Energieverteilungskurven erklärt sich die Bedeutung der ersten und zweiten Ionisierungsenergie, die photoelektrisch an EuO gemessen wurde (siehe a). Bei sehr stark erhöhter Meßempfindlichkeit zeigt sich in den Energieverteilungskurven von EuO ein Ausläufer besetzter Zustände mit der Breite von etwa 1 eV oberhalb der $4f$-Zustände. Diese sehr schwache Struktur ist Nichtstöchiometrie und Verunreinigungen zuzuschreiben. Erwartungsgemäß hängt diese Struktur vom Probenexemplar ab, während der Rest der Kurven reproduzierbar ist. Bandkrümmungseffekte scheinen keine wesentlichen Störungen zu verursachen. Die erste Ionisierungsenergie von 1,7 eV stellt also den Abstand des oberen Randes dieser Zustände vom Vakuumniveau dar. Entsprechend bestimmt die zweite Ionisierungsenergie von 2,7 eV die Lage des oberen Randes der 4f-Zustände in bezug auf das Vakuumniveau.

e) Schlußfolgerungen

Es zeigt sich, daß für die Europiumchalkogenide mit Hilfe der Photoemissionsspektroskopie Resultate erhalten werden, die, wo vergleichbar, in Übereinstimmung mit optischen Messungen sind. Dies deutet darauf hin, daß der Emissionsprozeß vor allem durch Volumeneigenschaften bestimmt wird. Im betrachteten Photonenergiebereich können die Energieverteilungskurven im Wesentlichen durch die Zustandsdichte erklärt werden. Die Photoemission kann also, zumindest im Falle der Europiumchalkogenide, als Ergänzung zu rein optischen Messungen detailliertere, zusätzliche Information liefern.

Zum Schluß möchten wir Herrn Prof. Dr. G. Busch, dem Vorsteher des Laboratoriums für Festkörperphysik der ETHZ, für die Anregung zu dieser Arbeit und für viele helfende Hinweise herzlich danken. Die Arbeit wurde durch den „Schweizerischen Nationalfonds zur Förderung der Wissenschaftlichen Forschung" finanziell unterstützt.

Literatur

1. Wachter, P.: Crit. Rev. in Solid State Sci. **3**, 189 (1972)
2. Busch, G., Cotti, P., Munz, P.: Solid State Commun. **7**, 705 (1969)
3. Easman, D. E., Holtzberg, F., Methfessel, S.: Phys. Rev. Letters **23**, 226 (1969)
4. Methfessel, S.: Z. angew. Phys. **18**, 414 (1965)
5. Schaich, W., Ashcroft, N. W.: Phys. Rev. **3 B**, 2452 (1971)
6. Berglund, C. N., Spicer, W. E.: Phys. Rev. **136 A**, 1030 (1964)
   Spicer, W. E.: Phys. Rev. **154**, 385 (1967)
7. Kane, E. O.: Phys. Rev. **175**, 1039 (1968).
8. DiStefano, T. H., Pierce, D. T.: Rev. Sci. Instrum. **41**, 180 (1970)
9. Koyama, R. Y., Hughey, L. R.: Phys. Rev. Letters **29**, 1518 (1972)
10. Spicer, W. E., Berglund, C. N.: Rev. Sci. Instrum. **35**, 1665 (1964)
11. Methfessel, S., Mattis, D. C.: Handbuch der Physik, XVIII (1968)
12. McClure: Solid State Physics 9, 399 (1959)
13. Brodén, G., Hagström, S. B. M., Norris, C.: Phys. kondens. Materie **15**, 327 (1973)
14. Hedén, P. O., Löfgren, H., Hagström, S. B. M.: Phys. status solidi (b) **49**, 721 (1972)
    Baer, Y.: private Mitteilung
15. Busch, G., Campagna, M., Siegmann, H. C.: Solid State Commun. **7**, 775, (1969)
    Sattler, K., Siegmann, H. C.: Phys. Rev. Letters **29**, 1565 (1972)
16. Eastman, D. E., Kuznietz, M.: J. appl. Phys. **42**, 1396 (1971)
17. Lendi, K.: erscheint in Phys. kondens. Materie.
18. Güntherodt, G., Wachter, P.: Phys. kondens. Materie **12**, 292 (1971)
19. Cho, S. J.: Phys. Rev. 1 B, 4589 (1970)
20. Güntherodt, G., Wachter, P.: AIP Conference Proceeedings 10, 1284 (1973)
21. Brance, U. R., Craig, R. D.: Vacuum 16, 647 (1966)

P. Munz
Laboratorium für Festkörperphysik
Eidgenössische Technische Hochschule Zürich
Hönggerberg
CH-8049 Zürich, Schweiz